D1187003

COMPUTERS IN NUMBER THEORY

COMPUTERS IN NUMBER THEORY

Proceedings of the Science Research Council Atlas Symposium No. 2 held at Oxford, from 18 - 23 August, 1969

Edited by

A. O. L. ATKIN

Mathematics Department,
Brown University, Providence, Rhode Island, U.S.A.

and

B. J. BIRCH

The Mathematical Institute,
University of Oxford, England

1971

ACADEMIC PRESS · LONDON AND NEW YORK

ACADEMIC PRESS INC. (LONDON) LTD.
Berkeley Square House,
Berkeley Square,
London, W1X 6BA

U.S. Edition published by
ACADEMIC PRESS INC.
111 Fifth Avenue,
New York, New York 10003

Library of Congress Catalog Card Number: 70–153520

ISBN: 0 12 065750 3

Printed in Great Britain by
ROYSTAN PRINTERS LIMITED
Spencer Court, 7 Chalcot Road
London N.W.1

Contributors

GEORGE E. ANDREWS, *Pennsylvania State University, University Park, Pennsylvania, U.S.A.*

P. BARRUCAND, *The Pascal Institute, Paris, France.*

P. T. BATEMAN, *University of Illinois, Urbana-Champaign, Illinois, U.S.A.*

J. BRILLHART, *University of Arizona, Tucson, Arizona, U.S.A.*

J. W. BROWN, *University of Illinois, Urbana-Champaign, Illinois, U.S.A.*

DONALD BURNELL, *Computer Science Department, Washington State University Pullman, Washington, U.S.A.*

STEFAN A. BURR, *Bell Telephone Laboratories, Inc., Whippany, New Jersey, U.S.A.*

DAVID G. CANTOR, *Department of Mathematics, University of California, Los Angeles, California, U.S.A.*

P. CARTIER, *Department of Mathematics, University of Strasbourg, Strasbourg, France.*

M. S. CHEEMA, *Mathematics Department, University of Arizona, Tucson, Arizona, U.S.A.*

R. F. CHURCHHOUSE, *Atlas Computer Laboratory, Chilton, Didcot, England.*

F. B. COGHLAN, *Department of Mathematics, University of Manchester, Manchester, England.*

A. M. COHEN, *Mathematics Department, University of Wales Institute of Science and Technology, Cathays Park, Cardiff, Wales.*

HARVEY COHN, *Mathematics Department, University of Arizona, Tucson, Arizona, U.S.A.*

P. ERDŐS, *Mathematical Institute, Academy of Science, Budapest, Hungary.*

AVIEZRI S. FRAENKEL, *Department of Applied Mathematics, Weizmann Institute of Science, Rehovot, Israel.*

R. A. GASKINS, *Virginia Polytechnic Institute, Blacksburg, Virginia, U.S.A.*

H. J. GODWIN, *Royal Holloway College, University of London, Egham, England.*

I. J. GOOD, *Virginia Polytechnic Institute, Blacksburg, Virginia, U.S.A.*

R. L. GRAHAM, *Bell Telephone Laboratories, Murray Hill, New Jersey, U.S.A.*

RICHARD K. GUY, *Department of Mathematics, University of Calgary, Calgary, Alberta, Canada.*

R. S. HALL, *Syracuse University, Syracuse, New York, U.S.A.*

MARSHALL HALL, *California Institute of Technology, Pasadena, California, U.S.A.*

OSKAR HERRMANN, *Department of Mathematics, University of Heidelberg, Heidelberg, Germany.*

LORNE HOUTEN, *Computer Science Department, Washington State University, Pullman, Washington, U.S.A.*

K. E. KLOSS, *National Bureau of Standards, Washington, D.C., U.S.A.*

RICHARD B. LAKEIN, *Department of Mathematics, State University of New York, Buffalo, New York, U.S.A.*

J. LARMOUTH, *Mathematical Laboratory, University of Cambridge, England.*

R. R. LAXTON, *Department of Mathematics, University of Nottingham, Nottingham, England.*

JOHN LEECH, *Department of Computing Science, University of Stirling, Stirling, Scotland.*

JOSEPH LEHNER, *Computer Science Center, Mathematics Department, University of Maryland, College Park, Maryland, U.S.A.*

D. H. LEHMER, *University of California, Berkeley, California, U.S.A.*

W. F. LUNNON, *Atlas Computer Laboratory, Chilton, Didcot, England.*

A. M. MACBEATH, *University of Birmingham, Birmingham, England.*

N. S. MENDELSOHN, *Mathematics Department, University of Manitoba, Winnipeg, Manitoba, Canada.*

N. METROPOLIS, *University of California, Los Alamos Scientific Laboratory, Los Alamos, New Mexico, U.S.A.*

J. C. P. MILLER, *Mathematical Laboratory, University of Cambridge, Cambridge, England.*

L. J. MORDELL, *St. John's College, Cambridge, England.*

JOSEPH B. MUSKAT, *Department of Mathematics, Bar Ilan University, Ramat Gan, Israel.*

MORRIS NEWMAN, *National Bureau of Standards, Washington, D.C., U.S.A.*

ALBRECHT PFISTER, *Mathematics Institute, The University, Mainz, Germany.*

P. A. B. PLEASANTS, *Department of Mathematics, University College, Cardiff, Wales.*

A. SCHINZEL, *Mathematics Institute PAN, Ul Sniadeckich* 8, *Warsaw, Poland.*

ERNST S. SELMER, *Institute of Mathematics, University of Bergen, Bergen, Norway.*

J. R. SMITH, *Royal Navy Engineering College, Manadon, Plymouth, England.*

ROBERT SPIRA, *Mathematics Department, Michigan State University, East Lansing, Michigan, U.S.A.*

H. M. STARK, *Department of Mathematics, Massachusetts Institute of Technology, Cambridge, Massachusetts, U.S.A.*

ROSEMARIE M. STEMMLER, *National Bureau of Standards, Washington, D.C., U.S.A.*

N. M. STEPHENS, *Pembroke College, Oxford, England.*

H. P. F. SWINNERTON-DYER, *Trinity College, Cambridge, England.*

OLGA TAUSSKY, *California Institute of Technology, Pasadena, California, U.S.A.*

J. TONASCIA, *Johns Hopkins University, Baltimore, Maryland, U.S.A.*

J. H. VAN LINT, *Mathematics Department, Eindhoven Technological University, Eindhoven, Netherlands.*

P. WEINBERGER, *University of California, Berkeley, California, U.S.A.*

M. C. WUNDERLICH, *Mathematics Department, Northern Illinois University, De Kalb, Illinois, U.S.A.*

Editors' Note

Several of the papers presented at the Symposium are published elsewhere.

Atkin, A. O. L. (Atlas Laboratory) and Swinnerton-Dyer, H. P. F. (Cambridge) (1971).
Modular forms on non-congruence subgroups. *In* "Proceedings of Symposia in Pure Mathematics, XIII". American Mathematical Society, Providence.

Berlekamp, E. R. (Bell Laboratories) (1970).
Factoring polynomials over large finite fields. *Math. Comp.* **24,** 713–735.

Birch, B. J. (Oxford) (1969).
K_2 of global fields. *In* "Proceedings of Symposia in Pure Mathematics, XX". Number Theory Institute. American Mathematical Society, Providence.

Fröberg, C.-E. (Lund) (1968).
On the prime zeta function. *BIT* **8,** 187–202.

Fröhlich, A. (King's College London) (1969).
On the classgroup of integral grouprings of finite abelian groups. *Mathematika* **16,** 143–152.

Good, I. J. and Churchhouse, R. F. (1968).
The Riemann hypothesis and pseudorandom features of the Möbius sequence. *Math. Comp.* **22,** 857–862.

Hasse, H. (Hamburg) and Liang, J. (1969).
Über den Klassenkörper zum quadratischen Zahlkörper mit der Diskriminante −47, *Acta Arithmetica* **16,** 89–98.

Riesel, H. (Stockholm) (1969).
Lucasian criteria for the primality of $N = h.2^n - 1$. *Math. Comp.* **23,** 869–875.

List of Participants Not Presenting Papers

Helen Alderson, Cambridge
Yvette Amice, Poitiers
I. Anderson, Glasgow
I. O. Angell, London
H. G. Apsimon, IBM Hursley Park
L. D. Baumert, Caltech
M. N. Bleicher, Madison
Lyliane Bouvier, Grenoble
D. A. Burgess, Nottingham
J. W. S. Cassels, Cambridge
J. H. H. Chalk, Toronto
J. H. E. Cohn, London
J. H. Conway, Cambridge
R. J. Cook, London
J. B. Cosgrave, London
A. L. Dulmage, Manitoba
V. Felsch, Kiel
H. Fredricksen, Caltech
J. Fresnel, Bordeaux
M. G. Gras, Grenoble
G. Greaves, Reading
J. M. Hammersley, Oxford
J.-C. Herz, IBM France
T. Hjelle, Bergen
J. Hunter, Glasgow
T. H. Jackson, York
M. Keates, Cardiff
K. E. Kloss, NBS Washington
T. Kløve, Bergen and Atlas
M. I. Knopp, Madison
Ø. Kølberg, Bergen
S. Kuroda, Karlsruhe and Maryland
Emma Lehmer, Berkeley
G. Ligozat, Paris
A. M. Macbeath, Birmingham
J. Martinet, Bordeaux
A. D. McGettrick, Cambridge
J. Merriman, Oxford
Margaret Millington, Reading
Marie-Nicole Montouchet, Grenoble
S. Mossige, Bergen
J. N. O'Brien, Exeter
T. R. Parkin, CDC Minneapolis
Jane Pitman, Adelaide
G. Poitou, Paris
R. A. Rankin, Glasgow
Ø. Rodseth, Atlas and Bergen
J. L. Selfridge, Illinois
J.-P. Serre, Paris
M. Sobel, Minnesota
J. Tate, Paris and Harvard
J. Todd, Caltech
H. Tverberg, Bergen

Preface

The Atlas Symposium No. 2, "Computers in Number Theory", was held at Oxford during the week of 18th to 23rd August, 1969. The Atlas Computer Laboratory, a part of the Science Research Council, was set up with the provision of a large scale computing service for British universities as its major purpose. Jack Howlett, the Director, recognized at an early stage that the Laboratory's active involvement, both in computer research itself and in other research which makes use of computers, would aid that purpose. Thus the Laboratory has employed internally research workers under fixed term contracts (commonly in conjunction with a Fellowship at an Oxford college), and has directly supported research at other centres. In addition, it has so far organized three international conferences devoted to the use of computers in specialized fields. The first, on "Computational Problems in Abstract Algebra", was held in the summer of 1967; the Proceedings, edited by John Leech, are published by Pergamon Press. The present volume contains the Proceedings of the second in the series. The third, on "Interdisciplinary Applications of Transport Theory", was held in the Summer of 1970; the Proceedings, edited by G. E. Hunt, will appear in 1971 in *Spectroscopy and Radiation Transfer,* **11**. The choice of fields has been made both on grounds of inherent interest, and on those of need. Thus it was felt that High Energy Physics, for instance, had already been adequately catered for elsewhere in this context.

The chief activity of the Symposium was the presentation of a number of papers, most of which are printed here. Some papers were presented which had already been accepted for publication elsewhere; these are listed on page ix. The papers illustrate all aspects of the use of computers in number theory: as an essential part of a proof, as an aid to discovery (Gauss would surely have approved), and negatively as a possible ally in doing what has not yet been done. The attitude sometimes maintained a few years ago, that a computer is a disreputable device in the context of "pure" mathematics, was noticeably absent at the Symposium. Computors and noncomputors alike were concerned with getting on with the job, rather than worrying about the relative reliability of computers and papers in mathematical journals.

There were in all about 120 invited participants, many of them accompanied by their families. However, not all of the participants presented talks.

About 60 invitations sent out were not accepted for a variety of reasons, mainly those of previous commitment. On Wednesday evening there was an open discussion on "Number Theoretic Subroutines and Tables". As might have been expected with so large a number of original thinkers no substantial agreement was reached, but the three following conclusions commanded the support of most of the 40 people present. First, that a complete list of what exists, in print and privately, would be useful. Second, that the subject is too various to lend itself to an agreed package of subroutines and a universal language. Third, that some facility should exist for making known the "results" of unsuccessful but substantial computation. The current tradition of mathematics does not look kindly on a statement in print that sulphuric acid was poured on a white powder without apparent reaction, but machine time can be expensive. A tape recording of the discussion is kept at the Atlas Computer Laboratory.

There were also some distractions not directly number theoretic. On Monday evening Professor Mordell gave us a fascinating account of his infinitely varied mathematical and personal experience, entitled "Reminiscences of an Octogenarian Mathematician". On Tuesday Dr. Howlett's sherry party at the Queen's College made serious thought impossible after 6 p.m. On Wednesday afternoon, there was an organized tour of the Atlas Laboratory in action. On Friday, the Symposium dinner at the Queen's College provided a memorable conclusion to the proceedings.

A large number of people and public bodies made the Symposium both possible and successful. First and foremost, Jack Howlett, without whose forethought and determination there would have been no background on which to build. The Science Research Council, through the Atlas and Rutherford Laboratories, provided the secretarial staff, and gave a substantial sum to cover the expenses of some of the speakers. International Computers Limited, the manufacturers of the Atlas computer, also contributed generously towards these expenses. Synolda Butler did all the hard secretarial work involved in planning and organizing the Symposium. The Programme Committee, consisting of the editors, R. F. Churchhouse, and H. P. F. Swinnerton-Dyer, had only to sign her letters, and we became increasingly aware of her exact attention to every detail.

At Oxford, most of the participants were housed in Jesus College, whose authorities gave us every cooperation. Their rooms were not big enough to accommodate the numbers at the sherry party and the dinner, and we were fortunate in our choice of the Queen's College for these occasions. The actual meetings were held at the Mathematics Institute, by kind permission of Professor Higman; he and the Institute staff were very helpful. The arrangements for travel were in the capable hands of "Robbie", Mr. C. L. Roberts of the Atlas Laboratory. He also attended to the important and easily forgotten details that are nobody's business in particular, dealing imperturbably with all emergencies.

Finally we must thank the authors for their cooperation, and Academic Press, particularly the production staff, for their frequent and ready advice during publication. The editors have confined themselves mainly to making small alterations of style or notation. The articles are either the original manuscripts given to us at the time of the Symposium or revised versions received after it; they are printed more or less in the same order as the talks were given, since we soon found that many of them would not fit in to any rigid scheme of classification.

A. O. L. ATKIN
B. J. BIRCH

May, 1971

Contents

The Economics of Number Theoretic Computation

D. H. LEHMER

University of California, Berkeley, California, U.S.A.

It is a privilege to be invited to open the Science Research Council's Atlas Symposium No. 2 on Computers in Number Theory. This general subject has been a concern of mine for nearly a half century. It would be a simple matter to impose on the many distinguished contributors to this Symposium a full hour of my own detailed opinions in this field. This, I have no intention of doing. Not only the number of the contributors, but also their geographic distribution attests to the fact that the subject of the Symposium is now of widespread interest. I believe that the time has come to look into the economics of our subject a little more closely and to consider what steps can be taken to offset, at least to some extent, certain developments that are unfavorable to us. It seems to me that to speak of this problem directly, though not in a minor key, might well be a suitable overture to the opera about to unfold here this week.

It is easy to cite reasons why a high speed digital computer is just the tool to apply to number theory. Less often does one read about the difficulties of bringing the machine to bear on a given problem. Of course there is the language barrier. There is also the fact that machines are usually built for applied mathematics. A little deeper there is the problem of timing. Between hand computing and automatic computing there is a huge gap in the times required to do arithmetic and symbol manipulation. Here the old phrase "a month a minute" comes to mind. On top of this is the infinite gap in speed between the fastest machine and a proof by induction, for example. Such considerations enter into the economics of our subject.

On the hardware side, over the past two decades, great forward strides in speed and reliability have been made, indeed more than in any other branch of science. More information can be processed in a single step and the steps are getting quicker. In order to speed up the approximate floating point operations the exact rational operations have been slighted by the designer to the detriment of the number theorist. Recent large machines show other disquieting trends. The IBM 360 has a smaller word size for its integers and this, as we shall see, hampers the use of multiprecision arithmetic. In the CDC

6000 type machines two divisions are required to obtain one quotient and one remainder.

On the software side, recent developments towards increasing throughput, providing graphic displays, accommodating large numbers of remote users, etc. though laudable in themselves, serve to reduce the time and space available to the poor number theorist.

On the purely economic side, the signs are also unfavorable. Although greater speeds have brought down the cost of a multiplication, there is a noticable hardening of the economy against the number theorist with a large problem. No longer does the establishment feel it useful to make free long exact machine runs in order to check the reliability of the system. Weekend and graveyard time is now being sold to medicine and psychology for long statistical runs involving mountains of data.

There are four different fronts on which the number theorist can fight back against these economic encroachments. These are:

I. Special Number Theory Subroutines,

II. Selective Use of the Central Processor,

III. Modification of Abandoned Equipment,

IV. Special Off-line Computers.

I hope to say more than a few words about topic I because it is of more general interest and to dwell on the other three topics only briefly.

The efficiency and flexibility of number theoretic programs can be greatly enhanced by the careful utilization of a library of well designed basic subroutines. Not only is it very important to reduce the time spent in each of the much used subroutines but it is desirable to have a measure of this time for economic considerations. If the time spent in one application of a subroutine is always the same, as for example with input subroutines, the problem is easy. More frequently the elapsed time depends upon a parameter n. If this time is less than a constant times $f(n)$ we say that the subroutine is of order $f(n)$, or less. Just the theoretical knowledge of f and the number of applications of the subroutine may deter us from embarking upon a too costly main program. We need not write and polish a subroutine to discover f. In some cases, however, a substantial saving in cost results from a clever design. In this case the constant implied by the relation

$$\text{Cost} = O(f(n))$$

is unusually small. Of course, in other cases it may be unusually large.

A surprising number of basic subroutines for number theory are of logarithmic order. The first such that comes to mind is the Euclidean algorithm.

It was Lamé (1844) who first observed that the number of divisions needed to find the greatest common divisor (m, n), $m < n$, does not exceed 5 times the number of decimal digits of m. By using the nearest integer algorithm this factor 5 can be lowered to 3 but we would say that either algorithm is of order $\log n$. Further reductions in cost result from exploiting the fact that most divisions give tiny quotients (41 % are equal to 1) and that subtraction is much cheaper than division. Of course if one is designing a subroutine for the G.C.D. only, one can ignore these quotients, and there is no need to store all the information generated by the algorithm, a fact that allows the whole operation to be carried out in the arithmetic registers of the modern central processor. Other closely related subroutines can be written to expand m/n in a regular, or semiregular, continued fraction and to solve the linear Diophantine equation

$$mx + ny = d \tag{1}$$

and thus to find the multiplicative inverse of n modulo m in case $d = 1$. In these cases one must be prepared to compute the convergents of the continued fraction. Gauss pointed out that by reversing the order of the partial quotients half of the work in solving (1) can be obviated. To achieve this saving, however, provision must be made for storing the partial quotients, in order to reverse them later. This is unpleasant in a subroutine. This illustrates the economic distinction between hand and machine computing.

I have gone into such detail about one class of subroutines merely to illustrate what considerations enter into their economic aspects. I promise not to do this again.

Another similar function of two variables is the Jacobi symbol (m/n), n odd. This we evaluate by the reciprocity law after shifting out the power of 2 dividing m. It is seen at once that this results in a subroutine of order $\log n$. The implied constant here can be lowered by various coding tricks depending on the machine's available instructions. This effort can be pretty rewarding if the subroutine is to be used to evaluate Dirichlet L-series, for example.

A very useful subroutine of order $\log n$ is the power algorithm that finds the number r_n in

$$b^n \equiv r_n \pmod{m},\ 0 \leqslant r_n < m$$

where b, n and m are given. If we were to use the recurrence

$$r_{k+1} \equiv br_k \pmod{m},$$

the cost would be of order n. Instead, we use

$$r_{2k} \equiv r_k^2 \pmod{m} \text{ or } r_{2k+1} \equiv br_k^2 \pmod{m}$$

and thus construct the nth power by successive squaring with an occasional multiplication by b, whenever n has a unit digit in its binary representation, using an obvious intentional overflow to control the process. To find b^n itself one has only to choose m large enough.

We can use this power subroutine to find primitive roots of a prime p, to determine power characters using Euler's criterion and to test large numbers p for primality by the various converses of Fermat's theorem, each problem being of order $\log p$. This last problem depends for its success upon an adequate knowledge of the factors of $p - 1$. When this is lacking one may try to factor $p + 1$ instead. In this case the simple power subroutine is replaced by another which calculates Lucas' function

$$V_n \equiv \alpha^n + \beta^n \pmod p$$

where α and β are quadratic integers with sum 1, as will be explained in Riesel's paper this morning. When $p + 1$ also resists factoring, one may try to factor $p^2 + p + 1$. Then one uses a still more elaborate subroutine to calculate

$$\Delta_n \equiv (1 - \alpha^n)(1 - \beta^n)(1 - \gamma^n) \pmod p$$

where α, β, γ are cubic integers satisfying

$$x^3 = kx^2 + kx + 1,$$

as will be explained in Selfridge's paper this morning. Although this last subroutine is 18 times as expensive as the simple power subroutine it is still of order $\log p$, since a duplication formula can be used. In contrast, the simpler test for primality based on Wilson's theorem is impractical, not because the factorial function is larger than the power function, but because the former function has no practical duplication formula and thus leads to a cost of order p.

Other important subroutines of order $\log p$ find solutions of quadratic congruences modulo p and represent p by binary quadratic forms. In spite of the text books, these problems can be solved via the power subroutine, Lucas' function V and continued fractions, with a cost of order $\log p$.

Before considering subroutines of higher order, it is well to point out that logarithmic order allows for (and by Parkinson's Law calls for) the use of very large values of the parameter, values occupying two or more machine words. For this reason we must prepare our logarithmic subroutines to operate in the environment of "multiprecise arithmetic", not just double precision or any fixed precision. That is, we must prepare subroutines for the rational operations too. Our logarithmic library now contains some score of subroutines such as the following

Addition
Subtraction
Multiplication
Division with remainder
Square root with remainder
Greatest common divisor
Power
Jacobi symbol
Linear equation solver
Decimal to binary input
Binary to decimal output

I have brought over the pole a "source deck" of such a library, prepared by Dr. Weinberger and myself, for the inspection and reproduction by any interested member of the Symposium. It is written in FORTRAN IV and therefore it is available at once to any computing system having a FORTRAN compiler. However it may not compete economically with a library prepared in machine language for a given machine.

The subroutines for multiplication, division, and square root are of order $(\log n)^2$. This increases the order of the subroutines which call for these operations. Thus if an algorithm, like the power algorithm, involving $O(\log n)$ steps calls on multiplication and division, one must realistically consider its order to be $O(\log n)^3$.

When we come to subroutines of order $\sqrt{n}$ we are thinking of fairly small numbers occupying only one or two machine words. The most obvious example of such a routine is that for factoring the integer n by trial divisions. Every computing establishment has its own "in house" subroutine for this problem if only for the edification of visitors. Such programs vary widely in efficiency and sincerity, depending mainly on the method used to generate the next trial divisor. If one's search extends not to $\sqrt{n}$ but some fixed limit, then the cost is really more like $O(\log n)$. This is often the case when a very large n is examined for possible small factors before testing for primality or embarking upon more sophisticated factorization routines.

Another useful algorithm to make into a subroutine of order $\sqrt{n}$ is the expansion of the regular continued fraction for the square root of n. In applications one is often forced to extend the expansion to its half period P and there are good heuristic reasons to support the inequality

$$P < 0.15\sqrt{n}\,\log n$$

so we bring this subroutine at this time. Since the subroutine is often used to solve the Pell equations

$$x^2 - ny^2 = \pm 1 \text{ or } \pm 4,$$

one must be prepared for very large values of x and y by immersing the subroutine in the multiprecision package.

As already mentioned, the solution of a quadratic congruence with a prime modulus p and the representation of p by a quadratic form are really problems of order $\log p$. If we replace p by a composite number m of unknown composition the price, in each case, goes up to $O(\sqrt{m})$. In fact, we can no longer fall back on Fermat's theorem

$$a^{p-1} \equiv 1 \pmod{p}$$

and the power subroutine. To be sure, we still have Euler's theorem

$$a^{\phi(m)} \equiv 1 \pmod{m}$$

but we don't know $\phi(m)$. If we had some way of approximating $\phi(m)$, a search procedure of order $\log m$ wolud be practical and we could reduce the cost accordingly back to $O(\log m)$. This simple truth has an analogue in the theory of composition of binary quadratic forms where ϕ is replaced by the class number function h. In this case we can approximate h by means of the Dirichlet L-function. This idea is exploited by Shanks in a paper once scheduled for Friday morning.

On the question of representing m by a given quadratic form, the distinction between definite and indefinite forms would, at first, seem crucial to the cost. However, by a theorem of Chebyshev, the cost in either case is $O(\sqrt{m})$. This rule suffers but one exception, namely when the form is $x^2 - y^2$, which is of course closely related to the factorization of n. Here the direct cost is of order m since the search for x or y involves one of the inequalities

$$\sqrt{m} < x < \frac{1}{2}\left(\frac{m}{N} + N\right) \qquad 0 < y < \frac{1}{2}\left(\frac{m}{N} - N\right)$$

where the constant N is the limit to which the direct search for prime factors of m has been pushed in vain. The popularity of this method of factorization is due to the fact that it is very effective when m is the product of two nearly equal factors so that it serves as a good supplement to the direct search for factors. Also when the prime factors of m are known to lie in the arithmetical progression $nt + 1, t = 1, 2, \ldots$, the value of x is restricted to one residue class modulo $n^2/2$. Hence this algorithm of order m competes strongly with others of order $\sqrt{m}$. In this connection R.S. Lehman has shown recently how to use many equations of the form

$$x^2 - y^2 = km$$

with different values of k to produce an algorithm to factor m in only $O(m^{1/3})$ steps. For numbers m with no particular properties this method could develop into an economic asset.

For problems whose cost is of order n or higher, one is economically limited to small values of n and relatively short runs, if one is to use the computer in the usual way. For example, if one is to look for primes $p \leqslant x$ for which the Fermat quotient

$$(a^p - a)/p$$

is divisible by p, a problem to be discussed by Brillhart on Friday afternoon, this will consume vast amounts of machine time for large x since the cost will be

$$O\left(\sum_{p \leqslant x} \log p\right) = O(x)$$

(according to the prime number theorem) no matter how many clever subroutines are used. To get a problem like this done when the weekend and graveyard time is not available one can still use the machine when it would otherwise stand idle provided it is a modern machine having a central processor surrounded by a number of peripheral processors. In this case it is possible to soak up every millisecond of time that the central processor is not occupied with its paying satellites, provided that one's own program occupies a very small amount of storage space and uses no input or output devices. Almost no one but a number theorist has problems of this sort. This technique has been used pretty steadily on the CDC 6400 machines at Berkeley during the past couple of years. A surprisingly large percentage of the time can be snatched from the queue of paying users in this way, something between 70% and 90%, depending on the number of short problems in the queue. The optimal type of problem here is one in which there is only one answer which may come up after running many hours. When the answer appears, we deliberately divide by zero. This causes an alarm and a dumpout of a section of the memory containing our answer in octal notation.

The advent of the new machine at one's computing establishment offers another economic opportunity to the number theorist. This is the possibility of using abandoned equipment. After a few years of service and maintenance many a machine, in better condition than its new replacement, is abandoned in favor of some more modern computer. The solid state components of those parts of the old system that don't move, namely the main frame and the core memory, need no regular maintenance and will operate reliably for months on end. Short programs can be read into storage either manually or by a cheap card reader. The machine can store away its answers for future display. A few minutes per day suffice to read out answers and to service the queue of future problems. In this way one has a fully committed computer working 24 hours per day on a whole stack of problems each with its own priority. The fact that the system solves problems 5 times slower than the new machine, that we cannot afford to use anyway, need not worry us too much. The real degradation consists in having to use machine language coding and octal representation of numbers. We have been operating an abandoned IBM 7094 at Berkeley in this way for a couple of years now. Types of problems that we run include

Elimination of small prime factors $< 2^{35}$ from very large numbers

Evaluation of Dirichlet L-functions

Search for primes p dividing $(a^p - a)/p$

Search for even numbers n dividing $2^n - 2$

Enumeration of special kinds of permutations

besides a host of other short problems that ran in less time than it would take to describe.

In conclusion, we come to the fourth way that the number theorist can try to overcome his economic difficulties, namely by using a special "off-line" computer. A piece of equipment, such as a card punch, is called off-line with respect to a computing system if it can be given an assignment by the main computer and be told to proceed with this assignment while the main computer goes about other business. An off-line computer would be a small machine independent of the main computer but depending on it for the preparation of input data and the processing of output.

The economic advantages to be gained with such a setup include the following: The off-line computer can be a special device and hence not only cheaper but also faster than the main computer's arithmetic unit. The off-line computer is also fully committed and hence under no political or economic pressure from the crowd of general users. The advent of integrated circuit technology brings the design of very fast special purpose equipment down to a level at which the prospective user can draw in his own requirements. Interrupting the main computer for a few seconds, via a remote console, is becoming commonplace for many computer systems. I forecast an interesting future for the off-line computer concept.

An example of an off-line computer is the Delay Line Sieve, DLS-127 that has been in constant operation at Berkeley since 1965. It runs without a budget on 100 watts of power and performs calculations which. if done in our all purpose computers (tying them up completely), would cost something like a million dollars per month. Since the sieve will not add 2 and 2 it is easily defended against intrusion by the general user. If the DLS will not add what will it do? In general it will divide a given number by 31 moduli $\leqslant 127$ and decide whether the set of 31 remainders is suitable or not in one microsecond. This means that it can find large solutions (x, y) of the general Diophantine equation $f(x, y) = 0$ where f is a polynomial with integer coefficients. For example solutions of Mordell's equations $x^3 = y^2 + k$ $(0 < k < 100)$ about which we are to hear later, were found with $x < 10^{10}$. Many other special Diophantine equations have been treated. Especially common is the problem of representing a given number by a binary quadratic form. Another class of problems has to do with the discovery of discriminants of forms with exceptionally small (or large) class numbers or having a single class in each genus.

The presentation of the problem to the sieve is via the CDC 6400 which also processes its output. Its services are available, to all number theorists, especially those at this Symposium, free of charge.

Linear Relations Connecting the Imaginary Parts of the Zeros of the Zeta Function

P. T. BATEMAN AND J. W. BROWN
University of Illinois, Urbana, Illinois, U.S.A.

R. S. HALL
Syracuse University, New York, U.S.A.

K. E. KLOSS AND ROSEMARIE M. STEMMLER
National Bureau of Standards, Washington, D.C., U.S.A.

1. Introduction

Let $\gamma_1, \gamma_2, \ldots$ be the imaginary parts of the zeros of the Riemann zeta function above the real axis, arranged in order of magnitude. In this paper we report on a numerical investigation of sums of the following three types

(A) $\sum_{n=1}^{N} c_n \gamma_n$ $\quad (c_n = -1, 0, 1$, not all c_n zero)

(B) $\sum_{n=1}^{N} c_n \gamma_n$ $\quad (c_n = -2, -1, 0, 1, 2$, not all c_n zero, at most one $|c_n|$ has the value 2)

(C) $\sum_{n=1}^{N} c_n \gamma_n$ $\quad$ (c_n integers with greatest common divisor 1).

The interest in these sums stems from the following theorem of Ingham (1942).

INGHAM'S THEOREM: *Let λ and μ denote the arithmetical functions of Liouville and Möbius respectively and put*

$$L(x) = \sum_{n \leqslant x} \lambda(n), \qquad M(x) = \sum_{n \leqslant x} \mu(n).$$

If at most a finite number of sums of type (C) *are zero then, when* $x \to \infty$,

$$\underline{\lim}\, x^{-\frac{1}{2}} L(x) = -\infty, \quad \overline{\lim}\, x^{-\frac{1}{2}} L(x) = +\infty, \tag{1}$$

$$\underline{\lim}\, x^{-\frac{1}{2}} M(x) = -\infty, \quad \overline{\lim}\, x^{-\frac{1}{2}} M(x) = +\infty. \tag{2}$$

For fixed N there are infinitely many sums of type (C) and so it is worthwhile to modify Ingham's theorem as follows.

VARIANT FORM OF INGHAM'S THEOREM. *If at most a finite number of sums of type* (B) *are zero, then* (1) *and* (2) *hold.*

We shall give a proof of this variant of Ingham's theorem in the last section of the paper. Naturally it would also be desirable to prove a similar variant involving sums of type (A), but this appears to be much more difficult if not impossible.

For each value of N from 1 to 20 we have determined the smallest sums of types (A) and (B), where we use the word "smallest" (as we shall throughout) to mean "smallest in absolute value". Needless to say no sums of types (A) or (B) were found to be zero. The minimal values found are given in Table I.

TABLE I. Values of Smallest Sums of Types (A) and (B)

	Smallest Sum of Type (A) Involving $\gamma_1, \ldots, \gamma_N$		Smallest Sum of Type (B) Involving $\gamma_1, \ldots, \gamma_N$	
	Predicted Value (a_N)	Actual Value	Predicted Value (b_N)	Actual Value
1	9·642	14·1347251417	10·02	14·1347251417
2	5·760	6·8873144970	3·847	6·8873144970
3	2·698	3·9888179413	1·306	2·8984965556
4	1·183	1·4732959513	0·4458	0·6821298910
5	0·4821	1·0368895105	0·1480	0·2884822691
6	0·1922	0·0807720835	0·04964	0·0807720835
7	0·07465	0·0807720835	0·01660	0·0594534168
8	0·02831	0·0210591095	0·005524	0·0210591095
9	0·01067	0·0210591095	0·001853	0·0003058342
10	0·003955	0·0017781898	0·0006180	0·0003058342
11	0·001453	0·0017781898	0·0002063	0·0001332060
12	0·0005310	0·0002796754	0·00006907	0·0000593242
13	0·0001927	0·0000286426	0·00002311	0·0000012150
14	0·00006931	0·0000286426	0·000007714	0·0000012150
15	0·00002490	0·0000117743	0·000002584	0·0000004140
16	0·000008892	0·0000033333	0·0000008642	0·0000004140
17	0·000003162	0·0000001925	0·0000002889	0·0000000458
18	0·000001120	0·0000001925	0·0000000966	0·0000000293
19	0·0000003966	0·0000001925	0·0000000323	0·0000000293
20	0·0000001397	0·0000000297	0·0000000108	0·0000000094

Once a computer program was set up to determine the smallest sum of type (A) for each N, all that was needed to find the smallest sum of type (B) was to make 20 further runs of the same program, in each of which one of the γ_i was replaced by $2\gamma_i$. For this reason the discussion of programs which follows will be limited to sums of type (A).

Two different programs were used for computing the smallest sum of type (A) for a given value of N. One was a backtrack program which simply computed all possible sums of type (A) involving the first N zeros and printed out the coefficients giving the smallest sum for each N. (In order to avoid duplication we computed only those sums of type (A) in which the first non-zero coefficient was $+1$.) Since this program was rather time-consuming, we used it only for values of N from 1 to 16. The approximate linear relations found in this way are listed in Tables II and III.

TABLE II. Coefficients of Small Sums of Type (A)

γ_1	$= 14{\cdot}13472\ 51417$
$-\gamma_1+\gamma_2$	$= 6{\cdot}88731\ 44970$
$-\gamma_2+\gamma_3$	$= 3{\cdot}98881\ 79413$
$-\gamma_1+\gamma_2+\gamma_3-\gamma_4$	$= 1{\cdot}47329\ 59513$
$\gamma_1-\gamma_2-\gamma_3+\gamma_5$	$= 1{\cdot}03688\ 95105$
$\gamma_1+\gamma_2-\gamma_4+\gamma_5-\gamma_6$	$= 0{\cdot}08077\ 20835$
$-\gamma_1-\gamma_2+\gamma_6+\gamma_7-\gamma_8$	$= 0{\cdot}02105\ 91095$
$\gamma_1+\gamma_2+\gamma_3+\gamma_5-\gamma_8-\gamma_{10}$	$= 0{\cdot}00177\ 81898$
$\gamma_1-\gamma_3+\gamma_5-\gamma_6-\gamma_7+\gamma_{12}$	$= 0{\cdot}00027\ 96754$
$\gamma_2-\gamma_4-\gamma_5+\gamma_7+\gamma_8-\gamma_9-\gamma_{10}+\gamma_{11}-\gamma_{12}$ $+\gamma_{13}$	$= 0{\cdot}00002\ 86426$
$-\gamma_1-\gamma_3+\gamma_4+\gamma_5-\gamma_7-\gamma_8-\gamma_9+\gamma_{11}+\gamma_{13}$ $+\gamma_{14}-\gamma_{15}$	$= 0{\cdot}00001\ 17743$
$\gamma_1-\gamma_3+\gamma_4+\gamma_5-\gamma_6-\gamma_7+\gamma_8+\gamma_{10}-\gamma_{16}$	$= 0{\cdot}00000\ 33333$
$-\gamma_3+\gamma_5-\gamma_7+\gamma_8+\gamma_9-\gamma_{11}+\gamma_{12}-\gamma_{13}+\gamma_{16}$ $-\gamma_{17}$	$= 0{\cdot}00000\ 01925$
$\gamma_1-\gamma_4-\gamma_5-\gamma_6+\gamma_7+\gamma_8-\gamma_9+\gamma_{10}+\gamma_{14}$ $-\gamma_{15}-\gamma_{18}+\gamma_{20}$	$= 0{\cdot}00000\ 00297$

TABLE III. Coefficients of Small Sums of Type (B)

γ_1	$= 14{\cdot}13472\ 51417$
$-\gamma_1+\gamma_2$	$= 6{\cdot}88731\ 44970$
$-\gamma_1+2\gamma_2-\gamma_3$	$= 2{\cdot}89849\ 65556$
$-\gamma_1-\gamma_2-\gamma_3+2\gamma_4$	$= 0{\cdot}68212\ 98910$
$-\gamma_1-\gamma_2-\gamma_4+2\gamma_5$	$= 0{\cdot}28848\ 22691$
$\gamma_1+\gamma_2-\gamma_4+\gamma_5-\gamma_6$	$= 0{\cdot}08077\ 20835$
$\gamma_3-2\gamma_5+\gamma_7$	$= 0{\cdot}05945\ 34168$
$-\gamma_1-\gamma_2+\gamma_6+\gamma_7-\gamma_8$	$= 0{\cdot}02105\ 91095$
$-2\gamma_1+\gamma_2+\gamma_3-\gamma_4-\gamma_5+\gamma_7-\gamma_8+\gamma_9$	$= 0{\cdot}00030\ 58342$
$\gamma_2+\gamma_4+\gamma_5+\gamma_6-2\gamma_7-\gamma_8-\gamma_{10}+\gamma_{11}$	$= 0{\cdot}00013\ 32060$
$\gamma_2+2\gamma_4-\gamma_5-\gamma_6+\gamma_8-\gamma_9+\gamma_{10}-\gamma_{12}$	$= 0{\cdot}00005\ 93242$
$\gamma_1+\gamma_5-\gamma_7+\gamma_8-2\gamma_{10}+\gamma_{11}+\gamma_{12}-\gamma_{13}$	$= 0{\cdot}00000\ 12150$
$-\gamma_1-\gamma_2-2\gamma_3+\gamma_4+\gamma_6-\gamma_7+\gamma_8+\gamma_9-\gamma_{10}$ $-\gamma_{11}-\gamma_{12}+\gamma_{14}+\gamma_{15}$	$= 0{\cdot}00000\ 04140$
$\gamma_1+2\gamma_2+\gamma_3+\gamma_4+\gamma_5-\gamma_9-\gamma_{10}-\gamma_{11}-\gamma_{12}$ $+\gamma_{15}+\gamma_{16}-\gamma_{17}$	$= 0{\cdot}00000\ 00458$
$-\gamma_2+\gamma_3+\gamma_4+\gamma_5-2\gamma_6+\gamma_7-\gamma_8-\gamma_9+\gamma_{10}$ $-\gamma_{11}+\gamma_{12}-\gamma_{16}+\gamma_{18}$	$= 0{\cdot}00000\ 00293$
$-\gamma_2-\gamma_4+\gamma_5-\gamma_6+\gamma_9+\gamma_{10}-2\gamma_{11}-\gamma_{12}+\gamma_{13}$ $+\gamma_{14}-\gamma_{15}+\gamma_{16}+\gamma_{19}-\gamma_{20}$	$= 0{\cdot}00000\ 00094$

The second program began by calculating all sums of type (A) involving $\gamma_1, \ldots, \gamma_{10}$ (and having first non-zero coefficient equal to $+1$). The absolute values of these sums were then sorted numerically. For each value of N from 11 to 20 each possible sum of type (A) involving $\gamma_{11}, \ldots, \gamma_N$ (and having its first non-zero coefficient equal to $+1$) was computed. For the absolute value of each sum a binary search was made in the previously obtained sorted list to determine the values immediately greater and immediately less. The lesser of the two differences thus found was then a candidate for smallest sum. The disadvantage of this second program was that the coefficients of only $\gamma_{11}, \ldots, \gamma_N$ were immediately recoverable. However, after the value of the smallest sum was found for a given value of N, we were then able to determine the other coefficients involved in the smallest sum by simply making further runs with the zeros permuted, Tables II and III give the coefficients producing the minimal sums in the various cases.

Consistent results were obtained in the range covered by both programs. Also, there was complete agreement between calculations made independently at the National Bureau of Standards using the now defunct "Pilot" computer and at the University of Illinois using the IBM 360 system.

We used values for $\gamma_1, \ldots, \gamma_{20}$ to fourteen decimal places taken from the tables of Bigg (1961) and Haselgrove (1961). We are grateful to the Royal Society for providing us with copies of the relevant portions of these tables. Spira (1964) also found values for $\gamma_1, \ldots, \gamma_{30}$ to sixteen places which are in agreement with those found by Bigg (1961) and Haselgrove (1961). In preparing the tables we have truncated our results to ten decimal places, since the subsequent figures can easily be reconstructed from Bigg's table and the data in Tables II and III.

In the next section we discuss the values we would expect to get for these smallest sums on the basis of simple probabilistic arguments. The actual values found for the minimal sums agree very well with these "predicted" values. In fact the smallest sums behave so regularly as to suggest that one would not be very likely to find a zero sum of type (A) or (B) within a finite amount of computer time. Even so, there is no harm in trying and it would certainly be worthwhile to extend our data to larger values of N.

2. Probabilistic Considerations

For fixed N consider the 3^N numbers we get by listing all sums of type (A) and in addition the trivial sum with all coefficients zero. These 3^N sums are distributed with mean zero and variance

$$\sigma_N^2 = \tfrac{2}{3}(\gamma_1^2 + \gamma_2^2 + \ldots + \gamma_N^2).$$

Further, since

$$\gamma_N = o(\{\tfrac{2}{3}(\gamma_1^2 + \gamma_2^2 + \dots + \gamma_N^2)\}^{\frac{1}{2}}) \qquad (N \to +\infty),$$

it follows from Lindeberg's version of the Central Limit Theorem (cf. Feller, 1966) that the distribution of our 3^N sums is approximately normal for large N. Thus the probability density of these numbers near the origin is about $\sigma_N^{-1}(2\pi)^{-\frac{1}{2}}$ and so the number per unit length near the origin is about $3^N \sigma_N^{-1}(2\pi)^{-\frac{1}{2}}$. Thus we would expect the smallest sum of type (A) to be about

$$a_N = 3^{-N} \sigma_N(2\pi)^{\frac{1}{2}} = 3^{-N}\{\tfrac{4}{3}\pi(\gamma_1^2 + \gamma_2^2 + \dots + \gamma_N^2)\}^{\frac{1}{2}}. \tag{3}$$

From the familiar asymptotic formula

$$\gamma_N \sim 2\pi N/\log N \qquad (N \to +\infty)$$

(which is proved, for example, by Ingham (1932)) it is easy to see that

$$a_N \sim \frac{4(\pi N)^{3/2}}{3^{N+1}\log N} \qquad (N \to +\infty).$$

Now consider for fixed N the $(2N + 3)3^{N-1}$ numbers we get by listing all sums of type (B) and in addition the trivial sum with all coefficients zero. These $(2N + 3)3^{N-1}$ sums have mean-value zero and variance τ_N^2 given by

$$\begin{aligned}(2N + 3)3^{N-1}\tau_N^2 &= 3^{N-1}(2\gamma_1^2 + 2\gamma_2^2 + \dots + 2\gamma_N^2)\\ &\quad + \sum_{j=1}^{N} \{3^{N-1} \cdot 8\gamma_j^2 + 2 \cdot 3^{N-2}(2\gamma_1^2 + \dots + 2\gamma_{j-1}^2\\ &\qquad + 2\gamma_{j+1}^2 + \dots + 2\gamma_N^2)\}\\ &= 3^{N-2}(4N + 26)(\gamma_1^2 + \gamma_2^2 + \dots + \gamma_N^2)\end{aligned}$$

or

$$\tau_N^2 = \frac{4N + 26}{6N + 9}(\gamma_1^2 + \gamma_2^2 + \dots + \gamma_N^2) = \frac{2N + 13}{2N + 3}\sigma_N^2.$$

As in the previous paragraph we would expect the smallest sum of type (B) to be about

$$b_N = \frac{\tau_N(2\pi)^{\frac{1}{2}}}{(2N + 3)3^{N-1}} = \frac{3(2N + 13)^{\frac{1}{2}}}{(2N + 3)^{3/2}} a_N. \tag{4}$$

Clearly

$$b_N \sim \frac{2\pi^{3/2} N^{1/2}}{3^N \log N} \, .$$

The values of a_N and b_N computed from formulas (3) and (4) are given in Table I, so that they can be compared with the actual values of the smallest sums of types (A) and (B).

3. Proof of the Variant Form of Ingham's Theorem

We now give the proof promised in Section 1 for the Variant Form of Ingham's Theorem.

As pointed out by Ingham (1942) the assertions (1) and (2) are immediate if ζ has a zero with real part greater than $\frac{1}{2}$ or a multiple zero with real part equal to $\frac{1}{2}$. Thus we may assume that the non-real zeros of ζ have real part $\frac{1}{2}$ and are simple. To prove (1) we put

$$A(u) = e^{-u/2}L(e^u), \qquad F(s) = \frac{\zeta(2s+1)}{(s+\frac{1}{2})\zeta(s+\frac{1}{2})}, \qquad \alpha_0 = \frac{1}{\zeta(\frac{1}{2})}.$$

To prove (2) we put

$$A(u) = e^{-u/2}M(e^u), \qquad F(s) = \frac{1}{(s+\frac{1}{2})\zeta(s+\frac{1}{2})}, \qquad \alpha_0 = 0.$$

In both cases

$$F(s) = \int_0^\infty A(u)e^{-su}\,du.$$

Ingham's idea was to consider the trigonometric sum

$$A_T^*(u) = \alpha_0 + 2\mathrm{Re} \sum_{0<\gamma_n<T} \left(1 - \frac{\gamma_n}{T}\right)\alpha_n \exp(^i\gamma_n u),$$

where α_n is the residue of $F(s)$ at $s = i\gamma_n$ $(n = 1, 2, \ldots)$ and α_0 is, as above, the residue of $F(s)$ at $s = 0$. Ingham proved that as $u \to +\infty$ we have

$$\underline{\lim}\, A(u) \leqslant \underline{\lim}\, A_T^*(u) \leqslant \overline{\lim}\, A_T^*(u) \leqslant \overline{\lim}\, A(u) \tag{5}$$

and that in both cases

$$\sum_{n=1}^{\infty} |\alpha_n| = +\infty. \tag{6}$$

If one uses Ingham's assumption that there are at most a finite number of zero sums of type (C), then (1) and (2) follow immediately from Ingham's results (5) and (6) and the Kronecker approximation theorem. (For a proof of Kronecker's theorem see Bohr and Jessen, 1932).

To get (1) and (2) from the weaker assumption that there are at most a finite number of zero sums of type (B), we argue as follows. If $\alpha_n = |\alpha_n| e^{i\phi_n}$, we may write $A_T^*(u)$ in the form

$$A_T^*(u) = \alpha_0 + 2 \sum_{0<\gamma_n<T} \left(1 - \frac{\gamma_n}{T}\right) |\alpha_n| \cos(\gamma_n u + \phi_n).$$

Now suppose M and N are positive integers satisfying the three conditions (i) $M < N$, (ii) any zero sum of type (B) involves only $\gamma_1, \gamma_2, \ldots, \gamma_{M-1}$, and (iii) $\gamma_N < T$. Let

$$Q(u) = \prod_{n=M}^{N} \{1 + \cos(\gamma_n u + \phi_n)\}$$

$$= \sum_{b_M,\ldots,b_N=-1}^{1} 2^{-|b_M|-\ldots-|b_N|} \cos\{b_M(\gamma_M u + \phi_M) + \ldots + b_N(\gamma_N u + \phi_N)\}.$$

We consider the mean-value of $A_T^*(u)\, Q(u)$, viz.

$$K = \lim_{U\to+\infty} \frac{1}{2U} \int_{-U}^{U} A_T^*(u)\, Q(u)\, du.$$

The limit defining K exists, since the limit

$$\lim_{U\to+\infty} \frac{1}{2U} \int_{-U}^{U} \cos(\beta u + \delta)\, du$$

exists for each pair of real numbers β and δ, the value of the latter limit being 0 or $\cos\delta$ according as $\beta \neq 0$ or $\beta = 0$.

On the one hand

$$K \leqslant \left\{\sup_{u \text{ real}} A_T^*(u)\right\} \cdot \lim_{U\to+\infty} \frac{1}{2U} \int_{-U}^{U} Q(u)\, du.$$

Now our assumption (ii) that any zero sum of type (B) involves only $\gamma_1, \ldots, \gamma_{M-1}$ guarantees in particular that any zero sum of type (A) involves only $\gamma_1, \ldots, \gamma_{M-1}$, i.e., there is no linear relation of the form

$$0 = b_M \gamma_M + \ldots + b_N \gamma_N \quad (b_j = -1, 0, 1)$$

except that in which $b_M = \ldots = b_N = 0$. Thus

$$\lim_{U \to +\infty} \frac{1}{2U} \int_{-U}^{U} Q(u) = 1,$$

so that

$$K \leqslant \sup_{u \text{ real}} A_T^*(u) = \overline{\lim_{u \to +\infty}} A_T^*(u). \tag{7}$$

The equality between the sup and $\overline{\lim}$ follows from the Dirichlet approximation theorem, which is proved, for example, in Hardy and Wright (1960), Section 11.12.

On the other hand our assumption (ii) that any zero sum of type (B) involves only $\gamma_1, \ldots, \gamma_{M-1}$ guarantees that there are no linear relations of the form

$$\pm \gamma_n = b_M \gamma_M + \ldots + b_N \gamma_N \quad (b_j = -1, 0, 1)$$

except for those in which $|b_M| + \ldots + |b_N| = 1$. It follows that the only terms in the product $A_T^*(u)\, Q(u)$ which contribute anything to its mean-value are the terms

$$\alpha_0 + 2 \sum_{n=M}^{N} \left(1 - \frac{\gamma_n}{T}\right) |\alpha_n| \cos^2(\gamma_n u + \phi_n).$$

Thus

$$K = \alpha_0 + \sum_{n=M}^{N} \left(1 - \frac{\gamma_n}{T}\right) |\alpha_n|. \tag{8}$$

Combining (7) and (8) and using (5) we obtain

$$\alpha_0 + \sum_{n=M}^{N} \left(1 - \frac{\gamma_n}{T}\right) |\alpha_n| = K \leqslant \overline{\lim}\, A_T^*(u) \leqslant \overline{\lim}\, A(u) \tag{9}$$

under conditions (i), (ii), and (iii) above.

Similarly by considering

$$H = \lim_{U \to \infty} \frac{1}{2U} \int_{-U}^{U} A_T^*(u)\, P(u)\, du,$$

where

$$P(u) = \prod_{n=M}^{N} \{1 - \cos(\gamma_n u + \phi_n)\},$$

we get

$$\underline{\lim}\, A(u) \leqslant \underline{\lim}\, A_T^*(u) \leqslant H = \alpha_0 - \sum_{n=M}^{N} \left(1 - \frac{\gamma_n}{T}\right) |\alpha_n| \tag{10}$$

under conditions (i), (ii), and (iii).

The device of using the kernels P and Q to derive (9) and (10) is a familiar technique in harmonic analysis going back at least to Bohr and Jessen (1932). For a good presentation of this technique see Katznelson (1968). We could have shortened the above argument somewhat by simply referring to Theorem 9.3 of Chapter VI of Katznelson's book.

Finally in (9) and (10) we may take T as large as desired while holding M and N fixed. Thus we have

$$\underline{\lim}\, A(u) \leqslant \alpha_0 - \sum_{n=M}^{N} |\alpha_n|$$

and

$$\overline{\lim}\, A(u) \geqslant \alpha_0 + \sum_{n=M}^{N} |\alpha_n|$$

for any values of M and N satisfying conditions (i) and (ii) above. In view of (6) we obtain the desired results

$$\underline{\lim}\, A(u) = -\infty, \overline{\lim}\, A(u) = +\infty$$

by letting $N \to +\infty$.

References

Bigg, M. D. (1961). The first 50 zeros of the Riemann zeta function to 50 decimal places. Royal Society Depository for Unpublished Mathematical Tables, Table No. 84.

Bohr, H. and Jessen, B. (1932). One more proof of Kronecker's Theorem. *J. London Math. Soc.* **7,** 274–275.

Feller, W. (1966). "An Introduction to Probability Theory and Its Applications", Volume II, John Wiley and Sons, Inc., New York–London–Sydney.

Hardy, G. H. and Wright, E. M. (1960). "An Introduction to the Theory of Numbers", The Clarendon Press, Oxford.

Haselgrove, C. B. (1961). The first 900 zeros of the Riemann zeta function to 12 decimal places. Royal Society Depository for Unpublished Mathematical Tables, Table No. 82.

Ingham, A. E. (1932). "The Distribution of Prime Numbers". The University Press, Cambridge.

Ingham, A. E. (1942). On two conjectures in the theory of numbers. *Amer. J. Math.* **64,** 313–319.

Katznelson, Y. (1968). "An Introduction to Harmonic Analysis". John Wiley and Sons, Inc., New York–London–Sydney–Toronto.

Spira, R. (1964). Table of the Riemann zeta function, reviewed in *Mathematics of Computation* **18,** 519–521 and deposited in the Unpublished Mathematical Tables File maintained in the editorial office of Mathematics of Computation.

An Explanation of Some Exotic Continued Fractions Found by Brillhart

H. M. STARK*

Department of Mathematics, M.I.T., Cambridge, Massachusetts, U.S.A. and The University of Michigan, Ann Arbor, Michigan, U.S.A.

1. Introduction

Late in 1964, John Brillhart embarked upon some extended computations of the continued fraction expansions of cubic irrationalities. Assisted by Michael Morrison, he was hoping that some kind of pattern would emerge—any such pattern would of course be of tremendous value if it could be proved to exist. No pattern was found but something equally unexpected occurred. The real root of the equation

$$x^3 - 8x - 10 = 0 \tag{1}$$

was found to have the continued fraction expansion:

$$x = [3, 3, 7, 4, 2, 30, 1, 8, 3, 1, 1, 1, 9, 2, 2, 1, 3, 22986, \\ 2, 1, 32, 8, 2, 1, 8, 55, 1, 5, 2, 28, 1, 5, 1, 1501790, \ldots].$$

Altogether 8 partial quotients over 10000 were found: if we write $x = [a_0, a_1, a_2, \ldots]$ then

$$a_{17} = 22986, \quad a_{33} = 1501790, \quad a_{59} = 35657, \quad a_{81} = 49405,$$
$$a_{103} = 53460, \quad a_{121} = 16467250, \quad a_{139} = 48120, \quad a_{161} = 325927.$$

This was brought to my attention by D. H. Lehmer who noted that the discriminant of (1) is $-4{\cdot}163$ and asked if the amazingly large partial quotients found were related to the fact that the class-number of $Q(\sqrt{-163})$ is one.

A. O. L. Atkin later brought to my attention a fact which greatly helped

* This research was supported in part by the N.S.F. under contract GP–13630

towards an affirmative answer to Lehmer's question. He noted that if we translate (1) by setting $x + 2 = f$, then we get the new equation

$$f^3 - 6f^2 + 4f - 2 = 0. \tag{2}$$

This equation may be found on p. 725 of Weber (1908). Its occurrence there relates to the quadratic field $Q(\sqrt{-163})$ and its form is due to the fact that the class-number of this field is one. Let us define the Schläfli modular function f by

$$f(z) = q^{-1/48} \prod_{n=1}^{\infty} (1 + q^{n-1/2}), \qquad q = e^{2\pi i z}, \qquad \operatorname{Im} z > 0. \tag{3}$$

Then $f(\sqrt{-163})$ is the real root of (2)!

2. Modular Functions and Quadratic Fields

Define $\gamma_2(z)$ by the equation

$$f(2z + 3)^{24} + \gamma_2(z) f(2z + 3)^{16} - 256 = 0 \tag{4}$$

and set

$$j(z) = \gamma_2(z)^3. \tag{5}$$

The function $j(z)$ is regular inside the upper halfplane and invariant under the full modular group, i.e.,

$$j\left(\frac{\alpha z + \beta}{\gamma z + \delta}\right) = j(z) \text{ if } \alpha\delta - \beta\gamma = 1 \text{ and } \alpha, \beta, \gamma, \delta \text{ are integers;}$$

this property and the first two terms of the expansion

$$j(z) = \frac{1}{q} + 744 + 196884q + 21493760q^2 + \ldots$$

completely determine $j(z)$.

Let $d < 0$ be the discriminant of the complex quadratic field $Q(\sqrt{d})$ and let $h(d)$ be its class-number. If 1 and ω form an integral basis of $Q(\sqrt{d})$ then the importance of $j(z)$ is illustrated by the fact that $j(\omega)$ generates the maximal unramified abelian extension of $Q(\sqrt{d})$ when it is adjoined to $Q(\sqrt{d})$. In fact $j(\omega)$ is an algebraic integer of degree exactly $h(d)$. For our purposes here, we will assume for the rest of this paper that

$$|d| \equiv 3 (\operatorname{mod} 8) \text{ and } 3 \nmid d. \tag{6}$$

In this case, we may take $\omega = \dfrac{-3+\sqrt{d}}{2}$. But now in fact $j(\omega)$ is a perfect cube; $\gamma_2(\omega)$ is also an algebraic integer of degree $h(d)$ (Weber, 1908).

Let

$$j = j\left(\frac{-3+\sqrt{d}}{2}\right), \quad \gamma = \gamma_2\left(\frac{-3+\sqrt{d}}{2}\right), \quad f = f(\sqrt{d}).$$

We see from (4) that

$$f^{24} + \gamma f^{16} - 256 = 0. \tag{7}$$

Thus f^8 is the root of a cubic equation over $Q(\gamma)$ [$= Q(j)$] and hence is an algebraic integer of degree $\leqslant 3h(d)$. In fact, with the restriction (6), f^8 is an algebraic integer of degree exactly $3h(d)$. But in fact we can do better. Weber (1908) conjectured that f is itself an algebraic integer of degree $3h(d)$. He verified this conjecture in many numerical cases, including $d = -163$, but the conjecture itself has only recently been proved by Birch (1968) (Weber did prove that f^2 has degree $3h(d)$).

For notational convenience we write the (unique) cubic equation for f^k over $Q(j)$ as

$$f^{3k} + B_k f^{2k} + A_k f^k - 2^k = 0 \tag{8}$$

where k is a positive integer. Here A_k and B_k are integers in $Q(j)$. When we deal with $d = -163$, we get $B_1 = -6$, $A_1 = 4$ which yields (2).

3. Our Basic Goal and some Numerical Tests Thereof

It is now clear that there is a relation between (1) and $Q(\sqrt{-163})$. Churchhouse and Muir (1969) have recently made some extended computations which related the successive large partial quotients of $f(\sqrt{-163})$ with the successive factors in (3). They also have a very nice discussion of what it means for several partial quotients of $f(\sqrt{-163})$ to be "larger than expected". Our aim is different from theirs. If indeed the theory of modular functions is responsible for the large partial quotients found, then we should be able to find modular functions that converge to the numerator and denominator of the corresponding convergents at $\frac{1}{2}(-3+\sqrt{-163})$ and the ratio of these modular functions as series in z should give exceptionally good approximations to $f(2z+3)$. Our goal is to find such functions.

If indeed this is possible then we would expect $f(\sqrt{d})$ to have some excellent approximations for other d satisfying (6) and $h(d) = 1$. There are four other fields which satisfy these conditions. Computations then reveal the following:

For $d = -67$, $f^3 - 2f^2 - 2f - 2 = 0$,

$$f = [2, 1, 11, 2, 3, 1, 23, 2, 3, 1, 1337, 2, 8, 3, 2, 1, 7, 4, 2, 2, 87431, \ldots].$$

For $d = -43$, $f^3 - 2f^2 - 2 = 0$,

$$f = [2, 2, 1, 3, 1, 1, 1, 1, 2, 1, 5, 456, 1, 30, 1, 3, 4, 29866, \ldots].$$

For $d = -19$, $f^3 - 2f - 2 = 0$,

$$f = [1, 1, 3, 2, 1, 95, 2, 1, 1, 2, 1, 127, \ldots].$$

For $d = -11$, $f^3 - 2f^2 + 2f - 2 = 0$,

$$f = [1, 1, 1, 5, 4, 2, 305, \ldots].$$

Note the tendency of the large a_n's to drift forwards and also to decrease. This seems a persistent enough pattern to support our goal. The modular functions that we get should converge to the numerator and denominator of the exceptional approximations in these cases also.

Next let us ask what turns out to be the key question. Why should just f have large partial quotients? Why not f^2 or f^4 or f^8, for example? In fact these other numbers also have large partial quotients. Consider the case of $d = -163$. We repeat some of the details of f for ease in comparison.

f is a root of $x^3 - 6x^2 + 4x - 2 = 0$, $a_0 = 5$,

$$a_{17} = 22986, \quad a_{33} = 1501790, \quad a_{59} = 35657,$$
$$a_{81} = 49405, \quad a_{103} = 53460, \quad a_{121} = 16467250.$$

f^2 is a root of $x^3 - 28x^2 - 8x - 4 = 0$, $a_0 = 28$,

$$a_{11} = 126425, \quad a_{31} = 8259853, \quad a_{49} = 1620,$$
$$a_{77} = 271730, \quad a_{99} = 294038, \quad a_{121} = 90569882.$$

f^4 is a root of $x^3 - 800x^2 - 160x - 16 = 0$, $a_0 = 800$,

$$a_4 = 3202800, \quad a_{12} = 209249628, \quad a_{32} = 41061,$$
$$a_{58} = 19068, \quad a_{78} = 20634, \quad a_{100} = 6355781.$$

f^8 is a root of $x^3 - 640320x^2 - 256 = 0$, $a_0 = 640320$,

$$a_1 = 1601600400, \quad a_3 = 2135467200, \quad a_7 = 20533337,$$
$$a_{19} = 9535605, \quad a_{35} = 10318433, \quad a_{55} = 3178287878.$$

Even f^{24} gets in on the act. For f^{24} we find

$$a_0 = 262537412640768767 \ (a_1 = 1), \quad a_2 = 1335334333499, \ldots.$$

Furthermore we can relate exceptionally good approximations to f^k with exceptionally good approximations to f^{2k}. This is because we may rewrite (8) as

$$f^{2k} = \frac{-A_k f^k + 2^k}{f^k + B_k} \tag{9}$$

The determinant of this linear fractional transformation (we will call it D_k) is

$$D_k = -(A_k B_k + 2^k)$$

and is generally not unity. Thus the continued fraction expansions of f^k and f^{2k} are not eventually the same but they are related. In fact since the expansions of f^k and $1/(f^k + B_k)$ are the same after the first term of the former and first two terms of the latter, and since

$$f^{2k} = -A_k - D_k \cdot \frac{1}{f^k + B_k},$$

we see that if p_n/q_n is a convergent of f^k and

$$a_{n+1} > 2|D_k|/g^2$$

where

$$g = (-A_k p_n + 2^k q_n, p_n + B_k q_n),$$

then

$$\frac{-A_k p_n + 2^k q_n}{p_n + B_k q_n}$$

is a convergent of f^{2k} (say the Nth) and a_{N+1} for f^{2k} is given by

$$a_{N+1} \approx \frac{a_{n+1}\, g^2}{|D_k|}.$$

For $d = -163$, we have $|D_1| = 22$, $|D_2| = 228$, $|D_4| = 128016$. Thus we see from the list given earlier that all of the convergents of f^k corresponding to the a_{n+1}'s listed ($k = 1, 2, 4$), except possibly those of f^4 corresponding to a_{32}, a_{58}, a_{78} yield convergents of f^{2k} and in fact the value of g is large enough in the three doubtful cases to enable them to work also. In fact the six values of a_{n+1} listed for f correspond to the six values given for f^2 which correspond to the six values given for f^4 which correspond to the six values given for f^8. This means that if we can find modular functions that converge to the numerators and denominators of the exceptional approximations to f^8, then we can do the same thing for f by using (9) inverted three times:

$$f^k = \frac{-B_k f^{2k} + 2^k}{f^{2k} + A_k} \tag{10}$$

It is interesting to note that for $d = -163$, the approximation p_{16}/q_{16} of f corresponds to the approximation $p_0/q_0 = a_0(= -\gamma)$ of f^8. If we were to pass to f^{24} by

$$f^{24} = \frac{256f^8}{f^8 + \gamma},$$

then we would completely lose this first spectacular approximation (or rather, it corresponds to $\infty = p_{-1}/q_{-1}$) while p_{32}/q_{32} of f moves up to the integer $p_1/q_1 = a_0 + 1 (= -j + 768)$ approximation to f^{24}.

Let me close this section with an acknowledgment. With the exception of Brillhart's expansion given after (1), all the computer calculations shown above were performed by Mr. Richard Schroeppel at M.I.T. (the accuracy was insufficient to trace numerically the last two of Brillhart's eight large partial quotients for f forward to f^8). In addition, Mr. Schroeppel provided me with equation (9) along with the calculations; he used (9) to find the continued fraction expansion of f^{2k} from that of f^k as a check on his calculations.

TABLE I. The continued fraction expansion of $f(\sqrt{d})^k$ for certain values of d and k. Shown are the first few values of a_n in $f(\sqrt{d})^k = [a_0, a_1, a_2, \ldots]$.

	0	1	2	3	4	5	6	7	8	9
0	1	1	3	2	1	95	2	1	1	2
10	1	127	2	2	32	1	4	1	35	1
20	1	7	3	1	1	5	9	7	4	1

$f(\sqrt{-19})$

	0	1	2	3	4	5	6	7	8	9
0	2	2	1	3	1	1	1	1	2	1
10	5	456	1	30	1	3	4	29866	2	5
20	1	3	1	7	1	3	2	5	2	2

$f(\sqrt{-43})$

	0	1	2	3	4	5	6	7	8	9
0	2	1	11	2	3	1	23	2	3	1
10	1337	2	8	3	2	1	7	4	2	2
20	87431	1	5	1	1	1	9	130	2075	1
30	158	1	5	2	1	1	4	1	22	4
40	7	1	2	1	1	2	1	3	1	1
50	3	7	1	1	8	122	1	4	15	2
60	3	1	4	1	15	1	1	7	1	1
70	1	9	1	1	4	1	2	1	4	1
80	608	155	1	3	1	2	5	1	5	1
90	5	5	17	1	7	8	3	14	3	1

$f(\sqrt{-67})$

	0	1	2	3	4	5	6	7	8	9
0	5	3	7	4	2	30	1	8	3	1
10	1	1	9	2	2	1	3	22986	2	1
20	32	8	2	1	8	55	1	5	2	28
30	1	5	1	1501790	1	2	1	7	6	1
40	1	5	2	1	6	2	2	1	2	1
50	1	3	1	3	1	2	4	3	1	35657
60	1	17	2	15	1	1	2	1	1	5
70	3	2	1	1	7	2	1	7	1	3
80	25	49405	1	1	3	1	1	4	1	2
90	15	1	2	83	1	162	2	1	1	1
100	2	2	1	53460	1	6	4	3	4	13
110	5	15	6	1	4	1	4	1	1	2
120	1	16467250	1	3	1	7	2	6	1	95

$f(\sqrt{-163})$

	0	1	2	3	4	5	6	7	8	9
0	28	3	2	9	3	18	1	6	14	13
10	2	126425	1	5	1	3	11	8	1	2
20	5	2	2	1	4	9	1	4	2	2
30	1	8259853	12	3	1	2	2	5	3	2
40	1	7	2	11	5	22	1	3	3	1620
50	1	1	5	3	5	1	1	2	2	1
60	1	1	3	1	2	1	30	2	1	3
70	3	1	4	2	1	1	4	271730	1	1
80	2	3	2	6	2	1	3	7	1	2
90	1	20	1	6	2	8	2	1	2	294038
100	4	5	1	2	2	1	1	1	1	5
110	1	5	1	3	1	2	2	3	2	4
120	2	90569882								

$f(\sqrt{-163})^2$

	0	1	2	3	4	5	6	7	8	9
0	800	5	1600	5	3202800	4	2000	2	1	1
10	570	1	209249628	2	1	1	20	5	2	1
20	1	1	2	4	5	2	1	2	1	1
30	4	8	41061	2	2	1	2	1	2	11
40	2	1	1	10	4	1	2	2	2	4
50	1	2	1	5	1	2	2	6	19068	1
60	2	17	3	22	1	12	2	1	3	4
70	1	4	1	3	1	1	5	38	20634	3
80	1	4	3	1	1	4	2	12	1	3
90	1	4	1	1	1	4	2	4	3	7
100	6355781									

$f(\sqrt{-163})^4$

	0	1	2	3	4	5	6	7	8	9
0	640320	1601600400	320160	2135467200	261949	10	1	20533337	1	1
10	4	3	1	3	1369	2	2	14	3	9535605
20	1	3	2	1	2	1	5	1	585	3
30	1	15	2	1	1	10318433	2	1	5	3
40	1	2	1	3	1	1	10	1	1	2
50	2	1	1	3	1	3178287878				

$f(\sqrt{-163})^8$

4. The Achievement of our Goal

The A_k and B_k are actually modular functions of z. However when expanded in series of fractional powers of $q = e^{2\pi iz}$, the coefficients are not always rational. This tends to be a nuisance. When we deal with f^8, we find that the coefficients of $B_8(z) = \gamma_2(z)$ are rational ($A_8(z) = 0$). Using $q = e^{2\pi iz}$, the expansion of $\gamma_2(z)$ begins,

$$\gamma_2(z) = q^{-1/3}(1 + 248q + 4124q^2 + 34752q^3 + \dots). \tag{11}$$

The expansion of $f(2z + 3)^8$ begins

$$f(2z + 3)^8 = q^{-1/3}(-1 + 8q - 28q^2 + 64q^3 - \dots). \tag{12}$$

Our object is to express $f(2z + 3)^8$ as the continued fraction

$$f(2z + 3)^8 = [a_0, a_1, a_2, \dots]$$

where the a_j are polynomials in $\gamma = \gamma_2(z)$ chosen so as to eliminate the negative (and zero) powers of q at each stage.

We set $\alpha_0 = f(2z + 3)^8$ and find α_{n+1} recursively from α_n and a_n by the usual continued fraction rule

$$\alpha_{n+1} = \frac{1}{\alpha_n - a_n}.$$

Comparing (11) and (12), we see that

$$a_0 = -\gamma$$

and thus

$$\begin{aligned}\alpha_1 &= \frac{1}{f(2z + 3)^8 + \gamma_2(z)} \\ &= \tfrac{1}{256} f(2z + 3)^{16} \\ &= \tfrac{1}{256} q^{-2/3}(1 - 16q + 120q^2 - 576q^3 + \dots)\end{aligned}$$

where it is convenient to use (1). From

$$\gamma_2(z)^2 = q^{-2/3}(1 + 496q + 69752q^2 + 2115008q^3 + \dots) \tag{13}$$

we see that we should take

$$a_1 = \tfrac{1}{256}\gamma^2$$

and now

$$\alpha_2 = \frac{1}{\alpha_1 - a_1} = -\tfrac{1}{2}q^{-1/3}(1 - 136q + 14364q^2 + \dots).$$

Thus we take

$$a_2 = -\tfrac{1}{2}\gamma$$

and hence

$$\alpha_3 = \frac{1}{\alpha_2 - a_2} = \tfrac{1}{192}q^{-2/3}(1 + \tfrac{80}{3}q + \dots).$$

Now

$$a_3 = \tfrac{1}{192}\gamma^2$$

which yields

$$\alpha_4 = \frac{1}{\alpha_3 - a_3} = -\tfrac{9}{22}q^{-1/3}(1 + \dots).$$

We now take

$$a_4 = -\tfrac{9}{22}\gamma$$

and at this point we have run out of accuracy. However, it is clear how we would proceed if we were willing to start with more terms in (11) and (12). It is also clear that a_5 will be at least of degree 2 in γ.

Let us assemble the convergents p_n/q_n which we find from the usual recursion relations. In order to get rid of needless powers of 2, we will find these in terms of

$$\Gamma = -\frac{\gamma}{8}.$$

We then find

$$p_0(\Gamma) = 8\Gamma, \quad q_0(\Gamma) = 1,$$

$$p_1(\Gamma) = 2\Gamma^3 + 1, \quad q_1(\Gamma) = \tfrac{1}{4}\Gamma^2,$$

$$p_2(\Gamma) = 8\Gamma^4 + 12\Gamma, \quad q_2(\Gamma) = \Gamma^3 + 1,$$

$$p_3(\Gamma) = \tfrac{8}{3}\Gamma^6 + 6\Gamma^3 + 1, \quad q_3(\Gamma) = \tfrac{1}{3}\Gamma^5 + \tfrac{7}{12}\Gamma^2$$

$$p_4(\Gamma) = \tfrac{96}{11}\Gamma^7 + \tfrac{304}{11}\Gamma^4 + \tfrac{168}{11}\Gamma, \quad q_4(\Gamma) = \tfrac{12}{11}\Gamma^6 + \tfrac{32}{11}\Gamma^3 + 1.$$

When we set

$$\gamma = \gamma_2\left(\frac{-3+\sqrt{d}}{2}\right),$$

we will get rational p_n and q_n if $h(d) = 1$, but not necessarily integral p_n and q_n. We can expect that the factor that we must multiply through by in order to get integers will increase with n; the approximation to f^8 compared to $q_n^{\ 2}$ is worsened by this denominator squared and thus only finitely many of the p_n/q_n should be convergents for the ordinary continued fraction expansion of $f(\sqrt{d})^8$. The number of really good approximations to $f(\sqrt{d})$ that we get will also be for this reason larger with larger $|\gamma|$ which in turn corresponds to larger $|d|$. In the five cases $d = -11, -19, -43, -67, -163$, we find that $8|\gamma$ which helps things very much. We find that $\Gamma = 4, 12, 120, 635, 80040$ respectively.

The three best approximations (of the five that we have found) to $f(2z+3)^8$ are given by

$$\frac{p_0(\Gamma)}{q_0(\Gamma)}, \qquad \frac{p_2(\Gamma)}{q_2(\Gamma)}, \qquad \frac{11p_4(\Gamma)}{11q_4(\Gamma)},$$

where in each case the numerators and denominators are (not necessarily relatively prime) integers when $d = -11, -19, -32, -67, -163$. When we trace these back to approximations to $f(2z+3)$ by applying (10), we find that for $d = -11$, the first two already give the convergents $p_3/q_3 (a_4 = 4)$ and $p_5/q_5 (a_6 = 305)$ of $f(\sqrt{-11})$ and for $d = -19, -43, -67, -163$, all three give convergents for $f(\sqrt{d})$. For each of these last four discriminants we get the first two spectacular approximations indicated in the expansions earlier but the third becomes spectacular only as $|d|$ grows. The third corresponds to p_{14}/q_{14} for $f(\sqrt{-19})$ $(a_{15} = 1)$, to p_{26}/q_{26} for $f(\sqrt{-43})$ $(a_{27} = 5)$, to p_{27}/q_{27} for $f(\sqrt{-67})$ $(a_{28} = 2075)$, and to p_{58}/q_{58} for $f(\sqrt{-163})$ $(a_{59} = 35657)$. It would be interesting to know if there is a ninth non-spectacular convergent to $f(\sqrt{-163})$ that comes from this process.

It is certainly possible to analyze all of this further to include a discussion of what multiples of the numerators and denominators we actually end up with. The numbers A_k and B_k are related to A_{2k} and B_{2k}. If we transpose the f^{2k} and constant terms of (8) to the other side of the equation and then square both sides, we find

$$B_{2k} = 2A_k - B_k^{\ 2}, \qquad A_{2k} = A_k^{\ 2} + 2^{k+1} B_k. \tag{14}$$

If we recall that $B_8 = \gamma$, $A_8 = 0$, then we find from (14) that we may set for $k = 1, 2, 4, 8$,

$$A_k = 2^k a_k, \qquad B_k = 2^{[k/2]+1} b_k, \tag{15}$$

and a_k and b_k are integers (in particular $4|\Gamma$). This conclusion is true for any discriminant d satisfying the restriction (6); we are of course then dealing with algebraic integers.

For any algebraic integer Γ, such that $2|\Gamma$, we find that

$$\big(p_0(\Gamma), q_0(\Gamma)\big) = 1, \qquad \big(p_2(\Gamma), q_2(\Gamma)\big) = 1, \qquad \big(11p_4(\Gamma), 11q_4(\Gamma)\big)|11;$$

if Γ is also rational then

$$\big(11p_4(\Gamma), 11q_4(\Gamma)\big) = \begin{cases} 1 \text{ if } 11 \nmid \Gamma \\ 11 \text{ if } 11|\Gamma. \end{cases}$$

In the cases of interest to us, this factor 11 occurs only for $d = -67$ but it makes a_{28} of $f(\sqrt{-67})$ about 121 times larger than it would otherwise have been.

Now when $4|\Gamma$ we see that

$$16|p_0(\Gamma), \qquad 16|p_2(\Gamma), \qquad 16|11p_4(\Gamma),$$
$$\big(q_0(\Gamma), 2\big) = \big(q_2(\Gamma), 2\big) = \big(11q_4(\Gamma), 2\big) = 1.$$

Suppose we have an approximation to f^8, $p(8)/q(8)$, where

$$16|p(8), (q(8), 2) = 1. \tag{16}$$

Then we get an approximation to f^4 given by (10),

$$\frac{p(4)}{q(4)} = \frac{-B_4\, p(8) + 16q(8)}{p(8) + A_4\, q(8)}.$$

Thanks to (15) and (16), we may remove a factor 16 from the numerator and denominator and take

$$p(4) = -\tfrac{1}{2}b_4\, p(8) + q(8), \qquad q(4) = \tfrac{1}{16}p(8) + a_4\, q(8)$$

(this enlarges the corresponding a_{n+1} by a factor of about 256). We see also that $\big(p(4), 2\big) = 1$. We now get an approximation to f^2,

$$\frac{p(2)}{q(2)} = \frac{-B_2\, p(4) + 4q(4)}{p(4) + A_2\, q(4)}.$$

Here we can't remove any factors of 2 and so we set

$$p(2) = -4b_2\, p(4) + 4q(4), \qquad q(2) = p(4) + 4a_2\, q(4).$$

Note that $(q(2), 2) = 1$ and $4|p(2)$. We are now ready to go to f,

$$\frac{p(1)}{q(1)} = \frac{-B_1 p(2) + 2q(2)}{p(2) + A_1 q(2)}.$$

Here we can save a factor 2 and so we take

$$p(1) = -b_1 p(2) + q(2), \qquad q(1) = \tfrac{1}{2}p(2) + a_1 q(2).$$

For $d = -19, -47, -67, -163$, we find that in each of the first three spectacular convergents, $(p(1), q(1))$ is divisible by 3 (but not 9). When $d = -11$, $(p(1), q(1))$ is not divisible by 3.

When we come to compare the first two spectacular convergents to $f(\sqrt{d})$, we note that the second is even more spectacular than the first. The reason is that although the same powers of 2 and 3 come out of $(p(1), q(1))$ for the first two convergents, the result is relatively prime for the first and usually not for the second. There are extra common factors

$$g = 7, 1, 7, 7, 7$$

in $p(1)$ and $q(1)$ for the second spectacular convergent with $d = -11, -19, -43, -67, -163$ respectively. The result is that the $a_{??+1}$ corresponding to the second spectacular convergent is about

$$\tfrac{256}{192} g^2 = \tfrac{4}{3} g^2$$

times as large as the $a_{?+1}$ corresponding to the first spectacular convergent.

5. Other Applications of our Results

While there are only five discriminants d satisfying (6) and $h(d) = 1$, there is no reason why we should restrict ourselves to these. If d satisfies (6) then we may give exceptionally good approximations to $f(\sqrt{d})$ by quotients of algebraic integers of degree $h(d)$. For example, we may find good approximations to $f(\sqrt{-427})$ (an algebraic integer of degree 6) by the quotients of two integers in $Q(\sqrt{61})$.

There is still another direction in which we may proceed. Consider the cubic equation

$$x^3 + tx^2 - 256 = 0, \tag{17}$$

where t is a negative integer, large in absolute value. Any such equation will have a unique real root and the continued fraction expansion of this root will have some partial quotients of the order of t^2. For if we let $y > \sqrt{3}$ be determined (uniquely) by

$$\gamma_2\left(\frac{-3 + iy}{2}\right) = t,$$

then the real root of (17) is $f(iy)^8$. We have even explicitly given in the last section three excellent convergents as quotients of polynomials in $\Gamma = -\frac{1}{8}t$.

When t is a large positive integer, (17) has three real roots (this is true for $t > 12$). In the last section, we have found good approximations to one of them. In fact if we determine $y > 1$ uniquely by

$$\gamma_2(iy) = t$$

then we have found good approximations to $f(3 + 2iy)^8$ which is one of the two negative roots of (17) (the one furthest removed from 0). The other two roots of (17) are of the order of $t^{-1/2}$ and hence not well approximable by polynomials in t.

Now let us consider the cubic equation

$$x^3 - 2s^2x^2 + 8sx - 16 = 0 \tag{18}$$

where s is an integer, large in absolute value. Any such equation will have a unique real root and the continued fraction expansion of this root will have some partial quotients on the order of s^5. In fact, if we determine $y > \sqrt{3}$ uniquely by

$$\gamma_2\left(\frac{-3 + iy}{2}\right) = t = 4s(4 - s^3),$$

the real root of (18) is $f(iy)^4$. We have $A_4 = 8s$, $B_4 = -2s^2$ and these numbers satisfy the relation (14) (with $k = 4$) where $B_8 = 4s(4 - s^3)$, $A_8 = 0$. Since

$$|D_4| = 16s^3 - 16,$$

and since we have approximations to $f(iy)^8$ with partial quotients of the order of at least t^2 which is of the order of s^8, we see that we get approximations to $f(iy)^4$ with partial quotients of the order of at least s^5.

An example of such an equation is given by $s = 60$ (the value of y is then transcendental and thus certainly not connected with quadratic fields). We then have the equation

$$x^3 - 7200x^2 + 480x - 16 = 0.$$

Since $2|x$, we may set $x = 2X$ and the equation for X is then,

$$X^3 - 3600X^2 + 120X - 2 = 0.$$

The people at Atlas very kindly furnished the continued fraction expansion of the real root of this equation. It is given in Table II. Besides the expected very large partial quotients we note some others in the neighborhood of 10000. These do not come from the expansion of $f(iy)^8$. We do get partial quotients for $f(iy)^4$ on the order of s from $f(iy)^8$ but the partial quotients on the order of 10000 for $f(iy)^4$ originate with $f(iy)^4$ and come from expanding $f(iy)^4$ in a continued fraction with partial quotients being polynomials in s.

TABLE II. The continued fraction expansion of the real root of $X^3 - 3600X^2 + 120X - 2 = 0$. Shown are the values of a_n in $X = [a_0, a_1, a_2, \dots]$.

	0	1	2	3	4	5	6	7	8	9
0	3599	1	28	1	7198	1	29	388787400	23	1
10	8998	1	13	1	10284	1	2	25400776804	1	1
20	3	4	93	3	1	2	11	1	9	1
30	99	1	3	1	3	9	1	603118914	1	1
40	2	24	1	1	3	2	1	1	2	2
50	1	1	26	1	8	1	18	1	2	2
60	1	2	1	1	3	9	3	2	1	2314761
70	6	1	2	5	5	61	1	1	4	1
80	1	5	1	22	1	4	2	1	1	1
90	9	2	1	1	2	1	2	2	1	1
100	12	1709319								

In view of all this, why did Brillhart come up with a cubic related to a quadratic field? The answer is that the magnitude of the coefficients involved in his search made sure that any spectacular approximations to $f(iy)^k$ covered by our discussion here that he might find would come with $k = 1$ (and conceivably 2). While there are infinitely many $y \geqslant \sqrt{3}$ such that $f(iy)^8$ is the root of a cubic equation of the form

$$x^3 + tx^2 - 256 = 0$$

with t an integer, and still infinitely many such y with the additional restriction that $f(iy)^4$ should generate a cubic extension of the rationals, there are only six values of y satisfying all this and having $f(iy)^2$ generate a cubic extension of the rationals. This is because the recursion relations (14) are very restrictive ($k = 2, 4$). They give a set of Diophantine equations with only six solutions (the corresponding y being $\sqrt{3}$, $\sqrt{11}$, $\sqrt{19}$, $\sqrt{43}$, $\sqrt{67}$, $\sqrt{163}$).

This is in fact the Heegner (1952) approach to proving that there are only nine values of d with $h(d) = 1$. [See also Birch (1968), Deuring (1968) and Stark (1969)].

We close by mentioning one more aspect of all this. Suppose $X = X(s)$, $Y = Y(s)$ are polynomials in s (with complex coefficients) of degrees n and $n - 2$ respectively. How small can we make the degree of

$$F(X, Y) = X^3 - 2s^2X^2Y + 8sXY^2 - 16Y^3?$$

Since there are $2n$ unknown coefficients we would expect that the degree of $F(X, Y)$ (in s) would be at least $n + 1$; if it were any smaller, we would have $2n$ homogeneous equations in the $2n$ coefficients and would expect the $2n$ unknowns to be all zero. In fact our expectations are wrong. For infinitely many n, there exist polynomials X and Y (with integral coefficients) such that the degree of $F(X, Y)$ in s is less than or equal to $n - 3$. For example, if

$$X(s) = 2s^6 - 8s^3 + 4, \qquad Y(s) = s^4 - 2s$$

then

$$F(X, Y) = -64(s^3 - 1).$$

I hope to say more about this in the future.

References

Birch, B. J. (1968). Diophantine analysis and modular functions, Proceedings of the conference on algebraic geometry, held at Tata Institute, Bombay, pp. 35–42.

Churchhouse, R. F. and Muir, S. T. E. (1969). Continued fractions, algebraic numbers and modular invariants, *J.I.M.A.* **5**, 518–328.

Deuring, M. (1968). Imaginäre Quadratische Zahlkörper mit der Klassenzahl Eins, *Inventiones Math.* **5**, 169–179.

Heegner, K. (1952). Diophantische Analysis und Modulfunktionen, *Math. Z.* **56**, 227–253.

Stark, H. M. (1969). On the "gap" in a theorem of Heegner, *J. Number Theory*, **1**, 16–27.

Weber, H. (1908). "Lehrbuch der Algebra", Vol. 3, reprinted 1961, Chelsea, New York.

Some Numerical Computations Relating to Automorphic Functions

P. CARTIER

University of Strasbourg, France

1. We denote by G the group of all 2 by 2 real matrices with determinant one and by Γ the subgroup consisting of the integral matrices in G. Then G is a three-dimensional real Lie group, Γ is a discrete subgroup of G, and the homogeneous space G/Γ is noncompact and of finite invariant measure. Let $\mathfrak{H}$ be the Hilbert space $L^2(G/\Gamma)$ of the square-integrable functions on G/Γ; we consider the elements of $\mathfrak{H}$ as functions f on G satisfying the relation $f(g\gamma) = f(g)$ for g in G and γ in Γ. We consider $\mathfrak{H}$ as the carrier of the left regular representation η of the group G given by $(\eta_g f)(g') = f(g^{-1} g')$.

One of the main problems in the theory of the automorphic functions is to decompose the representation $(\eta, \mathfrak{H})$ *of* G *into a direct sum and/or direct integral of irreducible unitary representations.* This problem has been dealt with in depth by Maass, Roelcke, Selberg, Gelfand, and Godement, and the best account of the known results is due to Godement. Before stating the results having relevance here, we remind the reader about the definition of the cuspidal functions and the classification of the irreducible unitary representations of G.

For every real number t we denote by u_t the matrix

$$\begin{pmatrix} 1 & t \\ 0 & 1 \end{pmatrix}$$

in G. From the equation $u_{t+t'} = u_t\, u_{t'}$ and the fact that u_n belongs to Γ for integral n, one deduces immediately that $f(gu_t)$ has period one in t for f in $\mathfrak{H}$ and g in G. Since any function f in $\mathfrak{H}$ is locally integrable on G, one sees from Fubini's theorem the existence of a Fourier expansion

$$f(gu_t) = \sum_{n=-\infty}^{\infty} c_n(g)\, e^{2\pi i n t}. \tag{1}$$

The function f in H is called *cuspidal* if $c_0(g) = 0$ for almost all g in G, which amounts to

$$\int_0^1 f(gu_t)\, dt = 0$$

for almost all g in G. The cuspidal functions form in $\mathfrak{H}$ a closed subspace $\mathfrak{H}_1$ invariant under $\eta(G)$. We denote by $\mathfrak{H}_2$ the orthogonal complement of $\mathfrak{H}_1$ in $\mathfrak{H}$, this space being also invariant under $\eta(G)$.

Let K be the rotation group in two variables consisting of the matrices

$$\begin{pmatrix} a & b \\ -b & a \end{pmatrix}$$

with a, b real and $a^2 + b^2 = 1$. An irreducible unitary representation of G is said to be of *class one* if there is in the representation space a nonzero vector invariant under K. The representations of class one form the *principal series* and the *complementary series* both depending on a continuous parameter, and the remaining representations form the *discrete series*. The only series of relevance to us is the principal series which we describe briefly now.

Let s be a positive real number. By $\mathfrak{K}_s$ we mean the space of all measurable functions f of two real variables subject to the conditions

$$f(tx, ty) = t^{-1+2is} f(x, y) \qquad (\text{for } t > 0) \tag{2}$$

$$2\, \|f\|^2 = \int_{-\infty}^{+\infty} |f(x, 1)|^2\, dx + \int_{-\infty}^{+\infty} |f(x, -1)|^2\, dx\theta. \tag{3}$$

This is a Hilbert space under the norm $\|f\|$ defined by (3). The space $\mathfrak{K}_s$ splits as the orthogonal sum of the subspaces $\mathfrak{K}_s^+$ and $\mathfrak{K}_s^-$ consisting respectively of the even and odd functions in $\mathfrak{K}_s$. If

$$g = \begin{pmatrix} a & b \\ c & d \end{pmatrix}$$

belongs to G, we define the unitary operator $\pi_s(g)$ in $\mathfrak{K}_s$ as transforming the function $f(x, y)$ into the function $f(ax + cy, bx + dy)$. We let $\pi_s^+(g)$ and $\pi_s^-(g)$ denote respectively the restrictions of $\pi_s(g)$ to $\mathfrak{K}_s^+$ and $\mathfrak{K}_s^-$. The representations $(\pi_s^+, \mathfrak{K}_s^+)$ and $(\pi_s^-, \mathfrak{K}_s^-)$ are unitary and irreducible for every real positive s, and when s varies they comprise the principal series.

We can now formulate the main results of the theory:

(a) *The representation η of G in the space $\mathfrak{H}_1$ of cuspidal functions is a direct sum of irreducible representations, a given irreducible representation having finite multiplicity.*

(b) *The representation* η *of* G *in* $\mathfrak{H}_2$ (the space orthogonal to the cuspidal functions) *is unitarily equivalent to the direct integral*

$$\oplus \int_{-\infty}^{+\infty} (\pi_s^+, \mathfrak{R}_s^+)\, ds,$$

the isomorphism being given explicitly by Eisenstein–Epstein series.

2. The representation of G in the space $\mathfrak{H}_2$ is well described by the isomorphism referred to in (b) above. The situation is less clear concerning the space of cuspidal functions. It has been shown by Gelfand how to relate precisely the classical (holomorphic) automorphic forms of integral dimensions to the occurence of the representations of the discrete series in the space of cuspidal functions. The major problems up to date can be roughly classified as follows:

(A) *Prove that the representations of the complementary series do not occur in the space of cuspidal functions.*

(B) *Prove that the multiplicity of a representation of the principal series in the space of cuspidal functions is at most one.*

(C) *Find the significance of the set of all real positive numbers such that* $(\pi_s^+, \mathfrak{R}_s^+)$ *does occur in the space of cuspidal functions* (the representation $(\pi_s^-, \mathfrak{R}_s^-)$ cannot occur for obvious reasons).

A conceptual proof of (A) and (B) seems to be out of sight at the time being and the numbers involved in (C) are highly mysterious. D. K. Faddeev suggested to me that I should try to get some numerical evidence about these problems. We report our results in some detail now. What has been obtained so far gives *some support to the validity of conjecture* (B), while conjecture (A) can be considered as having been "established" experimentally. Our present results are too imprecise to throw any light on problem (C) (see commentary at the end).

3. We shall first transform the problem into a form better suited for numerical computation. Let Π be the Poincaré half-plane consisting of the complex numbers $z = x + iy$ with $y > 0$. We let the group G act to the right on Π by

$$z \,.\, g = (az + c)/(bz + d) \quad \text{for} \quad g = \begin{pmatrix} a & b \\ c & d \end{pmatrix}.$$

The differential operator

$$\Delta = -y^2 \left(\frac{\partial^2}{\partial x^2} + \frac{\partial^2}{\partial y^2} \right)$$

as well as the differential form $dxdy/y^2$ are invariant under G. The set

$$M = \{z = x + iy \mid |x| \leqslant 1/2, x^2 + y^2 \geqslant 1\}$$

is a fundamental domain for the action of the discrete subgroup Γ of G acting on Π.

We let also $\mathfrak{K}$ denote the Hilbert space of measurable functions f on Π such that $f(z \,.\, \gamma) = f(z)$ for z in Π and γ in Γ and

$$\|f\|^2 = \iint_M |f(x + iy)|^2 \, dx \, dy / y^2 < \infty. \tag{4}$$

Let $\mathfrak{K}_0$ be the dense subspace of $\mathfrak{K}$ consisting of the twice continuously differentiable functions vanishing in some half-plane of the form

$$\{z = x + iy \mid y \geqslant y_0 > 0\}.$$

Then Δ maps $\mathfrak{K}_0$ into $\mathfrak{K}$ and is an essentially self-adjoint operator in $\mathfrak{K}$ with that domain $\mathfrak{K}_0$, its spectrum being of the form $[0, 1/4] \cup \{\lambda_n \mid n \geqslant 1\}$ with

$$0 < \lambda_1 < \lambda_2 < \lambda_3 < \lambda_4 \ldots \quad \text{and} \quad \lim_{n \to \infty} \lambda_n = \infty.$$

We denote by $\mu(n)$ the multiplicity of the eigenvalue λ_n.

Easy group-theoretic arguments show the following:

(a) The above conjecture (A) is equivalent to $\lambda_1 > 1/4$.

(b) Let $\lambda_n > 1/4$ and put $\lambda_n = s_n{}^2 + 1/4$. Then the representation $(\pi_{s_n}{}^+, \mathfrak{K}_{s_n}{}^+)$ occurs in the space of cuspidal functions with multiplicity $\mu(n)$ and we get in this way all irreducible components of the principal series in $\mathfrak{H}_1$. (In the sequel we shall speak of the "eigenvalues" s_n.)

A further reduction of the problem is due to the fortunate existence of the symmetry σ given by $\sigma(x + iy) = -x + iy$ leaving invariant the set M as well as the differential operator Δ. We split therefore the Hilbert space $\mathfrak{K}$ as $\mathfrak{K}^+ \oplus \mathfrak{K}^-$ (resp. $\mathfrak{K}^-$) consists of the functions $f(x, y)$ in $\mathfrak{K}$ which are even (resp. odd) with respect to x. Let N be the right half of M given by

$$N = \{z = x + iy \mid 0 \leqslant x \leqslant 1/2, x^2 + y^2 \geqslant 1\}$$

whose boundary we call B (see Fig. 1). The set B consists of the vertical straight lines B_1 and B_3 and the circular line B_2. The corresponding Dirichlet problem asks for the pairs (f, λ) where $f \neq 0$ is a real function in N and λ a real number such that the following holds:

(D_1) The function f is twice continuously differentiable and satisfies (4).

(D_2) $\Delta f = \lambda f$.

(D_3) f vanishes at infinity in the sense $\lim_{y \to \infty} f(x + iy) = 0$ uniformly for x between 0 and 1/2.

(D_4) f vanishes identically on B.

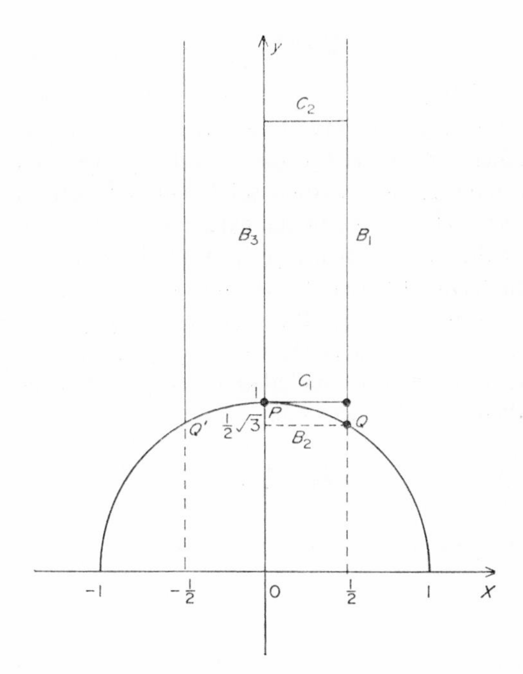

FIGURE 1.

The spectrum of Δ in $\mathfrak{R}^+$ is the set of corresponding eigenvalues λ. Similarly, the spectrum of Δ in $\mathfrak{R}^-$ is given by the Neumann problem (N) with conditions (N_1), (N_2) and (N_3) identical to (D_1), (D_2) and (D_3) respectively and (N_4) asserting that the normal derivative of f along B is identically 0. The sought-for spectrum of Δ in $\mathfrak{R}$ is the union of the spectra for the Dirichlet and Neumann problems.

4. Before going into the details of the computations, a few remarks are in order. The preceding Dirichlet and Neumann problems may seem to be of no particular scientific value, since similar boundary value problems for second-order elliptic differential operators are solved everyday in a routine way for technological or engineering purposes. But our demands were very drastic. Anything short of say 15 eigenvalues for each problem with a relative precision of 10^{-4} was worthless for our purposes whereas most practical methods of calculation give only a few eigenvalues with 10^{-2} or 10^{-3} accuracy.

Our domain N has two annoying features: the infinitely remote part of the boundary, and the acute angle at the vertex

$$Q = \frac{1 + i\sqrt{3}}{2}.$$

A preliminary investigation has shown that the boundary condition at infinity was rather innocuous for the lowest eigenvalues and that in the spectral region under study we could safely cut N by an horizontal line C_2 at the height 5 say. This is due to the existence of a strong damping factor going to 0 faster than $e^{-2\pi y}$ when y goes to infinity (see formula (8)).

In order to understand better the second difficulty, one has to remember the principle according to which the smaller the drum the higher the tone. More precisely, if one considers the Dirichlet problem for the operator Δ with two domains D, and D' included in D (under suitable smoothness assumptions), then the nth eigenvalue

$$\lambda_n = \tfrac{1}{4} + s_n{}^2$$

(counted including multiplicity) for D is majorized by the corresponding eigenvalue

$$\lambda'_n = \tfrac{1}{4} + s'_n{}^2$$

for D'. Moreover by the Weyl–Courant asymptotic formula, there are approximately St/π eigenvalues λ_n below t for large t where S is the hyperbolic area $\iint_D dxdy/y^2$ of D and a similar estimate holds for D' and its hyperbolic area S'. One expects therefore λ_n/λ_n' to be of the order of magnitude of S'/S and s_n/s_n' of the order of $(S'/S)^{\frac{1}{2}}$. Any method of numerical computation has to approximate in some way the domain N and the previous remarks show that for the required accuracy any unclever approximation may be disastrous. The factor $1/y^2$ in $S = \iint_D dxdy/y^2$ explains also why the eigenvalues respond wildly to any manipulation in the neighbourhood of Q whereas cutting the highest part of N is relatively harmless.

5. Our first method, a very crude one, consisted of replacing N by the rectangle R defined by the inequalities $0 \leqslant x \leqslant 1/2$, $1 \leqslant y \leqslant 5$. The method of separation of variables applies easily and gives eigenfunctions of the following form (with the self-explanatory indices D and N for the two boundary problems):

$$f(x+iy) = \sin 2\pi px \, . \, g(2\pi py) \tag{5}_D$$

$$f(x+iy) = \cos 2\pi px \, . \, g(2\pi py). \tag{5}_N$$

The function g is a solution of the Sturm–Liouville equation

$$g''(y) = \left(1 - \frac{s^2 + \frac{1}{4}}{y^2}\right) . \, g(y) \tag{6}$$

with the boundary conditions

$$g(2\pi p) = 0, \qquad g(10\pi p) = 0 \tag{7}_D$$

$$g'(2\pi p) = 0, \qquad g(10\pi p) = 0. \tag{7}_N$$

We may classify the eigenfunctions according to their mode, the above eigenfunction being of mode (p, q) in case the zeros of g divide the interval $[2\pi p, 10\pi p]$ into q subintervals.

The differential equation (6) is germane to the Bessel equation and its solutions can easily be expressed in terms of the Hankel functions $H_{iv}(ix)$ with parameter and argument being both purely imaginary. By an unfortunate oversight, these functions are not discussed in the classical text-books on Bessel functions. They play for hyperbolic equations the same role as the Bessel functions in the ordinary elliptic case and I suspect that some physicist working with relativistic elementary particles has met and computed them. But I have not been able to find a table of numerical values.

To solve the previous Sturm–Liouville problem, I have used the classical *shooting method*. Starting from the initial value $y = 2\pi p$ one computes the two solutions of (6) with initial values as in (7) by the Runge–Kutta method in the range $y = 2\pi p$ to $y = 10\pi p$ for different values of s. The final values for $y = 10\pi p$ are expressed as $G_D{}^p(s)$ and $G_N{}^p(s)$ in their dependence on s. The qth positive zero $s_{p,q}$ of the function $G_D{}^p$ gives the "eigenvalue" of type D and mode (p, q) and similarly for type N (the true eigenvalue is $s^2 + 1/4$). Any solution of (6) has an asymptotic expansion

$$g(y) \sim a \, . \, e^y + b \, . \, e^{-y} \qquad \text{for} \qquad y \to \infty. \tag{8}$$

Ideally, we are looking for the solution which vanishes at infinity, that is for which $a = 0$ and this special solution exhibits the damping factor mentioned before in Section 4. But the slightest numerical error will reintroduce in the true solution a term $a \,.\, e^y$ which will overwhelm the rest even for moderate values of y (and $10\pi p$ is quite large!). This well-known inherent instability makes $G_D{}^p(s)$ (or $G_N{}^p(s)$) vary very fast in the neighbourhood of an "eigenvalue", which is a very fortunate circumstance for the calculation of this "eigenvalue", provided one can rely on the numerical integration of (6). The phenomenon is clearly illustrated in Fig. 2 for the smallest "eigenvalue" of type D.

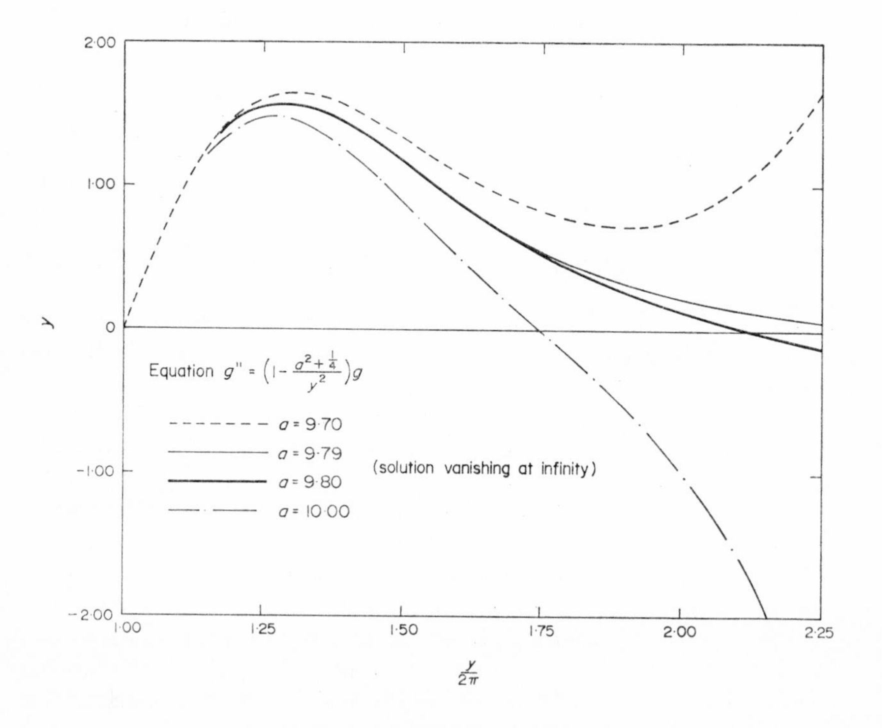

FIGURE 2.

We show in Table I the "eigenvalues" computed by this method. We recall that the "eigenvalues" of type D for the rectangle R majorize the "eigenvalues" of type D for the domain N; the square root of the ratio of the hyperbolic area of R to that of N is $(12/5\pi)^{\frac{1}{2}} = 0{\cdot}87 \ldots$.

TABLE I. "Eigenvalues" for a rectangle.

n	"Eigenvalue" s_n	mode	type	n	"Eigenvalue" s_n	mode	type
1	8·906	1,1	*N*	16	20·309	2,2	*D*
2	9·790	1,1	*D*	17	20·923	1,7	*N*
3	11·631	1,2	*N*	18	21·512	1,7	*D*
4	12·421	1,2	*D*	19	22·222	2,3	*N*
5	13·863	1,3	*N*	20	22·443	1,8	*N*
6	14·590	1,3	*D*	21	23·006	1,8	*D*
7	15·837	1,4	*N*	22	23·021	2,3	*D*
8	16·108	2,1	*N*	23	23·094	3,1	*N*
9	16·522	1,4	*D*	24	23·898	1,9	*N*
10	17·019	2,1	*D*	25	23·986	3,1	*D*
11	17·644	1,5	*N*	26	24·441	1,9	*D*
12	18·291	1,5	*D*	27	24·666	2,4	*N*
13	19·330	1,6	*N*	28	25·304	1,10	*N*
14	19·465	2,2	*N*	29	25·431	2,4	*D*
15	19·948	1,6	*D*				

6. The second method has been devised by my colleague P. Mignot, of the Centre de Calcul de l'Université de Strasbourg. Both the numerical procedure and the actual programming are due to him.

By a classical integration by parts, the differential equation $\Delta f = \lambda f$ is transformed into the integral form

$$\iint_N \left(\frac{\partial f}{\partial x}\cdot\frac{\partial g}{\partial x} + \frac{\partial f}{\partial y}\cdot\frac{\partial g}{\partial y}\right) dx\,dy = \lambda \iint_N fg\,dx\,dy/y^2 \tag{9}$$

where g runs over all continuously differentiable functions vanishing off a compact subset of the interior of N. In order to discretize the equation (9), we first choose a maximum ordinate Y and subdivide the intervals $[0,1/2]$ and $[\sqrt{3}/2, Y]$ by interpolation points

$$0 = x_0 < x_1 < \ldots < x_{m-1} < x_m = 1/2$$

$$\frac{\sqrt{3}}{2} < y_0 < y_1 < \ldots < y_{n-1} < y_n = Y.$$

We then consider the rectangular grid with vertices $M_{ij} = (x_i, y_j)$ for $0 \leqslant i \leqslant m$ and $0 \leqslant j \leqslant n$. To every vertex M_{ij} with $1 \leqslant i \leqslant m-1$ and $1 \leqslant j \leqslant n-1$ we associate the cross-shaped region C_{ij} which is the union of the rectangles

$$[\tfrac{1}{2}(x_{i-1} + x_i), \tfrac{1}{2}(x_i + x_{i+1})] \times [y_{j-1}, y_{j+1}]$$

and

$$[x_{i-1}, x_{i+1}] \times [\tfrac{1}{2}(y_{j-1} + y_j), \tfrac{1}{2}(y_j + y_{j+1})].$$

The vertices M_{ij} for which C_{ij} is contained in N form a finite set X and to every M_{ij} in X we associate the rectangle

$$R_{ij} = [\tfrac{1}{2}(x_{i-1} + x_i), \tfrac{1}{2}(x_i + x_{i+1})] \times [\tfrac{1}{2}(y_{j-1} + y_j), \tfrac{1}{2}(y_j + y_{j+1})].$$

Let V denote the vector space of all functions on N which are constant on each of the rectangles R_{ij} for M_{ij} in X and vanish outside the union of these rectangles. For any f in V, the approximate derivative $D_x f$ takes the constant value

$$\frac{f(x_i, y_j) - f(x_{i-1}, y_j)}{x_i - x_{i-1}}$$

on the rectangle $[x_{i-1}, x_i] \times [\tfrac{1}{2}(y_{j-1} + y_j), \tfrac{1}{2}(y_j + y_{j+1})]$ for $1 \leqslant i \leqslant m$ and $0 \leqslant j \leqslant n$ and vanishes outside the union of these rectangles. The approximate derivative $D_y f$ is defined similarly. Notice that the intersection of two among the considered rectangles is a null set and that the exact value of f, $D_x f$ or $D_y f$ on such a set does not matter. The approximate problem is to find a function $f \neq 0$ in V and a real number λ such that the equation

$$\textstyle\iint_N (D_x f \cdot D_x g + D_y f \cdot D_y g)\, dx\, dy = \lambda \iint_N fg\, dxdy/y^2 \qquad (10)$$

holds for all g in V.

In the actual computations, the horizontal steps $x_i - x_{i-1}$ were taken equal to the constant step $h = 1/2m$, whereas the vertical steps $h_j = y_j - y_{j-1}$ were variable. By tedious routine computations, one transforms the equation (10) into the linear system

$$\frac{1}{d_j^2}\left[\frac{h_j + h_{j+1}}{2h} + \frac{h}{h_{j+1}} + \frac{h}{h_j}\right] f_{ij} - \frac{1}{d_j^2}\frac{h_j + h_{j+1}}{2} \cdot f_{i+1,j}$$

$$- \frac{1}{d_j^2}\frac{h_j + h_{j+1}}{2} \cdot f_{i-1,j} - \frac{1}{d_j\, d_{j+1}}\frac{h}{h_{j+1}} \cdot f_{i,j+1}$$

$$- \frac{1}{d_{j-1} d_j}\frac{h}{h_j} \cdot f_{i,j-1} = \lambda f_{ij} \qquad (11)$$

where the unknowns $f_{i,j}$ are the values of the unknown function f at the vertices M_{ij} in X multiplied by d_j. The coefficient d_j is given by

$$d_j = \left[\frac{2h(h_j + h_{j+1})}{(y_{j-1} + y_j)(y_j + y_{j+1})}\right]^{\frac{1}{2}}. \qquad (12)$$

We enumerate the relevant pairs (i,j) in such a way that (i,j) precedes (i',j') whenever one has either $j < j'$ or $j = j'$ and $i < i'$. Then the system (11) is associated to the eigenvalue problem for a symmetric positive definite coefficient matrix B, and since any vertex in X has only four neighbours, B is a band matrix with only $2m - 1$ nonvanishing descending diagonals.

It is therefore advisable to keep the number m of horizontal steps within reasonable bounds.

The eigenvalue problem was solved by the Ruttishauser–Schwartz procedure. The following Table II gives the numerical results obtained for the Dirichlet problem in the one among our trials we consider the most accurate. In this trial, we had 1,903 vertices and the horizontal step $h = 0{\cdot}05$ (that is $m = 10$). We had $y_0 = 0{\cdot}925$ and $Y = 10$ and the following vertical steps:

43 steps of 0·0125 from $y = 0{\cdot}925$ to $y = 1{\cdot}5$

60 steps of 0·025 from $y = 1{\cdot}5$ to $y = 3$

80 steps of 0·05 from $y = 3$ to $y = 7$

30 steps of 0·1 from $y = 7$ to $y = 10$.

A comparison with other trails shows that decreasing Y from 10 to 7 by leaving aside 270 vertices does nothing to the first 9 digits of the eigenvalues, but decreasing h from 1/20 to 1/30 made a change of about 3×10^{-3}. In Table II, we consider the first three digits as reasonably significant, but not the fourth.

In Table III, we give the result of a similar computation for the Neumann problem, involving 2,026 vertices. In the output from the machine is included a bogus, very small, "eigenvalue", which arises from the discretization and has no significance.

TABLE II. "Eigenvalues" for the Dirichlet problem.

n	s_n	n	s_n
1	9·47	9	21·26
2	12·13	10	21·86
3	14·34	11	22·66
4	16·07	12	23·21
5	16·55	13	24·31
6	18·09	14	24·85
7	19·12	15	25·66
8	19·79		

TABLE III. "Eigenvalues" for the Neumann problem.

n	s_n	n	s_n
1	2·25	8	10·69
2	3·86	9	11·80
3	5·42	10	12·53
4	6·73	11	13·53
5	7·38	12	13·67
6	8·96	13	14·77
7	10·03	14	15·40

7. A comparison of Tables II and III with Table I show that the crude method was unexpectedly good for the Dirichlet problem (especially for the first 8 eigenvalues) but gave a completely distorted view of the Neumann problem. Also, the numerous clusters of eigenvalues for the rectangle disappear in the more exact results, and this gives a serious hope for conjecture (B).

Our plans for future calculations include revisiting the eigenvalues for the rectangle. By considering the inner rectangle $R = [0, 1/2] \times [1, \infty]$ and the outer rectangle $R' = [0, 1/2] \times [\sqrt{3}/2, \infty]$ we can get lower and upper bounds for each eigenvalue. To avoid the instability mentioned above, we shall integrate the differential equation (6) by a backwards Runge–Kutta method, using as a starting point an asymptotic expansion of $g(y)$ for y large. Sufficiently refining this method, we shall be able to offer a *numerical proof* for conjecture (A).

The second method has already given its best, until the advent of computers large enough to deal with 20,000 or more vertices. Future calculations will take into account the fact that *the eigenfunctions of* Δ *are also eigenfunctions of the Hecke operators* and have associated Dirichlet series with an Eulerian product. It is expected that calculations along these lines should be much more accurate.

Automorphic Integrals with Preassigned Period Polynomials and the Eichler Cohomology

JOSEPH LEHNER

University of Maryland and National Bureau of Standards, Washington, D.C., U.S.A.

1. The integrals of automorphic forms were studied systematically by Eichler (1957). He introduced a cohomology based on the period polynomials of the integrals and proved an interesting theorem relating the cohomology group to certain spaces of cusp forms. Further results were obtained by Gunning (1961) and new proofs of the Eichler–Gunning theorems were supplied by Husseini and Knopp (to be published).

In the present paper we offer another proof of these theorems. We use an elementary construction involving Poincaré series to prove the existence of an automorphic integral with preassigned periods. This construction was suggested and carried out in a special case by Eichler himself (1965). The existence theorem shows that a certain map is onto. The proof that it is 1–1 is the same as in the Husseini–Knopp paper and is made by application of the "supplementary function," a device introduced by Knopp (1962).

Professor Knopp has meanwhile generalized the above proof from automorphic forms of integral degree to forms of arbitrary real degree.

2. Let H be the upper half-plane $\{\tau; \operatorname{Im} \tau > 0\}$ and let Γ be a discrete group of linear-fractional transformations acting on H. We write the elements of Γ in the form $A\tau = (a\tau + b)/(c\tau + d)$ where a,b,c,d are real and $ad - bc = 1$. Then Γ is isomorphic to the matrix group $\{A = (ab \mid cd)\}$ if we identify A and $-A$.

We call a finitely-generated group Γ an H-group provided it contains translations and provided every real number is an accumulation point of images $A(\infty)$ with $A \in \Gamma$. The real points fixed by parabolic elements of Γ are called *cusps*.

A *multiplier system* on Γ is simply a complex character $v = v(A)$, $|v| = 1$, $A \in \Gamma$. Let s be an integer. We define a *stroke operator* with respect to Γ, s, and v:

$$(f|_s A)(\tau) \equiv f|_s A = \bar{v}(A)(c\tau + d)^s f(A\tau), \qquad A = (\,.\,.\mid cd) \in \Gamma \qquad (1)$$

where $\bar{v}$ denotes the complex conjugate. Note that

$$f|_s AB = (f|_s A)|_s B, \tag{2}$$

as a consequence of the fact that v is a character.

Let $f(\tau)$ be a function meromorphic in H and at the parabolic cusps and let f satisfy

$$f|_s A = f \tag{3}$$

for all $A \in \Gamma$. Then we say f is an automorphic form on Γ of degree s with multiplier system v, and we write

$$f \in \{\Gamma, s, v\}. \tag{4}$$

Meromorphicity at a cusp p means the expansion

$$e(-\kappa_p \tau/\lambda_p) f|_s V(\tau) = \sum_{m \geqslant M} a(m)\, e(m\tau/\lambda_p) \tag{5}$$

has only a finite number of terms with negative exponents (Lehner (1964), 272–273). Here

$$e(u) = \exp(2\pi i u),$$

$V = V_p \in SL(2, R)$, $V_{(\infty)} = p$, λ_p is the "width" at p, and κ_p is defined by

$$v(P) = e(\kappa_p), \qquad 0 \leqslant \kappa_p < 1$$

where P generates the stabilizer of p in Γ. The right member of (5) is called the Fourier series of f at p. When $p = i_\infty$ we have $P = (1\lambda \mid 01)$; we can choose $V = I$ and (5) becomes the usual Fourier series of a function with period λ.

The subspace of cusp forms, i.e., forms that vanish at each cusp, is denoted by $C^0(\Gamma, s, v)$.

When $f \in \{\Gamma, 0, v\}$, then $f' = df/d\tau \in \{\Gamma, -2, v\}$, since $A'(\tau) = (c\tau + d)^{-2}$. A generalization of this remark was made by G. Bol (1949). For any analytic F we have

$$\frac{d^{r+1}}{d\tau^{r+1}} \{(c\tau + d)^r F(A\tau)\} = (c\tau + d)^{-r-2} F^{(r+1)}(A\tau), \tag{6}$$

when r is a nonnegative integer. If in addition $F|_r A = F$, (6) gives

$$F \in \{\Gamma, r, v\} \quad \text{implies} \quad F^{(r+1)} \in \{\Gamma, -r-2, v\}. \tag{7}$$

Now we are ready to discuss *integrals*. Let r be a fixed positive integer. Suppose $\Phi(\tau)$ is meromorphic in H and at the cusps, and suppose

$$\Phi|_r A = \Phi + \omega_A, \qquad A \in \Gamma \tag{8}$$

ω_A being a polynomial (depending on A) of degree $\leqslant r$. We call Φ an *automorphic integral on* Γ *of degree* r *and multiplier system* v, and we write

$$\Phi \in I(\Gamma, r, v). \tag{9}$$

From (6) we deduce

$$\Phi \in I(\Gamma, r, v) \text{ if and only if } \Phi^{(r+1)} \in \{\Gamma, -r-2, v\}. \tag{10}$$

This explains the name "integral."

Again we need to define "meromorphic at the cusp p." With each Φ satisfying (8) we associate $\phi = \Phi^{(r+1)}$, which by (10) has a Fourier series of the form (5). Integrating ,we get

$$e(-\kappa_p\tau/\lambda_p)\,\Phi \mid V(\tau) = \sum_{m \geqslant u} b(m)\, e(m\tau/\lambda_p) + q_p(\tau), \tag{11}$$

with q_p a polynomial of degree $\leqslant r+1$. We demand as before that u be finite. The expansion of an automorphic integral at a cusp, then, consists of a Fourier series plus a polynomial.

The set of integrals $I(\Gamma, r, v)$ is a complex vector space. It includes $\{\Gamma, r, v\}$, as an automorphic form satisfies (8) with $\omega_A \equiv 0$. It also includes $\mathscr{P}_r$, the vector space of polynomials of degree $\leqslant r$, since $q \mid A - q \in \mathscr{P}_r$ if $q \in \mathscr{P}_r$.

The function Φ is called an integral of the third kind. If the degree of q_p is $\leqslant r$ for all cusps p, Φ is said to be of the second kind. (Φ is of the first kind when it is of the second kind and in addition is holomorphic in H and at the cusps.) Φ will be of the second kind if and only if the constant term in the expansion of ϕ is zero at each cusp.

The polynomial $\omega_A = \Phi|_r A - \Phi$ is called the period polynomial associated with A. For convenience we write

$$\Phi|_r(A-1)$$

in place of $\Phi|_r A - \Phi$. From (2) we derive

$$\omega_{AB} = \Phi|(AB-1) = \Phi|(A-1)|B + \Phi|(B-1),$$

or

$$\omega_{AB} = \omega_A|B + \omega_B\,; \qquad A, B \in \Gamma. \tag{12}$$

A set of polynomials $\{\omega_A, A \in \Gamma\} \subset \mathscr{P}_r$ satisfying (12) is called a *cocycle*; the set of all cocycles forms an abelian group. To each integral there is associated a unique cocycle, namely, the collection of its period polynomials. An automorphic form is associated to the cocycle zero. A polynomial $q \in \mathscr{P}_r$ is associated to the cocycle $\{q \mid (A-1); A \in \Gamma\}$ and this is called a *coboundary*. The group of cocycles modulo the subgroup of coboundaries is named the cohomology group $H_v^1(\Gamma, \mathscr{P}_r)$ with coefficient module $\mathscr{P}_r$. Adding an automorphic form to an integral does not change its cocycle. Adding $q \in \mathscr{P}_r$ to an integral changes its cocycle by a coboundary, i.e., does not change the cohomology class of the cocycle. Hence

a form $f \in \{\Gamma, -r-2, v\}$ determines a unique cohomology class in $H_v^1(\Gamma, \mathscr{P}_r)$. (13)

Eichler considered not H_v^1 but a subgroup $\tilde{H}_v^1$, in which the cocycles $\{\omega_A\}$ are restricted by the following condition. Let $P_1, \ldots, P_t$ generate the stabilizers of a complete set of inequivalent cusps $p_1, \ldots, p_t$; then there shall exist polynomials $q_i \in \mathscr{P}_r$ such that

$$\omega_p = q_i \mid (P_i - 1),\ i = 1, 2, \ldots, t. \tag{14}$$

It can be shown that this condition is independent of the choice of $\{P_i\}$; moreover, it is a *necessary* condition if there is to be an integral of the *second kind* with periods $\{\omega_A\}$ (cf. Lehner (1969), Section 1). The remark (13) can now be sharpened to:

a cusp form $f \in C^0(\Gamma, -r-2, v)$ determines a unique cohomology class in $\tilde{H}_v^1(\Gamma, \mathscr{P}_r)$. (15)

Our first object will be to prove an existence theorem (Section 3):

THEOREM 1. *Given a cocycle $\{\omega_A\}$ satisfying* (14), *there is an integral of the second kind Φ whose periods are exactly $\{\omega_A\}$. Furthermore ϕ is holomorphic in H and at the cusps except for a pole at i_∞.*

This theorem says that the necessary conditions (12) and (14) are also sufficient.

The theorems of Eichler (1957) and Gunning (1961), also proved by Husseini and Knopp (loc. cit.), are as follows:

THEOREM 2. $C^0(\Gamma, -r-2, v) \oplus C^0(\Gamma, -r-2, \bar{v}) \simeq \tilde{H}_v^1(\Gamma, \mathscr{P}_r)$. *Here $\bar{v}$ is the multiplier system obtained by replacing each $v(A)$ by its complex conjugate.*

THEOREM 3. $C^0(\Gamma, -r-2, \bar{v}) \oplus C^+(\Gamma, -r-2, v) \simeq H_v^1(\Gamma, \mathscr{P}_r)$, *where* $C^+(\Gamma, -r-2, v)$ *is the vector space of forms on* Γ *of degree* $-r-2$ *that are holomorphic in* H *and at the cusps.*

We shall see in Section 4 that Theorem 2 is a consequence of Theorem 1 and of the "supplementary function." The proof of Theorem 3, which we shall not give here, is very similar to that of Theorem 2.

3. Let $\{\omega_A\}$ be a cocycle satisfying (14). Let M be a system of right coset representatives of Γ_∞ (the stabilizer of i_∞) in Γ. Let E_α, $\alpha > 0$, be the region $|x| \leqslant 1/\alpha$, $y \geqslant \alpha$.

LEMMA. *Let* $A = (ab \mid cd) \in M$, $\tau \in E_\alpha$. *Then*

$$|\omega_A(\tau)| \leqslant m_1 |\tau|^r (c^2 + d^2)^{r/2} (\log (c^2 + d^2) + m_2) \tag{16}$$

with positive constants m_1, m_2.

The proof is given in Lehner (1969), Theorem 1, under the assumption $v \equiv 1$, which, however, makes no essential difference in the argument. The proof uses the iterative relation (12) together with an estimate for the length of the word representing an element of Γ in the generators of Γ (cf. Eichler (1965)).

In the last-mentioned paper Eichler proves, by means of Poincaré series, that there is an integral of the second kind with periods $\{\omega_A\}$, but only for Γ a finite-index subgroup of the modular group. This construction can be generalized to arbitrary H-groups, as is done in Lehner (1969), though the Poincaré series used there are of different type. We may define

$$\Psi(\tau) = - \sum_{A \in \Gamma} \omega_A(\tau)(c\tau + d)^{-r-4}(A\tau - i)^{-r-4}, \tag{17}$$

$$\psi(\tau) = \sum_{A \in \Gamma} (c\tau + d)^{-r-4}(A\tau - i)^{-r-4}, \qquad A = (\,.\,.\mid cd). \tag{18}$$

The absolute uniform convergence of (18) in regions E_α is classical, while that of (17) is proved by the estimate of the Lemma. It is then found without difficulty that

$$\Psi|_{-4} B(\tau) = \Psi(\tau) + \omega_B(\tau) \cdot \psi(\tau), \qquad B \in \Gamma$$

and of course $\psi|_{-r-2} B = \psi$, so that

$$\Phi_1(\tau) = \Psi(\tau)/\psi(\tau)$$

satisfies (8). From the behaviour of the series (17) and (18) at the cusps we deduce that Φ_1 is of the second kind.

The function Φ_1 may, however, have poles in H as well as at the cusps. But there exists an automorphic *form* Ω in $\{\Gamma, r, v\}$ that has the same principal parts as Φ_1 at points of H and at each cusp of Γ except i_∞, provided we allow a pole at i_∞ of sufficiently high order. This follows from the "principal parts" theorem of Petersson (1955), p. 384, (A). Since the periods of Ω vanish, the function $\Phi = \Phi_1 - \Omega$ satisfies the conclusion of Theorem 1.

4. For the proof of Theorem 2 we shall exhibit the isomorphism by an explicit map. First let us define the "supplementray function."

Let $g(\tau, \rho, \bar{v})$ be the Poincaré series belonging to $\{\Gamma, -r-2, \bar{v}\}$ with (integral) parameter ρ. Define κ by $\bar{v}(U) = e(\kappa)$, $0 \leqslant \kappa < 1$, where $U = (1\lambda|01)$ is the stabilizer of i_∞ in Γ. It is known (Lehner, 1964; 272 ff.) that when $\rho + \kappa < 0$, $g(\tau, \rho, \bar{v})$ has a pole of order $-(\rho + \kappa)$ at i_∞ and vanishes at all other inequivalent cusps; when $\rho + \kappa > 0$, it is a cusp form. The set of Poincaré series with positive parameters spans the space of cusp forms. Let $g = C^0(\Gamma, -r-2, \bar{v})$,

$$g(\tau) = \sum_{i=1}^{s} c_i \, g(\tau, \rho_i, \bar{v}), \qquad \rho_i > 0 \tag{19}$$

where $\{g(\tau, \rho_i, \bar{v}), i = 1, \ldots, s\}$ is a fixed but arbitrary basis for $C^0(\Gamma, -r-2, \bar{v})$. Let

$$\rho' = -\rho \quad \text{if} \quad \kappa = 0, \qquad \rho' = -1-\rho \quad \text{if} \quad \kappa > 0$$

$$\kappa' = 0 \quad \text{if} \quad \kappa = 0, \qquad \kappa' = 1-\kappa \quad \text{if} \quad \kappa' > 0.$$

Note that $v(U) = e(\kappa')$, $0 \leqslant \kappa' < 1$ and that $\rho' + \kappa' = -\rho - \kappa$. Then define

$$g^*(\tau) = \sum_{i=1}^{s} \bar{c}_i \, g(\tau, \rho_i', v).$$

We see that $g^* \in \{\Gamma, -r-2, v\}$, has a pole at i_∞, and vanishes at all other cusps. Let G^* be that $(r+1)$st primitive of g^* whose expansion at i_∞ is a pure Fourier series (no additive polynomial). We say G^* is *supplementary* to g. Clearly $G^* \in I(\Gamma, r, v)$ and its cocycle lies in the cohomology class determined by g^*.

The significance of the supplementary function is exhibited by the following result.

THEOREM 4. *Let r be a positive integer. Let $g \in C^0(\Gamma, -r-2, \bar{v})$ and let G^* be supplementary to g. Then G^* is an automorphic form, i.e., $G^* \in \{\Gamma, r, v\}$, if and only if $g \equiv 0$.*

This theorem was first proved by Petersson (1955). Another proof appears in Husseini and Knopp.

We are now ready to construct the map of Theorem 2, following the procedure* of Husseini and Knopp. Let $f \in C^0(\Gamma, -r-2, v)$ and let $\beta(f)$ be the cohomology class in $\tilde{H}_v^1(\Gamma, \mathscr{P}_r)$ determined by f. Let $g \in C^0(\Gamma, -r-2, \bar{v})$, let G^* be supplementary to g, and let $\alpha(g)$ be the cohomology class determined by g^* (i.e., the class of the cocycle of periods of G^*). Then the map

$$\mu:(g,f) \to \mu(g,f) = \alpha(g) + \beta(f)$$

is clearly a homomorphism of $C^0(\Gamma, -r-2, v) \oplus C^0(\Gamma, -r-2, \bar{v})$ into $\tilde{H}_v^1$. We must show it is onto and 1–1.

That μ is onto follows directly from Theorem 1. In fact, suppose $\langle \omega_A \rangle$ is a cohomology class in $\tilde{H}_v^1$, and suppose Φ is an integral of the second kind with period cocycle lying in $\langle \omega_A \rangle$. Set $\phi = d^{r+1}\Phi/d\tau^{r+1}$; then $\phi \in \{\Gamma, -r-2, v\}$ and determines the class $\langle \omega_A \rangle$. From Theorem 1 we can say that ϕ is holomorphic in H and at all inequivalent cusps except i_∞ where it has a pole, and it has no constant term in its Fourier series at any cusp. Hence with certain $v_j < 0$ we have

$$\phi(\tau) = \sum_{j=1}^{h} a_j\, g(\tau, v_j, v) + f(\tau) = h(\tau) + f(\tau), \tag{20}$$

where plainly $f \in C^0(\Gamma, -r-2, v)$. Choose

$$g = \sum_{j=1}^{h} \bar{a}_j\, g(\tau, v_j', \bar{v}) \in C^0(\Gamma, -r-2, \bar{v}).$$

We have $g^* = h(\tau)$. It follows that $\mu(g,f)$ is the cohomology class determined by $h + f = \phi$, i.e., $\langle \omega_A \rangle$. Hence ϕ is onto.

The proof that μ is 1–1 can be found in Husseini and Knopp, Section 3, for the case appropriate to Theorem 3. We have to show that the kernel of μ is (0, 0). So suppose $\mu(g,f) = 0$. Then any primitive of $f + g^*$ is associated to the class $\langle 0 \rangle$; in other words there is a primitive F of f and a $q \in \mathscr{P}_r$ such that $F + G^* + q$ has zero periods, or $F + G^* + q \in \{\Gamma, r, v\}$. Thus the expansion at i_∞ of $F + G^* + q$ is a pure Fourier series, and since the same is true of G^* by definition, it must be the case also with $F + q$. The Fourier series of $F + q$, however, has no terms with negative exponents since $f = d^{r+1}(F+q)/d\tau^{r+1}$ is a cusp form. Hence the principal parts of $F + G^* + q$ and G^* agree at i_∞. At the other cusps both functions are holomorphic, so the principal parts are all zero.

Since $F + G^* + q$ is now a form, its Fourier coefficients can be calculated by well-known formulas (cf., e.g., Lehner (1964), ch. IX) and it is seen that they depend linearly on the coefficients of the principal part. The coefficients of G^* can be calculated by integration from those of g^*, the latter being known because g^* is a linear combination of Poincaré series (Lehner, 1964; 298). Comparison of the two sets of coefficients shows that

$$F + G^* + q = G^*. \tag{21}$$

Hence $F = -q$, which implies $f = 0$. Moreover, (21) shows that $G^* \in \{\Gamma, r, v\}$, from which we deduce $g \equiv 0$ by Theorem 4. This completes the proof that μ is 1–1 and the proof of Theorem 2 .

References

Bol, G. (1949). Invarianten linearer Differentialgleichungen. *Abh. Math. Seminar Hamburgischen Universität* **16,** 1–28.

Eichler, M. (1957). Eine Verallgemeinerung der abelschen Integrale. *Math. Zeitschrift* **67,** 267–298.

Eichler, M. (1965). Grenzkreisgruppen und kettenbruchartige Algorithmen. *Acta Arithmetica* **11,** 169–180.

Gunning, R. C. (1961). The Eichler cohomology groups and automorphic forms. *Trans. Amer. Math. Soc.* **100,** 44–62.

Husseini, S. and Knopp, M. I. Eichler cohomology and automorphic forms. To be published.

Knopp, M. I. (1962). Construction of automorphic forms on *H*-groups and supplementary Fourier series. *Trans. Amer. Math. Soc.* **103,** 168–188.

Lehner, Joseph (1964). Discontinuous groups and automorphic functions. *Math. Surveys No.* 8, *Amer. Math. Soc.*, Providence.

Lehner, Joseph (1969). Automorphic integrals with preassigned periods, *J. Research Nat. Bur. Standards* **73B,** 153–161.

Petersson, H. (1955). Uber automorphe Formen mit Singularitäten im Diskontinuitätsgebiet, *Math. Ann.* **129,** 370–390.

The Number of Conjugacy Classes of Certain Finite Matrix Groups

MORRIS NEWMAN

National Bureau of Standards, Washington, D.C., U.S.A.

Let q be a prime power. As usual, let $GF(q)$ denote the finite field with q elements, $GL(n,q)$ the multiplicative group of non-singular $n \times n$ matrices over $GF(q)$, $SL(n,q)$ the subgroup of $GL(n,q)$ consisting of all $n \times n$ matrices over $GF(q)$ of determinant 1, and $PSL(n,q)$ the group $SL(n,q)$ modulo its center C. Then $SL(n,q)$ is a normal subgroup of index $q-1$ of $GL(n,q)$, C consists of all scalar matrices of $SL(n,q)$, and the orders of these groups are given by

$$\begin{aligned} o(GL(n,q)) &= (q^n-1)(q^n-q)\dots(q^n-q^{n-1}), \\ o(SL(n,q)) &= (q-1)^{-1}\,o(GL(n,q)), \\ o(PSL(n,q)) &= o(C)^{-1}\,o(SL(n,q)), \\ o(C) &= (n,q-1). \end{aligned}$$

We are interested in finding the number of conjugacy classes of $GL(n,q)$, $SL(n,q)$, and $PSL(n,q)$, which we denote by $k(n,q)$, $k_1(n,q)$, and $k_2(n,q)$, respectively. In his comprehensive paper on the characters of $GL(n,q)$, J. A. Green (1955) states (with an obvious slip corrected) that $k(n,q)$ is the coefficient of z^n in the infinite product

$$\phi(z)\cdot\prod_{k=1}^{\infty}\phi(z^k)^{-w(k,q)},$$

where $\phi(z)$ is the Euler product

$$\phi(z) = \prod_{l=1}^{\infty}(1-z^l),$$

and

$$w(k,q) = \frac{1}{k}\sum_{d|k}\mu(d)\,q^{k/d}$$

is the number of monic irreducible polynomials of degree k over $GF(q)$. Since the form of the generating function is not as simple as one might wish, and since the result is stated without proof, we prove here

THEOREM 1. *The number $k(n,q)$ of conjugacy classes of $GL(n,q)$ is the coefficient of z^n in the infinite product*

$$\prod_{k=1}^{\infty} \frac{1-z^k}{1-qz^k}.$$

Thus $k(n,q)$ is a monic polynomial in q of degree n with integral coefficients; and if we write

$$k(n,q) = \sum_{r=0}^{n} a_r(n) q^r,$$

then $a_r(n)$ is the coefficient of z^n in the infinite product

$$z^r(1-z^{r+1})(1-z^{r+2})\ldots.$$

We first prove

LEMMA 1. *Let*

$$w(k,q) = \frac{1}{k}\sum_{d|k} \mu(d) q^{k/d}$$

be the number of monic irreducible polynomials of degree k over $GF(q)$. Then the following formal power series expansion is valid:

$$\prod_{k=1}^{\infty} (1-z^k)^{w(k,q)} = 1-qz.$$

Proof. By the Möbius inversion formula,

$$\sum_{d|k} dw(d,q) = q^k.$$

Put

$$w = \prod_{k=1}^{\infty} (1-z^k)^{w(k,q)}.$$

Then

$$zw'/w = -\sum_{k=1}^{\infty} kw(k,q)\frac{z^k}{1-z^k} = -\sum_{k=1}^{\infty}\sum_{d|k} dw(d,q)\, z^k$$

$$= -\sum_{k=1}^{\infty} q^k z^k = -\frac{qz}{1-qz}.$$

Hence $w' = (1 - qz)'$, and the conclusion follows.

We turn now to the proof of Theorem 1. For this purpose, we note that if A, B are any two $n \times n$ matrices over a field F, then A and B are conjugate over $GL(n, F)$ if and only if $A - xI$ and $B - xI$ are equivalent over $GL(n, F[x])$, x an indeterminate. It follows easily that $k(n, q)$ is just the total number of different Smith normal forms of the matrices $A - xI$, $A \in GL(n, q)$.

Let $p_1, p_2, \ldots, p_t$ denote the non-constant monic irreducible polynomials in x over $GF(q)$ of degree not exceeding n, exclusive of the polynomial x. Put $d_j = \deg p_j$, $1 \leqslant j \leqslant t$. Then $k(n, q)$ is just the number of expressions

$$h_1 = p_1^{x_{11}} p_2^{x_{12}} \ldots p_t^{x_{1t}},$$

$$h_2 = p_1^{x_{21}} p_2^{x_{22}} \ldots p_t^{x_{2t}},$$

$$\ldots$$

$$h_n = p_1^{x_{n1}} p_2^{x_{n2}} \ldots p_t^{x_{nt}},$$

where the x_{ij} are non-negative integers satisfying

$$x_{1j} \leqslant x_{2j} \leqslant \ldots \leqslant x_{nj}, \qquad 1 \leqslant j \leqslant t,$$

and

$$\sum_{j=1}^{t} \sum_{i=1}^{n} x_{ij} d_j = n.$$

The quantities $h_1, h_2, \ldots, h_n$ are the invariant factors of $A - xI$, and x must be omitted since the matrices A are non-singular.

Put

$$x_{ij} = y_{1j} + y_{2j} + \ldots + y_{ij}, \qquad 1 \leqslant i \leqslant n, \qquad 1 \leqslant j \leqslant t.$$

Then the new variables y_{ij} are non-negative and must satisfy

$$\sum_{j=1}^{t} \sum_{i=1}^{n} \{y_{1j} + y_{2j} + \ldots + y_{ij}\} d_j = n,$$

which may be rewritten as

$$\sum_{j=1}^{t} \sum_{i=1}^{n} (n - i + 1) y_{ij} d_j = n.$$

A moment's consideration now shows that the required number is just the coefficient of z^n in the product

$$\prod_{i=1}^{t} \prod_{k=1}^{n} (1 - z^{kd_i})^{-1},$$

which may be replaced by

$$\prod_{i=1}^{t} \prod_{k=1}^{\infty} (1 - z^{kd_i})^{-1}$$

without changing the coefficient of z^n. Since the polynomials $p_1, p_2, \ldots, p_t$ comprise $q-1$ of degree 1, and $w(k,q)$ of degree k for $2 \leqslant k \leqslant n$, the expression above becomes

$$\phi(z)^{-(q-1)} \prod_{k=2}^{n} \phi(z^k)^{-w(k,q)} =$$

$$\phi(z) \prod_{k=1}^{n} \phi(z^k)^{-w(k,q)};$$

and this may be replaced by

$$\phi(z) \prod_{k=1}^{\infty} \phi(z^k)^{-w(k,q)}$$

without changing the coefficient of z^n.

We now make use of Lemma 1 to get that $k(n,q)$ is the coefficient of z^n in the infinite product

$$\phi(z) \prod_{k=1}^{\infty} (1 - qz^k)^{-1},$$

which is the first part of Theorem 1. The second part of Theorem 1 follows from the expansion

$$\prod_{k=1}^{\infty} \frac{1-z^k}{1-qz^k} = \sum_{r=0}^{\infty} q^r z^r (1-z^{r+1})(1-z^{r+2}) \ldots,$$

obtained by treating q as an independent variable, replacing q by zq, and determining the coefficients by recurrence. This completes the proof of the theorem.

The results of Theorem 1 make the polynomials $k(n,q)$ especially easy to compute. Table 1 gives them for $1 \leqslant n \leqslant 10$, and the author has computed them for $1 \leqslant n \leqslant 100$ with the aid of the computer of the National Bureau of Standards.

The corresponding problems for $SL(n,q)$ and $PSL(n,q)$ are much more difficult, and not completely solved. There is one case, however, which admits a complete answer: namely, when $(n, q-1) = 1$. For then the center C of

TABLE I.

n	$k(n,q)$
1	$q-1$
2	q^2-1
3	q^3-q
4	q^4-q
5	q^5-q^2-q+1
6	q^6-q^2
7	$q^7-q^3-q^2+1$
8	$q^8-q^3-q^2+q$
9	$q^9-q^4-q^3+q$
10	$q^{10}-q^4-q^3+q$

$SL(n,q)$ is trivial, so that $SL(n,q) \cong PSL(n,q)$, and an element $c \in GF(q)$ exists such that $c^n = \theta$ is a primitive element of $GF(q)$. It follows that

$$GL(n,q) = \{cI\} \times SL(n,q),$$

the direct product of the cyclic group $\{cI\}$ of order $q-1$ and $SL(n,q)$, and hence the number of conjugacy classes of $GL(n,q)$ must be equal to the number of conjugacy classes of $\{cI\}$ multiplied by the number of conjugacy classes of $SL(n,q)$. Since the number of conjugacy classes of an abelian group is equal to its order, we have proved the following theorem:

THEOREM 2. *Suppose that* $(n,q-1)=1$. *Then*

$$k_1(n,q) = k_2(n,q) = k(n,q)/(q-1).$$

Table 2 gives $k(n,q)/(q-1)$ for $1 \leqslant n \leqslant 10$.

The case $n=3$ can be treated directly, and it is easy to show:

THEOREM 3. *The number of conjugacy classes* $k_1(3,q)$ *of* $SL(3,q)$ *is given by*

$$k_1(3,q) = \begin{cases} q^2+q+8 & q \equiv 1 \bmod 3, \\ q^2+q & \textit{otherwise}. \end{cases}$$

TABLE II.

n	$k_1(n) = k_2(n) = k(n)/(q-1),\ (n, q-1) = 1.$
1	1
2	$q + 1$
3	$q^2 + q$
4	$q^3 + q^2 + q$
5	$q^4 + q^3 + q^2 - 1$
6	$q^5 + q^4 + q^3 + q^2$
7	$q^6 + q^5 + q^4 + q^3 - q - 1$
8	$q^7 + q^6 + q^5 + q^4 + q^3 - q$
9	$q^8 + q^7 + q^6 + q^5 + q^4 - q^2 - q$
10	$q^9 + q^8 + q^7 + q^6 + q^5 + q^4 - q^2 - q$

Proof. We need only consider the case $q \equiv 1 \bmod 3$. Let θ be a primitive element of $GF(q)$, and put $\alpha = \theta^{(q-1)/3}$. Then a complete set of representatives of the conjugacy classes of $SL(n, q)$ (together with the number of each type) is as follows:

$\mathrm{diag}(\beta, \beta, \beta);\ \beta = 1, \alpha, \alpha^2$ $\qquad$ 3

$\begin{pmatrix} \beta & 1 \\ 0 & \beta \end{pmatrix} \dot{+} (\beta^{-2});\ \beta \neq 0$ $\qquad$ $q - 4$

$\begin{pmatrix} \beta & \gamma & 0 \\ 0 & \beta & 1 \\ 0 & 0 & \beta \end{pmatrix};\ \beta = 1, \alpha, \alpha^2, \gamma = 1, \theta, \theta^2 \cdot$ $\qquad$ 9

$\mathrm{diag}\left(\beta, \beta, \frac{1}{\beta^2}\right);\ \beta \neq 0, 1, \alpha, \alpha^2$ $\qquad$ $q - 4$

$\mathrm{diag}(\beta, \gamma, \delta);\ \beta\gamma\delta = 1, \beta \neq \gamma, \beta \neq \delta, \gamma \neq \delta,$ the sets $\{\beta, \gamma, \delta\}$ distinct. $\qquad$ $\frac{1}{6}(q^2 - 5q + 10)$

$B \dot{+} (1/\det B);\ \det(B - xI)$ irreducible $\qquad$ $\frac{1}{2}(q^2 - q)$

$A;\ \det A = 1,\ \det(A - xI)$ irreducible $\qquad$ $\frac{1}{3}(q^2 + q - 2).$

It is readily verified that every conjugacy class occurs once and once only in the enumeration above. The most difficult part is counting the number of the last type. More generally, it can be shown that the number $w_1(n,q)$ of monic irreducible polynomials over $GF(q)$ of degree n and with constant term 1 is given by

$$w_1(n,q) = \frac{1}{n} \sum_{d|n} \mu(d)\,(q-1,d)\,\frac{q^{n/d}-1}{q-1}.$$

A final result of the same type, whose proof we omit, is the following:

THEOREM 4. *Let $\lambda_1, \lambda_2, \ldots, \lambda_r$ be r distinct non-zero elements of a field F. Let S be the totality of $n \times n$ matrices over F whose eigenvalues belong to the set $\{\lambda_1, \lambda_2, \ldots, \lambda_r\}$ (repetitions allowed). Then S is normalized by $GL(n,F)$ and the number of conjugacy classes of S over $GL(n,F)$ is $p_{-r}(n)$, the coefficient of x^n in the infinite product*

$$\phi(x)^{-r} = \prod_{k=1}^{\infty} (1-x^k)^{-r}.$$

The result is of interest in that the numbers $p_{-r}(n)$ have been given a natural combinatorial interpretation. Related remarks on this question may be found in Petersson (1954) and Newman (1959).

Note added in proof: The generating function (1) was given by Feit, W. and Fine, N. J. (1960). Pairs of commuting matrices over a finite field. *Duke Math. J.* **27**, 91–94. I am indebted to J. A. Green for this reference.

References

Green, J. A. (1955). The characters of the finite linear groups. *Trans. Amer. Math. Soc.* **10**, 402–447.

Newman, M. (1959). Weighted restriction partitions. *Acta Arith.* **5**, 371–380.

Petersson, H. (1954). Uber Modulfunktionen und Partitionenprobleme. *Abh. Deutsch. Akad. Wiss. Berlin. Kl. Math. Allg. Nat.* **2**, 1–59.

Hilbert's Theorem 94†

OLGA TAUSSKY

California Institute of Technology, Pasadena, California, U.S.A.

1. Introduction, Formulation of Theorem 94, Discussion of Various Proofs, Examples

Hilbert's Theorem 94 is concerned with the absolute or Hilbert class field of an algebraic number field. This is the largest extension field which is both abelian and unramified (i.e. all prime ideals split into different factors in the extension). Hilbert had conjectured that all ideals of the original field (call it F) become principal in the absolute class field. Much later this 'principal ideal theorem' was proved by Furtwängler (1930) using Artin's reciprocity law (1927). However, Hilbert's Theorem 94 is already a genuine part of the principal ideal theorem and is also in so far a starting point of class field theory as it connects a fact concerning the field F with a fact concerning the existence of an extension with prescribed properties. The extension here is an unramified cyclic field K of odd prime degree p (though 2 also qualifies if the ideal class group in F is considered in the narrow sense only). Theorem 94 states that an ideal in F belonging to a class of order p becomes principal in K. Hence the existence of K implies that the class number of F is divisible by p.

Apart from Hilbert's proof there are proofs implicit in the works of Brumer and Rosen (1963) and Rosen (1966) alone. Earlier Iwasawa (1956) connected principal ideal questions with cohomological facts. The author suggested the application of Artin's methods to this theorem to M. Hall who then gave a proof on these lines in 1952. Later Browkin and Kisilevsky found similar proofs, all entirely in terms of finite p-groups.

The theorem is an existence theorem. It does not tell which class of F becomes principal, nor how many. Since all classes of F become principal in the absolute class field and since the investigation of the possibilities in its subfields are far from easy, interest in these problems has dwindled. Yet the problems are there and recently also Hasse (1967) has stressed them again.

† This work was carried out (in part) under National Science Foundations Grant 3909 and 11236. Thanks are due to A. Brumer and H. Kisilevsky for helpful discussions.

To show their complexity three numerical examples will now be discussed. The first two are taken from Furtwängler (1916), the third from Scholz and Taussky (1934).

I. *The field* $Q(\sqrt{-21})$.

This field has a 2-class group of type (2, 2), hence three quadratic unramified extensions generated by $\sqrt{-7}, \sqrt{-3}, \sqrt{21}$. In each of them all classes of F become principal.

II. *The field* $Q(\sqrt{-195})$.

This field again has a 2-class group of type (2, 2). The three unramified quadratic extensions are generated by $\sqrt{-3}, \sqrt{13}, \sqrt{5}$. In each of them exactly one class of F becomes principal and it is a different class each time.

III. *The field* $Q(\sqrt{-4027})$.

This field has a 3-class group of type (3, 3). This is the quadratic field $Q(\sqrt{-m})$, with m a square free positive integer, and m minimal for this type.

In each of the four cubic unramified extensions exactly one class and its square becomes principal, but in three of them a different class does it while the fourth field copies one of the earlier three fields. Hence there exists a class in F which does not become principal in a cubic unramified extension, only in the whole Hilbert class field.

At present it is not possible to make a guess at the laws which govern the behaviour in general. Hence the computation of numerical examples ought to be continued and programing may be envisaged.† Another approach is again via p-groups. As in the proof of the principal ideal theorem or in the proof of Theorem 94 the explicit isomorphism, proved by Artin, between the class group in F and the Galois group of the extension is used. Next the class field F_2 of the class field F_1 of F is brought into the picture. It is a normal extension of F with a non abelian Galois group G, which, however, has an abelian commutator subgroup. Since only the p-part of all the ideal class groups is considered, G is a p-group and the study of this group is the main tool. The difficulties in this approach are of two types:

(1) It is not known which groups G can occur as Galois groups of a second class field, particularly since one does not even know what class groups occur. This is not important when one proves that a certain property cannot occur. However, if one proves that a certain property can occur for a certain G one has to add "provided this G can occur in numerical cases".

† The method used in Scholz and Taussky (1934) for the case of cyclic extensions uses rational operations only.

(2) The proofs have to be carried out for a given G/G' with no knowledge of G' and of the linkage between G/G' and G' apart from the necessary group theoretic relations. Hence the proofs have many times to be carried out for all possible structures of G.

The tower of fields involved is as follows:

$$\begin{array}{c} F_2 \\ | \\ K_1 \\ | \\ F_1 \\ | \\ K \\ | \\ F \end{array}$$

In this tower every field is contained in all fields above it and the larger field is normal with respect to the smaller one. The field K_1 is the first class field of K. Denote the Galois group of F_2 with respect to K by S. Then S' is isomorphic with the Galois group of F_2 with respect to K_1 and S/S' with the Galois group of K_1 with respect to K. Further G/S is isomorphic with the Galois group of K with respect to F and is hence cyclic of order p. Finally, the Galois group of F_2 with respect to F_1 is G' and G/G' is isomorphic with the Galois group of F_1 with respect to F, and hence with the ideal class group in F.

The p-group proof of Theorem 94 consists in showing that the kernel of the transfer map from G to S mod S' is not empty.

2. A Division into Two Different Cases

By the laws of class field theory the extension field K of F induces a division of the ideal classes of F into two types which can be described in a simple way: The classes which contain prime ideals which factorize in K into p different factors and those which remain prime. The former make up a subgroup H of the class group of index p while the latter lie in the cosets. Hence the fields K belong to one of the two following types

A. An element of H becomes principal in K.

B. No element of H becomes principal in K.

These conditions can be given a cohomological interpretation in terms of the group G and its subgroup S:

A. The cohomology of S/S' is non trivial when viewed as a module over G/S.

B. The cohomology of S/S' is trivial when viewed as a module over G/S.

To use number theoretic language one has to replace the module S/S' by the ideal class group in K and the quotient group G/S by the Galois group of K over F.

This interpretation is a special case of a theorem obtained by Kisilevsky (1968):

Let K be a cyclic unramified extension of F. Let j be the transfer map from G to S mod S', or, in number theoretic language, the map of the ideal class group of F into that of K by extension of ideals. Let $N_{K/F}$ be the map $g^{1+x+\ldots+x^{p-1}}$ where $g \in S/S'$, x a generating element of G/S and $g^x = x^{-1}gx$; in number theoretic language N is the map induced by the norm map of the ideal classes of K into the classes of F. Then $|\ker j \cap H| = |H^0(G/S, S/S')|$. He further shows:

$$H^0(G/S, S/S') = 0$$

if and only if

$$G/G' = \ker j \times N_{K/F}(S/S').$$

3. Some Results Concerning the Occurrence of Cases A and B for Class Groups of Type (p, p)

In the examples mentioned earlier the distribution is as follows: $Q(\sqrt{-21})$: all classes become principal in all extensions K.

$Q(\sqrt{-195})$: the extension field generated by $\sqrt{5}$ is of type A; both the extension fields generated by $\sqrt{13}$ and $\sqrt{-3}$ belong to type B, the subgroups H belonging to either of them becoming principal in the other.

$Q(\sqrt{-4027})$: only one of the four fields belongs to type A.

For type (3, 3) the following result was obtained earlier, in Scholz and Taussky (1934):

Assume that only one class (and its powers) of F becomes principal in K (this is, e.g., the case for an imaginary quadratic F, because of the lack of units in F). It then follows that not all of the four relatively cubic extension fields can belong to type A, nor can all belong to type B. Further it cannot happen that exactly one of them belongs to type B.

The author showed recently that for $p > 3$ all these possibilities can occur for the $p + 1$ fields K, at least from the group theoretic point of view. That all can belong to B is shown in Taussky (1970); an alternative proof for this will be presented here as well as proofs for the two other facts.

The alternative proof uses the fact that in the example of a group G for which all the K's belong to type B all the groups S/S' are elementary abelian and of order p^3. Further S/S' is a G/S-module. If $H^0(G/S, S/S') = 0$ held, then it would follow that S/S' is a free G/S-module (cf. Serre, 1962;

p. 150). From this it would follow (see Kisilevsky, 1970; p. 203) that S/S' has at least the order p^p. This is alright for $p = 3$, but not for $p > 3$.

Before summarising this result some notation for the p-group G will be introduced. The group G has an abelian commutator subgroup G' and G/G' is of type (p, p). Let S_1, S_2 be the generators of G. The group G' is generated by the commutator

$$T = S_1^{-1} S_2^{-1} S_1 S_2$$

and all $T^{f(S_1, S_2)}$, where f is a polynomial and T^A stands for $A^{-1}TA$. In what follows the polynomials will be expressed as polynomials in Δ_1, Δ_2 where

$$\Delta_i = S_i - 1.$$

The group G is completely determined if the ideal $\mathfrak{M}$ of all polynomials $f(\Delta_1, \Delta_2)$ for which $T^f = 1$ is known. Among these are the relations which were found by Schreier (1926) and which are the only necessary relations.

THEOREM 1. *Let G be a p-group, $p > 3$, with G/G' of type (p, p) and assume that the ideal $\mathfrak{M}$ is generated by p, Δ_1^2, $\Delta_1 \Delta_2$, Δ_2^2. Assume further that $S_1^p = T^{\Delta_1}$, $S_2^p = T^{\Delta_2}$. This group is of order p^5. If this G is the Galois group with respect to F of the second class field of F then all cyclic unramified extensions of degree p are of type* A *and only one class of F becomes principal in each of them.*

Because of the preceding remarks it is sufficient to show that the $p + 1$ subgroups S are all elementary abelian of order p^3. For this purpose we state some further facts about the above G, the proofs of which are contained in Taussky (1970).

Since the Schreier relations are implied by the assumptions made there is no relation between T, T^{Δ_1}, T^{Δ_2}. The groups S can be defined by $\{S_2, G'\}$ and $\{S_1 S_2^r, G'\}$, $r = 0, \ldots, p-1$. The corresponding groups S' are then $\{T^{\Delta_2}\}$, $\{T^{\Delta_1 + r\Delta_2}\}$. Since the latter are of order p we have that S/S' is of order p^3. Further, S/S' is elementary abelian in all cases, for it is $\{S_2, T, T^{\Delta_1}\}S'$, resp. $\{S_1 S_2^r, T, T^{\Delta_2}\}S'$ and since S_2 has order p with respect to $\{T^{\Delta_2}\}$, while $S_1 S_2^r$ has order p with respect to $\{T^{\Delta_1 + r\Delta_2}\}$. It is easy to check that only one class (and its powers) becomes principal.

We next construct an example of a group G which corresponds to a field F for which all K's belong to type B.

THEOREM 2. *Let G be a p-group with G/G' of type (p, p) and assume that the ideal $\mathfrak{M}$ is generated by p, Δ_1^2, $\Delta_1\Delta_2$, Δ_2^2; assume further that*

$$S_1^p = T^{\Delta_1 + \mu\Delta_2}, \qquad S_2^p = T^{p\Delta_1 + \Delta_2}$$

where $\mu \not\equiv 0(p)$, $\rho \not\equiv 0(p)$, μ/ρ a quadratic non residue mod p. *This group is of order p^5. If this G is the second class field of F then all cyclic unramified extensions of degree p are of type* B *and only one class of F becomes principal in each of them.*

Proof. As in the preceding theorem the group G is of order p^5 and is generated by $S_1, S_2, T, T^{\Delta_1}, T^{\Delta_2}$. The Schreier relations are satisfied automatically. For the group $S = \{S_2, G'\}$ the transfer map does not have S_2 in its kernel. For the groups $S = \{S_1 S_2^r, G'\}$ the transfer map does not have $S_1 S_2^r$ in its kernel. For, the transfer is

$$T^{\Delta_1 + \mu\Delta_2 + r\rho\Delta_1 + r\Delta_2} = T^{\Delta_1(1+r\rho)+\Delta_2(\mu+r)}.$$

If this were to lie in $\{T^{\Delta_1 + r\Delta_2}\}$ then

$$\frac{1+r\rho}{\mu+r} = \frac{1}{r}.$$

This implies

$$r(1+r\rho) \equiv \mu + r(p)$$

or

$$r^2\rho \equiv \mu(p),$$

which is impossible by the assumptions on ρ and μ.

Note that the groups S/S' are not elementary abelian here.

We next show that for $p > 3$ it is possible to exclude exactly one field K for type B.

THEOREM 3. *Let G be a p-group with G/G' of type (p, p) and assume that the ideal $\mathfrak{M}$ is generated by $p, \Delta_1^2, \Delta_1\Delta_2, \Delta_2^2$. Assume further that*

$$S_1^p = T^{\lambda\Delta_1 + \mu\Delta_2}, \qquad S_2^p = T^{\lambda\Delta_2}, \qquad \lambda \not\equiv 0, \quad \mu \not\equiv 0(p).$$

If this G is the Galois group with respect to F of the second class field of F then all but one of the cyclic unramified extensions of degree p are of type B *and only one class of F becomes principal in each of them.*

Proof. Again, the Schreier relations are satisfied. Again S is either $\{S_2, G'\}$ or $\{S_1 S_2^r, G'\}$ and the corresponding groups S' are as before. We then see that the transfer map of S_2 brings it into the corresponding S_1', hence we have here a case of A. However, all the $S_1 S_2^r$, $r = 0, \dots, p-1$, have a transfer map $T^{\lambda\Delta_1 + \mu\Delta_2 + r\lambda\Delta_2} \notin S'$ for the latter is $\{T^{\Delta_1 + r\Delta_2}\}$ and $\mu \not\equiv 0(p)$. This completes the proof.

References

Artin, E. (1927). Beweis des allgemeinen Reziprozitätsgesetzes. *Hamb. Sem. Abh.* **5**, 353–363.

Artin, E. (1929). Idealklassen in Oberkörpern und allgemeines Reziprozitätsgesetz. *Hamb. Sem. Abh.* **7**, 46–51.

Browkin, J. (1964). Unpublished.

Brumer, A. and Rosen, M. (1963). Class number and ramification in number fields. *Nagoya Math. J.* **23**, 97–101.

Furtwängler, P. (1916). Über das Verhalten der Ideale des Grundkörpers im Klassenkörper. *Monatshefte f. Math. u. Physik* **27**, 1–15.

Furtwängler, P. (1930). Beweis des Hauptidealsatzes für den Klassenkörper algebraischer Zahlkörper. *Hamb. Sem. Abh.* **7**, 14–36.

Hall, M. (1952). Unpublished.

Hasse, H. (1967). History of classfield theory. *In* "Algebraic Number Theory", (Cassels, J. W. S. and Fröhlich, A., Eds.) 266–277. Academic Press, London and New York.

Hilbert, D. (1897). Die Theorie der algebraischen Zahlkörper. *In* "Gesammelte Abhandlungen", I (1932) 69–363, Springer, Berlin, in particular, 155–156.

Iwasawa, K. (1956). A note on the group of units of an algebraic number field. *J. Math. Pure Appl.* (9) **35**, 189–192.

Kisilevsky, H. (1969). Unpublished.

Kisilevsky, H. (1970). Some results related to Hilbert's Theorem 94, *J. of Number Theory.* 199–206.

Rosen, M. (1966). Two theorems on Galois cohomology. *Proc. Amer. Math. Soc.* **17**, 1183–1185.

Scholz, A. and Taussky, O. (1934). Die Hauptideale der kubischen Klassenkörper imaginär-quadratischer Zahlkörper. *J. f. d. reine und angew. Mathematik* **171**, 19–41.

Schreier, O. (1926a). Über die Erweiterung von Gruppen I. *Monatsh. f. Math. und Physik*, **34**, 165–190.

Schreier, O. (1926b). Über die Erweiterung von Gruppen II. *Hamb. Sem. Abh.* **4**, 321–346.

Serre, J. P. (1962). "Corps Locaux", Hermann, Paris.

Taussky, O. (1932). Über eine Verschärfung des Hauptidealsatzes für algebraische Zahlkörper. *J. f. d. reine und angew. Math.* **168**, 194–210.

Taussky, O. (1970). A remark concerning Hilbert's Theorem 94, *J. f. d. reine und angew. Math.* **239/240**, 435–438.

Reducibility of Polynomials

A. SCHINZEL

Mathematics Institute PAN, Ul Sniadeckich, Warsaw, Poland

I have proved recently (Schinzel, 1970) the following:

THEOREM. *For any non-zero integers A, B and any polynomial $f(x)$ with integer coefficients such that $f(0) \neq 0$ and $f(1) \neq -A-B$ there exist infinitely many irreducible polynomials of the form $Ax^m + Bx^n + f(x)$.*

One stage in the proof of this theorem is the following

LEMMA. *Under the assumptions of the theorem there exist integers a, b, d such that for $m \equiv a$, $n \equiv b \bmod d$, $Ax^m + Bx^n + f(x)$ has no cyclotomic factors.*

If $f(x) \neq A\xi x^p + B\eta x^q$, where $\xi = \pm 1$, $\eta = \pm 1$ then, indeed, there exists d such that every cyclotomic factor of $Ax^m + Bx^n + f(x)$ divides $x^d - 1$ (l.c. Lemma 2). Therefore in order to prove the above lemma it is enough to establish the existence of integers a, b such that

$$(Ax^a + Bx^b + f(x),\, x^d - 1) = 1.$$

This is done by an enumerative argument, one counts the number of pairs m, n such that $0 \leqslant m < d$, $0 \leqslant n < d$;

$$(Ax^m + Bx^n + f(x),\, x^d - 1) \neq 1$$

and one finds that it is less than d^2. The count involves some computations, which are particularly cumbersome if $A = B$; then the success of the method depends upon the inequality:

$$\frac{15}{14}\prod\left(1 - \frac{2(p^2-1)}{p(p^3-p^2-3p+1)}\right) + \sum\frac{p^2-1}{p(p^3-p^2-3p+1)} + \frac{127}{30}$$
$$> 4\prod\left(1 + \frac{p}{p^3-p^2-3p+1}\right) - 2\sum\frac{p}{p^3-p^2-3p+1}, \quad (1)$$

where p runs over all odd primes.

In the case $A = B = 1$ the problem considered has the following analogue in number theory: do there exist integers k, d such that

$$(2^m + 2^n + k, 2^d - 1) > 1 \tag{2}$$

for all $m, n \geqslant 0$. This question is of interest in connection with the well known problem of representing integers as sums of a prime and of powers of two (see Sierpinski, 1964; Chapter XII, Section 5). Erdős (1950) found k such that

$$(2^m + k, 2^{24} - 1) > 1$$

for all $m \geqslant 0$. Two students of the University of Warsaw (K. Rukat-Hoinska in 1962 and K. Gozdek in 1963) tried to satisfy inequality (2) with $d = 144$ and 420, respectively. Their efforts were unsuccessful; with the best choice of k there remained about $4{\cdot}5d$ and $22d$ pairs (m, n) with $m < d$, $n < d$ not satisfying (2). However the comparison with the problem concerning polynomials shows two differences:

1. The cyclotomic polynomial $\phi_l(x)$ is irreducible, while $\phi_l(2)$ is not always a prime.

2. The congruence $x^n + x^m \equiv x^p + x^q \bmod \phi_l(x)$ implies $(n, m) \equiv (p, q)$ or $(q, p) \bmod l$, while $2^n + 2^m \equiv 2^p + 2^q \bmod \phi_l(2)$ does not always imply the same.

These phenomena indicate that the situation for numbers may differ from that for polynomials and if one takes into account the small margin by which (1) holds, it seems that finding d and k satisfying (2) may not be impossible for a computer.

Here are two problems directly concerning reducibility of polynomials which may also be treated by a computer.

1. Does there exist a constant K such that every trinomial with integral coefficients has an irreducible factor with at most K terms?

2. Does there exist a reducible trinomial of the form

$$x^n - ax^m + 1 \text{ where } n > m > 0,\ n \neq 2m,\ |a| \text{ integer} > 2?$$

The origin of the first problem is discussed in Schinzel (1963). The following example due to H. Smyczek shows that K if it exists is at least 6.

$$x^{10} - 12x^2 - 196 = (x^5 + 2x^4 + 2x^3 - 4x^2 - 10x - 14)$$
$$\times (x^5 - 2x^4 + 2x^3 + 4x^2 - 10x + 14)$$

(in the quoted paper, the name Smyczek is misspelled).

The second problem was a subject of M.A. dissertation of T. Karwowska, who proved that there is no example of reducibility for $n \leqslant 8$. A similar work by Z. Łutczyk on the trinomials $x^n - ax^m - 1$ revealed for $n \leqslant 8$ only the following example:

$$x^8 + 3x^3 - 1 = (x^3 + x - 1)(x^5 - x^3 + x^2 + x + 1)$$

and its trivial derivatives.

References

Erdős, P. (1950). On integers of the form $2^k + p$ and some related problems. *Summa Brasil. Math.* **2,** 113–123.

Schinzel, A. (1963). Some unsolved problems on polynomials, *Matematička Biblioteka* **25,** 63–70.

Schinzel, A. (1970). Reducibility of lacunary polynomials II, *Acta. Arith.* **16,** 371–392.

Sierpinski, W. (1964). "Elementary Theory of Numbers", Warszawa.

Sums of Squares in the Function Field $\mathbb{R}(x, y)$

ALBRECHT PFISTER

University Mathematical Institute, Mainz, Germany

1. The problem of representing rational functions as sums of squares was initiated by Hilbert (1888). In (1893) he showed that every positive definite rational function in two variables over the reals is a sum of 4 squares. For simplifications of the proof and generalisations to n variables we refer to papers by Artin (1927), Witt, Ax (1966), and Pfister (1967). In this talk I will restrict myself to the original 2-variable case and will make some comments on the following problem (unsolved at the time of speaking):

Is every positive definite function $f \in \mathbb{R}(x, y)$ representable as a sum of 3 *squares in $\mathbb{R}(x, y)$?*

The number 3 is the best possible since Cassels (1964) has shown that the element

$$1 + x^2 + y^2 \in \mathbb{R}(x, y)$$

is not a sum of 2 squares.

2. I will show that "locally" the answer to the problem is yes. For this let $k = \mathbb{R}(x)$ and consider $K = \mathbb{R}(x, y) = k(y)$ as a function field in one variable over k. Let $\mathfrak{p}$ run through a set of inequivalent valuations of K/k and denote by $K_\mathfrak{p}$ the completion of K with respect to $\mathfrak{p}$, by $k_\mathfrak{p}$ the residue class field of $\mathfrak{p}$. If $t \in K$ is a prime element for $\mathfrak{p}$ then $K_\mathfrak{p} = k_\mathfrak{p}((t))$ is the field of formal power series in t over $k_\mathfrak{p}$. We have

Proposition 1. In $K_\mathfrak{p}$ every sum of squares is a sum of 3 squares.

Proof. $k_\mathfrak{p}$ is a finite algebraic extension of k, hence a field of transcendence degree 1 over $\mathbb{R}$. This implies (Witt, 1934; Pfister, 1967) that in $k_\mathfrak{p}$ every sum of squares is a sum of 2 squares. We now distinguish two cases:

(a) $k_\mathfrak{p}$ is a non-real field, i.e. -1 is a sum of squares in $k_\mathfrak{p}$. Then $-1 = a^2 + b^2$ with $a, b \in k_\mathfrak{p}$. This implies

$$f = \left(\frac{f+1}{2}\right)^2 - \left(\frac{f-1}{2}\right)^2 = \left(\frac{f+1}{2}\right)^2 + \left(\frac{a(f-1)}{2}\right)^2 + \left(\frac{b(f-1)}{2}\right)^2$$

for any $f \in K_\mathfrak{p}$.

(b) $k_\mathfrak{p}$ is a (formally) real field. Then the same holds for $K_\mathfrak{p}$: $f \neq 0$ being a sum of squares in $K_\mathfrak{p}$ it then follows that f has a power series expansion

$$f = f_0 t^{2r} + f_1 t^{2r+1} + \ldots$$

with $r \in \mathbb{Z}$, $f_i \in k_\mathfrak{p}$, $f_0 \neq 0$. If now $f_0 = a^2 + b^2$ in $k_\mathfrak{p}$, $a \neq 0$, we see that $a^2 t^{2r} + f_1 t^{2r+1} + \ldots$ is a square g^2 in $K_\mathfrak{p}$, hence $f = g^2 + (bt^r)^2$ is a sum of 2 squares and *a fortiori* a sum of 3 squares.

3. The original problem is therefore equivalent to the question whether the Local–Global Principle of Hasse holds for sums of 3 squares in $k(y)$. In this respect it is of interest that the principle does hold for sums of 1, 2 or 4 squares.

The first case is easy: Let $f = c\Pi p_i^{r_i}$ be the prime decomposition of $f \in k(y)$. If f is a square at the prime spot $\mathfrak{p}_i$ corresponding to p_i then, since p_i is not a square in $k_{\mathfrak{p}_i}((p_i))$, r_i must be even. Hence we may suppose $f = c \in k$, and c is a square in all completions of $k(y)$, for instance in $k((y))$. But this implies that c is a square in k.

If now f is a sum of 2 squares everywhere locally then a similar argument shows that r_i must be even for all prime spots $\mathfrak{p}_i$ with $k_{\mathfrak{p}_i}$ real or $k_{\mathfrak{p}_i}$ non-real and -1 not a square in $k_{\mathfrak{p}_i}$. For the remaining $\mathfrak{p}_i = \mathfrak{p}$ we have:

$$-1 = a^2 \text{ in } k_\mathfrak{p} = k[y]/p,$$

$$1 + g(y)^2 = p(y)\, q(y)$$

for some polynomials $g(y), q(y) \in k[y]$ of degree less than $\deg p$. From this it is easily deduced by induction on the degree of p that p is a sum of 2 squares in $k(y)$, and then the same result follows for f since sums of 2 squares are closed under multiplication (for details see Pfister, 1967). Note that the above arguments for sums of 1 or 2 squares work over an arbitrary field k, not only for $k = \mathbb{R}(x)$.

If finally f is a sum of 4 squares everywhere locally then as above r_i must be even for all $\mathfrak{p}_i$ with real residue class field $k_{\mathfrak{p}_i}$. For the other $\mathfrak{p}_i$ we know from the proof of proposition 1 that -1 is a sum of 1 or 2 squares in $k_{\mathfrak{p}_i}$ and this in turn implies that the corresponding prime polynomial p_i is a sum of 2 or 4 squares in $k(y)$. Since also the constant c must be a sum of squares (in fact of at most two squares) in $k = \mathbb{R}(x)$, and since sums of 4 squares are closed under multiplication, the result for f follows. (Note that the assumption $k = \mathbb{R}(x)$ becomes essential in this case.)

The discussion above suggests on one hand that the Local–Global Principle should also be true for sums of 3 squares and perhaps for arbitrary quadratic forms over $\mathbb{R}(x, y)$. But on the other hand, this would imply that sums of 3 squares are closed under multiplication, which would be rather surprising since the corresponding result is not even true for the field $\mathbb{Q}$ of rational numbers (where the Local–Global Principle is known to hold).

4. For positive definite functions f of a special type the following results are known:

(a) If f is a positive definite polynomial in $\mathbb{R}[x, y]$ of total degree $\leqslant 4$, then f is a sum of 3 squares in $\mathbb{R}[x, y]$ (Hilbert, 1888).

(b) If f is a positive definite polynomial in $\mathbb{R}[x, y]$ of degree $\leqslant 2$ with respect to y, then f is a sum of 3 squares in $\mathbb{R}(x)[y]$. This may be seen as follows:

We can suppose that

$$f(x, y) = g(x)y^2 + h(x) \text{ where } g = g_1{}^2 + g_2{}^2, h = h_1{}^2 + h_2{}^2 \text{ in } \mathbb{R}[x].$$

Put

$$f = (g_1 y + g_2 \eta)^2 + (g_2 y - g_1 \eta)^2 + \xi^2$$

with some functions $\xi, \eta \in \mathbb{R}(x)$. Then the condition on ξ, η is

$$\xi^2 + g\eta^2 = h.$$

This equation is soluble since a quadratic form of shape $(1, g)$ represents all totally positive elements of the field $\mathbb{R}(x)$ (Pfister, 1967).

5. The next case which has been studied to some extent is the case where f has the form

$$f(x, y) = 1 + g(x)y^2 + h(x)y^4 \quad \text{with} \quad g, h \in \mathbb{R}[x].$$

Here f is positive definite whenever

$$h(r) \geqslant 0 \text{ for all } r \in \mathbb{R}$$

and

$$4h(r) - g(r)^2 \geqslant 0 \text{ for all } r \in \mathbb{R} \text{ with } g(r) < 0.$$

Suppose that f is a sum of 3 squares in $\mathbb{R}(x, y)$ and therefore (by Cassels, 1964) in $\mathbb{R}(x)[y]$. Then

$$f = \sum_1^3 (a_i + b_i y + c_i y^2)^2 \quad \text{with} \quad a_i, b_i, c_i \in \mathbb{R}(x).$$

After an orthogonal transformation over $\mathbb{R}(x)$ we may assume that $a_1 = 1$, $a_2 = a_3 = 0$, which implies $b_1 = 0$. The remaining conditions are

$$b_2{}^2 + b_3{}^2 = g - 2c_1, \qquad b_2c_2 + b_3c_3 = 0, \qquad c_2{}^2 + c_3{}^2 = h - c_1{}^2.$$

This implies

$$(g - 2c_1)(h - c_1{}^2) = (b_2c_3 - b_3c_2)^2$$

or

$$(g - \xi)(4h - \xi^2) = \eta^2 \tag{1}$$

for $\xi = 2c_1, \eta = 2(b_2c_3 - b_3c_2) \in \mathbb{R}(x)$.

A necessary and sufficient condition for f to be a sum of 3 squares is therefore that the elliptic curve (1) has a "rational point" (ξ, η) over $\mathbb{R}(x)$ for which $g - \xi$ and $4h - \xi^2$ are positive definite. The trivial point $\xi = g$, $\eta = 0$ does not in general satisfy the side conditions since $4h - g^2$ is not necessarily positive definite.

From the theory of elliptic curves over function fields (see Ogg, 1962), it seems likely that (1) has no non-trivial rational points for suitable choice of g and h, even though locally (i.e. in the completions of $\mathbb{R}(x)$ with respect to its valuations) such points exist. A particularly promising example is given by

$$g(x) = x^2(x^2 - 3), \qquad h(x) = x^2. \tag{2}$$

This example has been found by Motzkin (1965) and gives the simplest known positive definite polynomial, namely

$$f(x, y) = 1 - 3x^2y^2 + x^4y^2 + x^2y^4,$$

which is not a sum of squares of polynomials in $\mathbb{R}[x, y]$. An expression as a sum of 4 squares in $\mathbb{R}(x)[y]$ can easily be deduced from

$$f(x, y) = \frac{(1 - x^2y^2)^2 + x^2(1 - y^2)^2 + x^2(1 - x^2)^2y^2}{1 + x^2}.$$

Unfortunately however, even for the special functions g, h given by (2), it seems to be very hard to find rational points (ξ, η) on (1) or to disprove the existence of such points.

6. In conclusion I should like to suggest that one might search on a computer for rational points on (1) in the case (2). If the degree of the denominator $C(x)$ of a hypothetical point

$$(\xi, \eta) = \left(\frac{A(x)}{C^2(x)}, \frac{B(x)}{C^3(x)}\right)$$

on (1) (where $A(x)$, $B(x)$, $C(x) \in \mathbb{R}[x]$) is of moderate size, this should be possible. The coefficients of A, B, C, though *a priori* real numbers, will lie in some (presumably "small") algebraic number field.

Editorial note. We understand that Pfister and Cassels have independently dealt with (1) in the case (2). It follows that Motzkin's polynomial is not the sum of 3 squares. The answer to the problem is 'No!'.

References

Artin, E. (1927). Über die Zerlegung definiter Funktionen in Quadrate. *Hamb. Abh.* **5,** 100–115. (Collected Papers 273–288).

Ax, J. (1966). On ternary definite rational functions. (Unpublished).

Cassels, J. W. S. (1964). On the representation of rational functions as sums of squares. *Acta Arithmetica* **9,** 79–82.

Hilbert, D. (1888). Über die Darstellung definiter Formen als Summe von Formenquadraten. *Math. Ann.* **32,** 342–350. (*Ges. Abh.* II, 154–161).

Hilbert, D. (1893). Über ternäre definite Formen. *Acta Math.* **17,** 169–197. (*Ges. Abh.* II, 345–366).

Motzkin, T. S. (1967). "The Arithmetic–Geometric Inequality". *In* "Inequalities" (Shisha, O., Ed.) 205–224. Academic Press, London and New York.

Ogg, A. P. (1962). Cohomology of abelian varieties over function fields. *Ann. of Math.* **76,** 185–212.

Pfister, A. (1967). Zur Darstellung definiter Funktionen als Summe von Quadraten. *Invent. Math.* **4,** 229–237.

Witt, E. (1934). Zerlegung reeller algebraischer Funktionen in Quadrate. Schiefkörper über reellem Funktionenkörper. *J. reine angew. Math.* **171,** 4–11.

The Location of Four Squares in an Arithmetic Progression, with some Applications

John Leech

Department of Computing Science, University of Stirling, Scotland

1. Introduction

We consider arithmetic progressions of terms $t_n = t_0 + nd$, where t_0 is the *origin* of the progression, d the *difference*, and n is the *location* of the term t_n. The progression is *trivial* if $d = 0$, otherwise *non-trivial*. All variables except angles range over rational numbers; our interest in results relating to integers will arise only in contexts where multiplication by common denominators is acceptable because of homogeneity, though locally it will often be assumed that variables are integers. The number of variables used is so large that it is not convenient to avoid casual re-use of letters used elsewhere with different connotations.

The main problems considered in this paper are of the following form. Given a finite set of integers $0 = n_1, n_2, \ldots$, can we choose t_0 and $d \neq 0$ such that the terms in locations $n_1, n_2, \ldots$ of the arithmetic progression are all perfect squares? (Clearly $n_1 = 0$ is no restriction—the choice of origin of the progression is arbitrary.) It turns out that for sets of three locations this is always possible, and indeed a general parametric solution can be given, but for sets of four locations the state of affairs is more complicated. We shall obtain solutions in all cases where the locations do not fall into either of two exceptional patterns. In the first of these, the locations are in an arrangement of the form $0, a, b, a + b$; solutions are possible for many values of a, b and are known to be impossible for many others. In the second, the locations are in an arrangement of the form $0, a^2, b^2, (a + b)^2$. In this case there is always the trivial solution with $t_0 = 0$; the question of interest is whether there are others. Solutions are found for certain values of a, b and are known to be impossible for $a = 1, b = 2$; they are probably impossible for many other values of a, b.

A *perfect rational cuboid* is a rectangular parallelepiped whose edges, face diagonals and body diagonal are all integers. It is not known whether any exists. If one omits the condition on the body diagonal, or on one of the face diagonals, the problem reduces to that of finding three squares whose sums in pairs are all squares or whose differences are all squares respectively. These problems are both related to the first exceptional pattern for squares in an arithmetic progression above; the second is related to the problem of finding right-angled spherical triangles whose sides and angles are such that all their trigonometrical functions have rational values.

Another problem considered is that of finding sets of unequal integers such that the sum of their quotients, taken over all distinct pairs, is zero. This has no solution for sets of fewer than four integers; the problem for sets of four integers is closely related to that of the second exceptional pattern for squares in an arithmetic progression above.

Many references in this area are of considerable antiquity. To save space listing details which few readers will pursue, I list all references earlier than the present century by their reference numbers in Dickson (1920), for example Euler (XVI^{81}) refers to footnote 81 of Chapter XVI of Dickson.

2. Three and Four Squares, General Case

Of three squares, we may take any two to be $(x \pm y)^2$; then the third is $x^2 + 4\lambda xy + y^2$ for some value of λ (the factor 4 is introduced for later convenience), so we have to make $x^2 + 4\lambda xy + y^2$ square. The general solution of this is $x = c(u^2 - 1)$, $y = c(2u + 4\lambda)$, which gives $x^2 + 4\lambda xy + y^2 = c^2(u^2 + 4\lambda u + 1)^2$; this solution is complete and needs no further comment.

Of four squares, we take two to be $(x \pm y)^2$ as above, and the others to be $x^2 + 4\lambda xy + y^2$ and $x^2 + 4\mu xy + y^2$. We make $x^2 + 4\lambda xy + y^2$ square as above by putting $x = u^2 - 1$, $y = 2u + 4\lambda$. Then

$$\begin{aligned} x^2 + 4\mu xy + y^2 &= u^4 - 2u^2 + 1 + 4u^2 + 16\lambda u + 16\lambda^2 + 8\mu u^3 \\ &\qquad + 16\lambda\mu u^2 - 8\mu u - 16\lambda\mu \\ &= u^4 + 8\mu u^3 + (2 + 16\lambda\mu)u^2 + (16\lambda - 8\mu)u \\ &\qquad + (1 + 16\lambda^2 - 16\lambda\mu), \end{aligned}$$

which it is required to make square, say $(u^2 + 4\mu u + v)^2$. Equating these expressions, we obtain

$$(2 + 16\lambda\mu - 16\mu^2 - 2v)u^2 + (16\lambda - 8\mu - 8\mu v)u + (1 + 16\lambda^2 - 16\lambda\mu - v^2) = 0,$$

or, rearranged in powers of v,

$$v^2 + (8\mu u + 2u^2)v - \{(2 + 16\lambda\mu - 16\mu^2)u^2 + (16\lambda - 8\mu)u + (1 + 16\lambda^2 - 16\lambda\mu)\} = 0.$$

This equation is quadratic in each of u and v, and we seek rational pairs u, v

which satisfy it. In general, each suitable value of u corresponds to two values of v and vice versa. A trivial solution is obtained by equating to zero the coefficient of u^2, which corresponds to u infinite or $y = 0$ (since $y/x \to 0$ as $u \to \infty$). This gives $v_1 = 1 + 8\mu(\lambda - \mu)$, and this in turn gives (apart from u infinite) $u_1 = \mu - \lambda$. With this value, we have

$$x_1 = (\lambda - \mu)^2 - 1, \qquad y_1 = 2(\lambda + \mu),$$

and the squares are

$$(x_1 \pm y_1)^2,\ \{(\lambda - \mu)(3\lambda + \mu) - 1\}^2,\ \{(\mu - \lambda)(\lambda + 3\mu) - 1\}^2.$$

Call this the *first solution*; it is non-trivial provided that $x_1 y_1 \neq 0$, i.e. $\lambda + \mu \neq 0$ and $|\lambda - \mu| \neq 1$. The exceptional cases all correspond to the required squares being in a set of locations of the form $0, a, b, a + b$, i.e. two of the squares have the same sum as the other two. Fermat (XV^{11}) noted that his method of "triple equations" fails in just these circumstances, although there may be solutions, e.g. for $a = 5, b = 16$ ($t_0 = 1, d = 3$). This, the *first exceptional case*, is discussed in Section 3 below.

To obtain a second solution in the general case, we set $u_1 = \mu - \lambda$ and obtain a quadratic in v whose roots are $v_1 = 1 + 8\mu(\lambda - \mu)$ and $v_2 = -1 - 2(\lambda - \mu)^2$. This value for v_2 gives a quadratic in u whose roots are

$$u_1 = \mu - \lambda \quad \text{and} \quad u_2 = \frac{(\lambda - \mu)^3 - (3\lambda + \mu)}{(\lambda - \mu)(\lambda + 3\mu) + 1}.$$

It may be shown that the solutions given by u_1 and u_2 are not equivalent unless both are trivial, but the *second solution*, that given by u_2, may be trivial when the first is not. In this case it is found that the first solution has $t_0 = 0$ and the locations of the squares are a set of the form $0, a^2, b^2, (a + b)^2$. Although the arithmetic progression is not trivial in our sense, it is clear that this solution is to be regarded as trivial. This, the *second exceptional case*, is discussed in Section 8 below.

If neither exceptional case obtains, the first two solutions are distinct and non-trivial, and further solutions may be obtained by repeating the construction. Thus non-trivial solutions exist in all such cases. Whether or not an exceptional case obtains, this construction may not give all solutions. For fixed values of n_1, n_2, n_3, n_4 the problem is one of finding rational points on a cubic curve of genus 1. These are well known to form a finitely generated group. The solutions constructed by the present method are those generated by a certain set of generators which may or may not be the complete set. In the first exceptional case, the trivial solutions form a finite group of order 8, and in the second exceptional case a group of order 12. If in the first exceptional case a, b, $a + b$ are all squares, forming a Pythagorean triple, then the points corresponding to $t_0 = 0$ form with the trivial solutions a group of order 16. If neither of these conditions obtains, the solutions involved in the present construction do not form a finite group.

3. First Exceptional Case

In this case the required squares are in locations $0, a, b, a + b$, and we may therefore (using a different notation) take the squares to be

$$(x \pm y)^2 = e \pm f, \qquad (z \pm t)^2 = e \pm g,$$

and we have to investigate for what ratios $f : g$ these equations can be solved.

We have

$$e = x^2 + y^2 = z^2 + t^2,$$

from which the factorization

$$x^2 + y^2 = z^2 + t^2 = (\alpha^2 + \beta^2)(\gamma^2 + \delta^2)$$

gives

$$x = \alpha\gamma + \beta\delta, \qquad z = \alpha\gamma - \beta\delta,$$
$$y = \alpha\delta - \beta\gamma, \qquad t = \alpha\delta + \beta\gamma.$$

Now $f = 2xy$ and $g = 2zt$, whence

$$\frac{b}{a} = \frac{g + f}{g - f} = \frac{(\alpha\gamma - \beta\delta)(\alpha\delta + \beta\gamma) + (\alpha\gamma + \beta\delta)(\alpha\delta - \beta\gamma)}{(\alpha\gamma - \beta\delta)(\alpha\delta + \beta\gamma) - (\alpha\gamma + \beta\delta)(\alpha\delta - \beta\gamma)}$$

$$= \frac{\alpha^2 - \beta^2}{2\alpha\beta} \cdot \frac{2\gamma\delta}{\gamma^2 - \delta^2}.$$

Write $P(h, k) = (h^2 - k^2)/2hk$; then $P(h, k)$ is the ratio of the perpendicular sides of a rational right-angled triangle, since $(h^2 - k^2)^2 + (2hk)^2 = (h^2 + k^2)^2$. Let us call $P(h, k)$ the *Pythagorean ratio* (or P-ratio for short) formed from h, k. ($P(h, k)$ is $\cot\theta$, where $\tan\frac{1}{2}\theta = k/h$.) In this terminology we may express our conclusion above that solutions are possible in this first exceptional case whenever b/a is expressible as the product or quotient of two P-ratios. (We need not distinguish between products and quotients since $P(h + k, h - k) = 1/P(h, k)$, and it will often be more convenient to talk of products.) Possible values of b/a are readily constructed, and we may seek general classes of values. Among integers we have, for example, $n(4n + 3) = P(4n + 2, 1)/P(2n + 1, 2n)$ and $n(4n - 3) = P(4n - 2, 1)/P(2n, 2n - 1)$, which include the integers 7, 10, 22, 27, 45, 52, ...

A certain amount of negative information is also obtainable. Fermat (XV^{11}) knew that $b/a = 2$ is impossible; this is the result that four consecutive terms of an arithmetic progression cannot all be squares. Genocchi (XIV^{44}) extended this to the case $b/a = p$, solutions being impossible whenever p is a prime of the form $8n \pm 3$ such that $p^2 - 1$ has no prime factor of the form $4n + 1$. This excludes the values $p = 3, 5, 13, 37, 43, \ldots$. The proof is by infinite descent. Similar arguments show that we have no solutions for $b/a = 4p$ if p is a prime such that $16p^2 - 1$ has no prime factor of the form $4n + 1$, or for $b/a = \frac{1}{4}p$ if p is a prime such that $p^2 - 16$ has no prime factor of the form $4n + 1$. The former enables us to exclude $4p = 8, 20, 68, \ldots$, and the latter to exclude $\frac{1}{4}p$ for $p = 2, 3, 5, 7, 23, \ldots$. These and some similar results (Leech, unpublished) allow us to conclude that the smallest possible integer values of b/a are 7, 10, 11, 12, 14, and that of rational values having $a + b \leqslant 10$, the only possible values are $b/a = 7$ above and $b/a = 5/2 = P(4, 1)/P(2, 1)$. The squares in these two cases are $1^2, 11^2, 29^2, 31^2$ and $1^2, 7^2, 11^2, 13^2$ respectively.

4. First Exceptional Case, *b*/*a* Square

Because of its applications in Section 6, we give special consideration to the case where b/a is the square of an integer or rational number. In this case we require $P(\alpha, \beta)P(\gamma, \delta)$, and hence also $P(\alpha, \beta)/P(\gamma, \delta)$, to be square, and the product and quotient of $\alpha\beta(\alpha^2 - \beta^2)$ and $\gamma\delta(\gamma^2 - \delta^2)$ will be square also. This problem has arisen in several contexts (see XVI^{81} and cross-references there cited). The earliest reference is Diophantus (Heath, 1910), who, in Lemma 2 to Book 5 Prop. 7, gave a triple solution which may be expressed as follows. If $\xi^2 + \xi\eta + \eta^2$ is a perfect square ζ^2, then the triangles [having P-ratios] formed from ζ, ξ, from ζ, η and from $\zeta, \xi + \eta$ have equal areas. In our terminology, the product and quotient of any two of the P-ratios $P(\zeta, \xi)$, $P(\zeta, \eta)$, $P(\zeta, \xi + \eta)$ are squares. To make $\xi^2 + \xi\eta + \eta^2$ square, we put $l + m + n = 0$ and set $\xi = l^2 - m^2, \eta = m^2 - n^2, \xi + \eta = l^2 - n^2, \zeta = \frac{1}{2}(l^2 + m^2 + n^2)$. The six values of the products and quotients of these P-ratios are then just the squares of the ratios of pairs of the four numbers $\xi, \eta, \xi + \eta, 2\zeta$.

Many other solutions exist, and a computer search for them is easily organised. The values of $\frac{1}{3}\alpha\beta(\alpha^2 - \beta^2)$ are calculated and their squared factors removed (values of α, β which are both odd or are not coprime are omitted). The list is then sorted into increasing order, and equal values, which give suitable pairs of P-ratios, are now adjacent. I have a list obtained thus for all $\alpha, \beta \leqslant 100$, which takes only a few minutes, and the range could be extended easily.

Among results of interest found from this table are the following. As remarked in Section 2, if a, b, $a + b$ are all squares, forming a Pythagorean triple, there is a finite group of 16 points on the cubic curve. The table gave the following non-trivial solutions in which b/a is the square of a P-ratio:

$$P(13, 8)P(14, 1) = \{P(4, 1)\}^2,$$
$$P(19, 8)P(22, 3) = \{P(4, 1)\}^2,$$
$$P(50, 23)P(96, 73) = \{P(8, 5)\}^2.$$

The first of these admits generalization. It is a case of the Diophantine solution given above with $\eta = 8$, $\xi + \eta = 15$, depending on the fact that the integers 8, 15 satisfy both $m^2 + n^2 =$ square and $m^2 - mn + n^2 =$ square. Now this is precisely the case $\lambda = 0$, $\mu = -\frac{1}{4}$ of the general problem as dealt with in Section 2, and indeed this is the first solution for these values. The second solution gives

$$P(3637, 1768)P(3026, 611) = \{P(52, 17)\}^2,$$

and further solutions may be found by the same method.

One may enquire what higher powers of rational numbers are representable as products of P-ratios. The table is here less fruitful than that of Section 6, from which is deduced the example

$$P(1313, 703)/P(1924, 1919) = 256.$$

Other powers of integers represented include 3^3, 3^4, 11^2, 13^2, 18^2, 21^2, 22^2, 31^2, all but the first being deducible from the table of Section 6. Powers of rational numbers represented include $(3/4)^4$, $(2/5)^4$ and $(3/14)^4$.

5. Three Squares whose Sums in Pairs are Squares

This is the classical rational cuboid problem (XIX^{1-30}) and has been treated extensively by Kraitchik (1947). Suppose the sums to be made square are $y^2 + z^2$, $z^2 + x^2$, $x^2 + y^2$, so that y/z, z/x, x/y are P-ratios. Then the problem is that of finding triads of P-ratios whose product is 1, or equivalently finding pairs of P-ratios whose product (or quotient) is also a P-ratio. When x, y, z have had any common factor removed, we see that one of them is odd and the other two are even. Kraitchik gives a list of 240 solutions in which the odd term is smaller than 10^6, and gives also the parameters α, β from which the P-ratios $P(\alpha, \beta)$ are formed. Lal and Blundon (1966) give a list of solutions corresponding to pairs of P-ratios whose parameters do not exceed 70 and whose quotients are also P-ratios. Their list was made by computer search, and includes eighteen solutions within the range of Kraitchik's list but absent from it.

Analysis of these lists of solutions shows that none of them admit combination so as to give sets of four squares whose sums in pairs are all squares. However, any two solutions involving a common P-ratio can be combined to give sets of four squares such that five of the sums of the pairs are squares; a simple way of constructing such sets is to take any set x, y, z, as above and adjoin t such that $t/z = y/x$, which makes $t^2 + z^2$ and $t^2 + y^2$ square, though it seems highly unlikely that $t^2 + x^2$ will be square for any such set.

This analysis shows that many P-ratios do not appear in solutions. A result stated in Section 3 is that the product of two P-ratios cannot be $\frac{3}{4}$, so $P(2, 1)$ cannot occur in any solution, but this seems to be the only negative result known. A result given by Kraitchik is that for two (or possibly all three) of the P-ratios involved in any solution the product $\alpha\beta(\alpha^2 - \beta^2)$ is divisible by 11. So in only one of them can we have $\alpha + \beta < 11$. The only P-ratios formed from parameters α, β with sum smaller than 11 occurring in any of the listed solutions are $P(5, 2)$ and $P(4, 3)$, which occur in the solutions $P(5, 2)P(18, 7) = P(8, 3)$ and $P(4, 3)P(10, 1) = P(16, 5)$. The solution with the smallest squares is 44, 117, 240, corresponding to $P(6, 5)P(11, 2) = P(8, 5)$; this is also the earliest solution on record (given by Halcke in 1719, XIX[1]).

Another problem which has received attention is that of the perfect rational cuboid, in which we require $x^2 + y^2 + z^2$ to be square also. No solution is known, but attempts to prove that none exists (XIX[25-29]) are not valid. Lal and Blundon remark that no solution in their list satisfies this further condition; Kraitchik does not comment on the possibility. Lal and Blundon give a solution of the modified problem in which the sum of all three squares is to be square but the sum of one of the pairs is not, having evidently not identified this problem with that of the next section.

6. Three Squares whose Differences are Squares

Many writers, notably Euler (see Heath, 1910 or Dickson, 1920), have discussed two related problems. The first is that of finding sets of three integers, all pairs of which have their sums and differences squares (XV[28]) and cross-references there cited); the other is that of finding three squares whose differences are all squares (XIX[40-44]). Clearly the sums of the pairs of integers in the first problem are squares satisfying the requirements of the second, and any set of squares satisfying these requirements can, after doubling their sides if necessary, be made the sums of pairs of three integers satisfying the former conditions (perhaps with changes of sign if positive integers are insisted on). We consider here the problem in the latter form.

Suppose we have a solution of the set of equations $x^2 + y_1{}^2 = z_1{}^2$, $x^2 + y_2{}^2 = z_2{}^2$, $x^2 + y_1{}^2 + y_2{}^2 = t^2$. Then the squares $t^2, z_1{}^2, y_1{}^2$ are such that the difference between any two of them is square, since $t^2 - z_1^2 = y_2^2$, $t^2 - y_1^2 = z_2^2$ and $z_1^2 - y_1^2 = x^2$, and so similarly are the squares t^2, z_2^2 and y_2^2. Thus solutions of this problem go in pairs, and it is convenient to study the problem in the symmetrical form stated, a solution of which will be referred to as "a solution". If we could solve also the equation $y_1^2 + y_2^2 =$ square, we should have a perfect rational cuboid. Thus if any such exists it would give three solutions to our present problem; none has been found among the solutions examined.

To find solutions, we note that $x^2 + t^2 = z_1^2 + z_2^2$, so x^2, z_1^2, z_2^2, t^2 are four squares in an arithmetic progression, in the first case of exception (Section 3), and further we have $b/a = y_2^2/y_1^2$, so we have the special case considered in Section 4. Also we need the progression to have difference $d = 1$, which requires not merely that $P(\alpha, \beta)P(\gamma, \delta)$ be square, but that the numerator and denominator of this product, namely $(\alpha^2 - \beta^2)(\gamma^2 - \delta^2)$ and $4\alpha\beta\gamma\delta$, be squares separately. This has many solutions. For example if $\alpha, \beta, \gamma, \delta$ are any integers satisfying $P(\alpha, \beta)P(\gamma, \delta) =$ square, and $A = (\alpha^2 + \beta^2)^2$, $B = 4\alpha\beta(\alpha^2 - \beta^2)$, $C = (\gamma^2 + \delta^2)^2$, $D = 4\gamma\delta(\gamma^2 - \delta^2)$, then both $(A^2 - B^2) \times (C^2 - D^2)$ and $4ABCD$ are squares. Euler (XV[28]) gave a solution equivalent to the following. Let $\alpha, \beta, \gamma, \delta$ be the squares of A, B, C, D. Then $4\alpha\beta\gamma\delta$ is clearly square, and we require to make $(A^4 - B^4)(C^4 - D^4)$ square. Euler listed values of differences of fourth powers, noting that $3^4 - 2^4$, $9^4 - 7^4$, $11^4 - 2^4$ are all square multiples of 65, so any pair will have their product square.

Another way to find solutions is to note that $t^2 = y_1^2 + z_2^2 = y_2^2 + z_1^2$, the sum of two squares in two different ways, so it is the product of two sums of two squares. If $y_1/x = P(\alpha, \beta)$ and $y_2/x = P(\gamma, \delta)$, this gives

$$t^2 = 4(\alpha^2\gamma^2 + \beta^2\delta^2)(\alpha^2\delta^2 + \beta^2\gamma^2),$$

so we require $\alpha, \beta, \gamma, \delta$ to satisfy $(\alpha^2\gamma^2 + \beta^2\delta^2)(\alpha^2\delta^2 + \beta^2\gamma^2) =$ square. Euler (XVI[81]) satisfied this condition by making each factor square separately, giving two Pythagorean conditions. Thus $\alpha\gamma/\beta\delta$ and $\alpha\delta/\beta\gamma$ are two P-ratios; their product and quotient are the squares of α/β and γ/δ, so we have solutions from any pairs of P-ratios whose product and quotient are squares. For example Diophantus's solution (Section 4) shows that, with $l+m+n=0$, we may take $\alpha, \beta, \gamma, \delta$ to be $l^2 + m^2 + n^2$, $l^2 - m^2$, $l^2 - n^2$, $m^2 - n^2$ in any order; thus each set l, m, n gives three distinct solutions, although replacing l, m, n by $m - n$, $n - l$, $l - m$ gives only the same set of solutions again.

We have just seen that if α^2/β^2 is the product of two P-ratios, then $P(\alpha, \beta)$ is a P-ratio occurring in a solution to this problem. The converse is also valid, namely for any P-ratio $P(\alpha, \beta)$ occurring in a solution of the problem, α^2/β^2 is the product (or, more immediately, quotient) of two P-ratios. We found above that the product $(\alpha^2\gamma^2 + \beta^2\delta^2)(\alpha^2\delta^2 + \beta^2\gamma^2)$ is square, and the case we have to consider is when the factors are not separately square. Their quotient is square, however, so we have $h^2(\alpha^2\gamma^2 + \beta^2\delta^2) = k^2(\alpha^2\delta^2 + \beta^2\gamma^2)$ for some integers h, k. This gives $\alpha^2(\gamma^2h^2 - \delta^2k^2) = \beta^2(\gamma^2k^2 - \delta^2h^2)$, from which it follows at once that $P(\gamma k, \delta h)/P(\gamma h, \delta k) = \alpha^2/\beta^2$. We thus see that a necessary and sufficient condition for a P-ratio $P(\alpha, \beta)$ to be a value of y/x occurring in solutions to the present problem is that α^2/β^2 be expressible as the product of two P-ratios. This condition, for fixed α, β, is one of finding rational points on the cubic curve $\alpha^2/\beta^2 = P(u, 1)/P(v, 1) = v(u^2 - 1)/(u(v^2 - 1))$, so if it has any non-trivial solution it will have infinitely many. Results stated in Section 3 include those that 4 and 9 are not the products of pairs of P-ratios. Thus $P(2, 1)$ and $P(3, 1)$ do not occur in the solutions of the problem of this section, and so the simplest and most familiar case of a rational right-angled triangle, with $3^2 + 4^2 = 5^2$, cannot occur in solutions to the problems of this or the preceding section.

Lyness 1961 observed that solutions of this problem come in sets of five related solutions, and was led to obtain the sequence u_i determined by two initial non-zero values and the relation $u_{i-1}u_{i+1} = 1 + u_i$, which is of period 5 whatever the initial values. In terms of arbitrary u_1 and u_2, we have

$$u_3 = \frac{1 + u_2}{u_1}, \qquad u_4 = \frac{1 + u_1 + u_2}{u_1 u_2}, \qquad u_5 = \frac{1 + u_1}{u_2}.$$

Suppose that $u_1 = y_1{}^2/x^2$ and $u_2 = y_2{}^2/x^2$ in any solution to the problem of this section. The requirements of the problem are that $u_1, u_2, 1 + u_1, 1 + u_2, 1 + u_1 + u_2$ be rational squares, so u_3, u_4, u_5 are squares also, and any two consecutive u_i can be used to solve our problem. Indeed if x is chosen so that x^2u_i is an integer square for each i, we have a solution of the fivefold composite problem to find integers $x, y_1, y_2, y_3, y_4, y_5$ such that $x^2 + y_i{}^2$ and $x^2 + y_i{}^2 + y_{i+1}{}^2$ are all squares (including $x^2 + y_5{}^2 + y_1{}^2$).

H. S. M. Coxeter (1971) has observed that this sequence u_i is the algebraic basis of Napier's rules for the solution of spherical triangles. Let a, b, c be the sides and A, B, C the angles of a spherical triangle, right-angled at A, and let $\theta_1 = b, \theta_2 = 90° - C, \theta_3 = 90° - a, \theta_4 = 90° - B, \theta_5 = c$, or in the polar triangle with $a = 90°$ let $\theta_1 = B, \theta_2 = 90° - c, \theta_3 = A - 90°$, $\theta_4 = 90° - b, \theta_5 = C$. Then in either case Napier's rules take the form $\sin\theta_i = \tan\theta_{i-1}\tan\theta_{i+1}$ and $\sin\theta_i = \cos\theta_{i-2}\cos\theta_{i+2}$, all subscripts being taken modulo 5. If we put $u_i = \cot^2\theta_i$, these rules take the form $u_{i-1}u_{i+1} = 1 + u_i$ and $u_{i-2}u_iu_{i+2} = 1 + u_{i-2} + u_{i+2}$. The result that all five u_i can be squares of rational numbers now admits the following interpretation. Each u_i is the square of a P-ratio, and may be expressed as $u_i = (P(\alpha, \beta))^2$ where $\beta/\alpha = \tan\frac{1}{2}\theta_i$. Then all the trigonometric functions of the θ_i are rational, and we have sets of five right-angled (or quadrant-sided) spherical triangles such that all the trigonometric functions of their sides and angles are rational quantities. Pairs of such right-angled spherical triangles can be combined to make non-right-angled spherical triangles with all the trigonometric functions of their sides and angles rational, for example we may abut two triangles with their right-angled vertices common and a common side of length θ_1 but the other sides adjacent to the right angle being θ_2 and θ_5.

I conclude this section by giving the simplest parametric set of values for α, β, and a table, obtained from a computer search, of the numerical values involving the smallest integers. The parametric solution is derived from Diophantus's solution of $P(\alpha, \beta)P(\gamma, \delta) =$ square, as given above but is more concisely stated in terms of parameters p, q, r satisfying $p^2 + 3q^2 = r^2$, so that the ξ, η, ζ of Section 4 are $p + q, q - p, r$ respectively. Then the u_i are the P-ratios corresponding to

$$\tan \tfrac{1}{2}\theta = \frac{\beta}{\alpha} = \frac{q-p}{q+p}, \quad \frac{q}{r}, \quad \frac{pr}{2q^2}, \quad \frac{pq}{p^2+2q^2}, \quad \frac{qr-(p^2+q^2)}{qr+(p^2+q^2)}.$$

A few remarks, based on inspection of the table of solutions and easily proved, follow. A value of $\tan \frac{1}{2}\theta_i$ is said to be *divisible* by an integer n when, being expressed as a fraction in its lowest terms, it has either its numerator or its denominator divisible by n. Of the five values in any cycle (always treated cyclically) two adjacent values are not divisible by 2, while of the other three, which are all divisible by 2, the index of the power of 2 dividing the middle value is one greater than the sum of the indices of the powers dividing the others. Similarly two adjacent values are not divisible by 3 while the middle of the others is divisible by a power of 3 whose index is the sum of the indices of the powers of 3 dividing the other values. Other primes either behave in the same way as 3, or they divide only two adjacent values to the same power, or they divide none of the values. In particular any prime dividing a value must divide at least one of the adjacent values. It follows from this that the product of all the numerators and denominators is twice a square; indeed the square root of eight times the product is the least value of x such that x and all the y_i are integers. The smallest such values are for the second cycle in the list, and are $x = 78624$, $y_i = 115668, 508032, 349440, 55432, 21645$. The smallest triad x, y_1, y_2, also from this cycle, are 104, 153, 672.

7. The Quest for the Perfect Rational Cuboid

As already remarked, it is not known whether any perfect rational cuboid exists, and part of my interest in the problem of Section 6 was the possibility that this area, less well worked than that of Section 5, might be more fruitful. So far it has not been so in this respect, but a survey of the prospects is opportune. We saw in Section 6 that the ratio $y_1{}^2/y_2{}^2$ has to be the product of two P-ratios in such a way that the products of the numerators and of the

TABLE. giving all cycles of parameter pairs α_i, β_i in which α_1, β_1, α_2, β_2 do not exceed 50.

α_1	β_1	α_2	β_2	α_3	β_3	α_4	β_4	α_5	β_5	notes
7	4	5	3	45	11	33	4	32	7	1,2
13	1	9	1	27	14	21	16	13	4	
14	3	8	5	55	34	187	39	91	9	1,2
14	5	8	3	54	19	171	55	77	25	1,2
15	7	13	4	143	32	153	44	189	85	1,2
19	9	14	3	49	8	22	7	33	19	
21	1	18	1	24	11	22	19	19	7	
21	5	19	8	247	128	297	104	385	81	1,2
21	16	11	10	275	37	555	7	196	9	1
21	17	20	9	240	19	418	29	319	119	1
23	7	14	3	32	9	24	11	23	11	
24	11	5	2	91	25	195	49	63	22	1
26	7	15	8	190	99	627	161	299	49	1,2,3
26	15	8	7	154	41	451	15	225	13	1,2
26	19	16	1	252	5	175	27	65	57	
28	3	13	9	39	31	62	11	176	7	1
29	3	23	2	64	23	21	16	63	29	
29	22	11	1	119	3	126	17	29	24	
31	1	22	1	64	33	161	144	217	69	
31	23	31	7	108	7	608	81	92	57	
33	19	16	9	273	64	98	13	77	19	1
35	11	31	12	713	288	817	376	1045	301	1,2
36	7	35	33	319	215	1247	34	272	3	1
37	9	17	12	272	161	46	7	37	3	
37	20	33	7	1309	171	969	260	800	481	1,2
38	5	21	16	396	301	1419	185	703	25	1,2
38	21	16	5	340	59	1003	231	441	209	1,2
38	21	27	23	207	59	118	5	280	19	
40	7	30	7	99	47	1551	851	74	23	1
43	19	33	7	693	124	93	32	76	43	
43	24	35	13	1729	335	1273	264	1152	473	1,2
45	17	16	5	217	64	539	186	297	119	1
47	16	13	4	279	91	217	81	47	18	
48	29	35	4	931	55	2717	651	522	403	1
49	11	47	21	171	94	152	51	119	22	

Notes: 1. These solutions are such that $\alpha_1{}^2\alpha_2{}^2 + \beta_1{}^2\beta_2{}^2$ and $\alpha_1{}^2\beta_2{}^2 + \beta_1{}^2\alpha_2{}^2$ are squares.

2. These solutions are of the parametric form given in the text.

3. In this solution, $\alpha_1{}^2\alpha_2{}^2 + \beta_1{}^2\beta_2{}^2$, $\alpha_1{}^2\beta_2{}^2 + \beta_1{}^2\alpha_2{}^2$, $\alpha_2{}^2\alpha_3{}^2 + \beta_2{}^2\beta_3{}^2$ and $\alpha_2{}^2\beta_3{}^2 + \beta_2{}^2\alpha_3{}^2$ are all squares. Since products such as $(\alpha_1{}^2\alpha_2{}^2 + \beta_1{}^2\beta_2{}^2)(\alpha_2{}^2\alpha_3{}^2 + \beta_2{}^2\beta_3{}^2)$ are square multiples of $\alpha_2{}^2 + \beta_2{}^2$, this is possible only when $\alpha_2{}^2 + \beta_2{}^2$ is square, of which this is the only example within the range of the table.

denominators have to be squares separately. To lead to a perfect rational cuboid, we need also that y_1/y_2 is itself a P-ratio. Combining these requirements, we require two P-ratios such that the products of their numerators and denominators are the squares of the numerator and denominator of another P-ratio. Alternatively expressed, by putting the P-ratios involved on a common denominator $2N$, we require an integer N which admits three decompositions into pairs of factors $N = x_1 y_1 = x_2 y_2 = x_3 y_3$ such that the differences of squares $x_1^2 - y_1^2, x_2^2 - y_2^2, x_3^2 - y_3^2$ form a geometric progression; this is equivalent to solving $(\alpha^2\gamma^2 - \beta^2\delta^2)(\alpha^2\delta^2 - \beta^2\gamma^2) = (\alpha^2\beta^2 - \gamma^2\delta^2)^2$ in integers. At the time of writing, I have not pursued this approach.

A few examples of products of P-ratios which are squares of P-ratios are given in Section 4. In each case the numerator and denominator of the product are non-square multiples of squares, the non-squares being 91, 627, 5037 and 15213571 respectively for the examples displayed. This can hardly be described as encouraging belief in the prospects of finding an example without a non-square factor. Some encouragement might be gleaned from the fact that two of the P-ratios whose squares are products of P-ratios are themselves products of P-ratios, as any example without a non-square factor would have to be, namely $P(8, 5) = P(6, 5)P(11, 2)$ and $P(52, 17) = P(61, 11) \times P(416, 255)$.

8. Second Exceptional Case

In this case we require squares of the forms

$$t^2, \quad u^2 = t^2 + a^2 d, \quad v^2 = t^2 + b^2 d, \quad w^2 = t^2 + (a+b)^2 d.$$

We have

$$t^2 + w^2 - u^2 - v^2 = 2abd,$$

whence

$$(t^2 + w^2 - u^2 - v^2)^2 = 4a^2b^2d^2 = 4(u^2 - t^2)(v^2 - t^2),$$

$$\text{i.e. } (u^2 + v^2 + w^2 - t^2)^2 = 2(u^4 + v^4 + w^4 - t^4).$$

This equation is conveniently studied by making the transformation

$$\begin{aligned} 2x_1 &= t + u + v + w, & 2t &= x_1 + x_2 + x_3 + x_4, \\ 2x_2 &= t - u + v - w, & 2u &= x_1 - x_2 + x_3 - x_4, \\ 2x_3 &= t + u - v - w, & 2v &= x_1 + x_2 - x_3 - x_4, \\ 2x_4 &= t - u - v + w, & 2w &= x_1 - x_2 - x_3 + x_4, \end{aligned}$$

under which it takes the symmetrical form

$$\sum_{i \neq j \neq k} x_i^2 x_j x_k = 0, \qquad \text{or} \qquad \sum_{i \neq j} \frac{x_i}{x_j} = 0.$$

We thus have to solve, for the case $N = 4$, the problem of finding sets of N integers whose quotients in pairs have their sum zero. Much of what follows can be applied to any $N \geqslant 4$; it is obvious that there is no solution for $N = 2$ since x_1/x_2 and x_2/x_1 have the same sign, and it can be proved by an infinite descent argument that there is no solution for $N = 3$.

Suppose that x_1, x_2, x_3 are chosen, and we are seeking values of x_4 to complete solutions. Let s_1, s_2, s_3 be the symmetric functions $x_1 + x_2 + x_3$, $x_2x_3 + x_3x_1 + x_1x_2$, and $x_1x_2x_3$. Then the equation $\Sigma x_i^2 x_j x_k = 0$ becomes

$$s_2 x_4^2 + (s_1 s_2 - 3s_3)x_4 + s_1 s_3 = 0,$$

and we seek sets of x_1, x_2, x_3 such that this has rational solutions.

First solution. Put $s_1 = 0$, or $x_1 = l, x_2 = m, x_3 = n$, say, where $l + m + n = 0$. Then $x_4 = 0$ or $x_4 = 3s_3/s_2 = 3lmn/(mn + nl + lm)$. Thus, in integers, $x_4 = -3lmn$ and $x_1, x_2, x_3 = (mn + nl + lm)(l, m, n)$, giving the solution $t = -\frac{3}{2}lmn, d = -(mn + nl + m), a = l^2 - m^2, b = m^2 - n^2, a + b = l^2 - n^2$ to our original problem, e.g. $9^2 + 7(3^2, 5^2, 8^2) = 12^2, 16^2, 23^2$. In this solution $a^2 + ab + b^2$ is a perfect square; l, m, n can be chosen such that a, b are any integers satisfying this condition.

Second solution. Put $s_2 = 0$. Then $x_4 = \frac{1}{3}s_1$, so, taking $l + m + n = 0$ again, we may put $x_1 = 3mn, x_2 = 3nl, x_3 = 3lm, x_4 = mn + nl + lm$, which leads to $t = -2(mn + nl + lm), d = -3$ and again $a = l^2 - m^2, b = m^2 - n^2$, $a + b = l^2 - n^2$, e.g. $14^2 - 3(3^2, 5^2, 8^2) = 13^2, 11^2, 2^2$.

$s_3 = 0$ leads only to $x_4 = 0$ (as well as one of x_1, x_2, x_3 being zero) or $x_4 = -s_1$, i.e. $t = 0$, which we have already discounted as trivial. This deals with all cases in which the quadratic for x_4 has a degenerate root and so also a rational root.

Third solution. Next we make the quadratic for x_4 a perfect square. Its discriminant is

$$(s_1s_2 - 3s_3)^2 - 4s_1s_2s_3 = (s_1s_2 - s_3)(s_1s_2 - 9s_3).$$

If $s_1s_2 = s_3$, then x_1, x_2, x_3 are roots of the cubic equation

$$x^3 - s_1x^2 + s_2x - s_1s_2 = 0, \quad \text{i.e. } (x - s_1)(x^2 + s_2) = 0,$$

and we have $x_1 = s_1, x_2 = -x_3 = \sqrt{(-s_2)}$, the quadratic for x_4 reducing to $s_2(x_4 - s_1)^2 = 0$; we thus have the trivial solution in which x_1 and x_4 have the same arbitrary value and x_2 and $-x_3$ share another arbitrary value.

If $s_1s_2 = 9s_3$, this equation may be regarded as a plane cubic curve with equation

$$(x_1 + x_2 + x_3)(x_2x_3 + x_3x_1 + x_1x_2) = 9x_1x_2x_3;$$

this has a double point at (1, 1, 1) which enables us to express it parametrically in terms of l, m, n, with $l + m + n = 0$, as

$$x_1 : x_2 : x_3 = mn(m - n) : nl(n - l) : lm(l - m).$$

The quadratic reduces to $s_2(x_4 + \frac{1}{3}s_1)^2 = 0$; here we have

$$s_1 = -(m - n)(n - l)(l - m),$$

so we may take

$$x_1 = \tfrac{3}{2}mn(m - n), \quad x_2 = \tfrac{3}{2}nl(n - l), \quad x_3 = \tfrac{3}{2}lm(l - m),$$
$$x_4 = \tfrac{1}{2}(m - n)(n - l)(l - m).$$

This gives $t = \frac{1}{2}(m - n)(n - l)(l - m)$, $d = -3(mn + nl + lm)$ and yet again $a, b, a + b$ are $l^2 - m^2, m^2 - n^2, l^2 - n^2$, e.g. $10^2 + 21(3^2, 5^2, 8^2) = 17^2, 25^2, 38^2$.

We see that these three solutions all give only values of a, b subject to $a^2 + ab + b^2$ square. They are in fact closely related. From any solution x_1, x_2, x_3, x_4 we can obtain others by replacing each x_i by its reciprocal, or replacing each x_i by $\frac{1}{2}(x_1 + x_2 + x_3 + x_4) - x_i$. Both transformations preserve the values of a and b; the latter preserves the moduli of t, u, v, w and so does not give significantly different solutions to our original problem. Performed alternately, these two transformations give a cycle of six solutions for the x_i, pairs of which give the same values to t^2, u^2, v^2, w^2. The solutions above with $a^2 + ab + b^2$ square are three of such a cycle; the other three may be derived from them by replacing l, m, n by $m - n, n - l, l - m$.

The foregoing solutions deal with all cases in which the quadratic for x_4 has coincident or degenerate roots. In all other cases the values for x_4 will be distinct, so we can obtain further solutions by forming the symmetric functions of any three of the x in a known solution and solving the quadratic to find the second value for the fourth x, which must be rational and distinct from the known value. For this purpose the known solution need not be non-trivial, and this construction is illustrated beginning with a completely trivial solution $m, m, n, -n$. Replacing an m leads to no new solution, as this is one of the cases in which the quadratic has equal roots, but replacing $x_4 = -n$ is more successful. Here $s_1 = 2m + n$, $s_2 = m^2 + 2mn$, $s_3 = m^2n$, and the quadratic reduces to

$$m(x_4 + n)\big((m + 2n)x_4 + m(2m + n)\big) = 0.$$

Expressing the new solution in integers, we get

$$x_1 = x_2 = m(m + 2n), \qquad x_3 = n(m + 2n), \qquad x_4 = -m(2m + n).$$

As x_1 and x_2 are still equal, we fix x_2, x_3, x_4 and find a new value for x_1.

This turns out to be

$$x_1 = \frac{-n(2m+n)(m^2-2mn-2n^2)}{2m^2+2mn-n^2},$$

so, expressing the new solution in integers, we get

$$x_1 = -n(2m+n)(m^2-2mn-2n^2)$$

and

$$x_2, x_3, x_4 = (2m^2+2mn-n^2)\big(m(m+2n),\ n(m+2n),\ -m(2m+n)\big).$$

These x_i are now all distinct, and we have a non-trivial solution having $t = -m(m+2n)(m^2-2mn-2n^2)$, $d = (m^2-n^2)(2m^2+2mn-n^2)$, and $a = 2(m^2+mn+n^2)$, $b = 3mn$, $a+b = (2m+n)(m+2n)$. Noticing that these values of $a, b, a+b$ are unaltered by exchange of m and n, we may suspect, and duly verify, that this exchange gives two of a cycle of three solutions related as above. The third solution has

$$t = 3(m^2-n^2)(m^2+mn+n^2) \text{ and } d = -(m^2-2mn-2n^2)(2m^2+2mn-n^2).$$

By repeating this process, we can find further parametric solutions, of increasing degree and complexity, but it does not seem worth while to do so explicitly. A short computer study of numerical cases involving small integers (not a search—new solutions were derived only by applying the foregoing methods to known solutions of the forms given above) was made, and showed the following two interesting examples.

The second solution above, with $l, m, n = 1, 4, -5$, gives $x_i = -20, -5, 4, -7$ (after removing a common factor 3). Fixing the last three and solving for a replacement for -20, we find the value $-56/13$, or in integers $x_i = 56, 65, -52, 91$ and $t = 80, d = -39, a = 4, b = 7$. Replacing these x_i by $\frac{1}{2}(x_1+x_2+x_3+x_4) - x_i$ we get $x_i = 24, 15, -11, 132$. Fixing the first three and solving for a replacement for 132, we get 280/23, or in integers $x_i = 552, 345, -253, 280$. This gives the interesting values $t = 462, d = -299$, $a = 9, b = 16, a+b = 25$, showing that $a, b, a+b$ can be a Pythagorean triad of squares. (It also shows the rate at which the size of the integers increases with successive operations.)

The second example may have been noticed before, but if so I have failed to trace it. The second solution above, with $l, m, n = 1, 3, -4$, gives $x_i = 9, -12, -36, 13$. Solving for a replacement for -12, we find the value -520, giving $x_i = 9, -520, -36, 13$, and $t = 240, d = 1, a = 44, b = 117$. This solution is of interest in that the numbers 240, 44, 117 also give the smallest solution to the rational cuboid problem (Section 5), so we have simultaneously made $x^2+y^2, x^2+z^2, y^2+z^2$ and $x^2+(y+z)^2$ squares, namely

$240^2 + 44^2 = 244^2$, $240^2 + 117^2 = 267^2$, $44^2 + 117^2 = 125^2$ and $240^2 + (44 + 117)^2 = 289^2$. One might regard this as being almost as unlikely as a solution to the problem of the perfect rational cuboid.

The feature $d = 1$ of this last example is not so unusual; it gives us solutions with $t^2 + a^2$, $t^2 + b^2$ and $t^2 + (a + b)^2$ all squares. Any solution in which d is square may be so written, an example being the first solution of this section in which $d = -(mn + nl + lm)$ which is square whenever l, m, n are of the form $\mu^2 - \nu^2$, $\nu^2 - \lambda^2$, $\lambda^2 - \mu^2$. In such cases a and b have a common factor. An example in which they do not is $60^2 + (11^2, 80^2, 91^2) = 61^2, 100^2, 109^2$; this is the solution in smallest integers.

Pocklington (1913) proved by an infinite descent argument that there is no non-trivial solution with $a = 1$, $b = 2$, but this seems to be the only negative result known. Determining possibility or impossibility for given values of a, b seems substantially more difficult in this second exceptional case than in the first exceptional case. Apart from Pocklington's paper, this case seems to have received no previous attention.

References

Coxeter, H. S. M. (1971). Frieze patterns. *Acta. Arith.* **18,** 297–310.

Dickson, L. E. (1920). "History of the Theory of Numbers" Vol. 2: "Diophantine Analysis". 256. Carnegie Inst. of Washington.

Heath, T. L. (1910). "Diophantus of Alexandria". Cambridge University Press.

Kraitchik, M. (1947). "Théorie des nombres", Vol. 3: "Analyse Diophantine et applications aux cuboides rationnels". Gauthier–Villars, Paris.

Lal, M., and Blundon, W. J. (1966). Solutions of the Diophantine equations $x^2 + y^2 = l^2, y^2 + z^2 = m^2, z^2 + x^2 = n^2$. *Math. Comp.* **20,** 144–147.

Lyness, R. C. (1961). Cycles. *Math. Gazette* **45,** 207–209. (Earlier announcements in (1942). *Math. Gazette* **26,** 62 and *Math. Gazette* (1945) **29,** 231–233).

Pocklington, H. C. (1913). Some Diophantine impossibilities. *Proc. Cambridge Phil. Soc.* **17,** 110–118.

Diophantine Equations Involving Generalized Triangular and Tetrahedral Numbers

AVIEZRI S. FRAENKEL*

The Weizmann Institute of Science, Rehovot, Israel

and

Bar Ilan University, Ramat Gan, Israel

1. Introduction

Let

$$T_x^m(\delta) = x(x+\delta)(x+2\delta)\dots\big(x+(m-1)\,\delta\big),$$

where m and δ are arbitrary but fixed positive integers, $m > 1$. We concern ourselves with the diophantine equation

$$T_x^m(\delta) + T_y^m(\delta) = T_z^m(\delta), \tag{1.1}$$

mainly for the cases $m = 2, 3$.

It is easy to see that (1.1) has infinitely many solutions for the case $m = 2$, $\delta = 1$, i.e., the sum of two triangular numbers is infinitely often a triangular number. An infinite class of solutions is given by the identity:

$$\left(\frac{n(n+1)}{2} - 1\right)\frac{n(n+1)}{2} + n(n+1) = \frac{n(n+1)}{2}\left(\frac{n(n+1)}{2} + 1\right).$$

See Sierpiński (1964). Three other infinite classes were given by Khatri (1955):

$$\left.\begin{aligned} 3k(3k+1) + (4k+1)(4k+2) &= (5k+1)(5k+2) \\ (5k+4)(5k+5) + (12k+9)(12k+10) &= (13k+10)(13k+11) \\ (15k+9)(15k+10) + (8k+4)(8k+5) &= (17k+10)(17k+11), \end{aligned}\right\} \tag{1.2}$$

$$k = 0, 1, 2, \dots.$$

* This work was sponsored, in part, by the U.S. National Bureau of Standards. Reproduction in whole or in part is permitted for any purpose of the U.S. Government.

In Section 2 we characterize all solutions of

$$T_x^2(\delta) + T_y^2(\delta) = T_z^2(\delta), \tag{1.3}$$

for any integer δ. For $\delta = 1$, the result subsumes those of Sierpiński and Khatri, and it also includes, for $\delta = 0$, the Pythagorean triplets case.

To begin with, any solution (x, y, z) of (1.3) for which $(x, y, z, \delta) = d$, induces a solution

$$x'(x' + \delta') + y'(y' + \delta') = z'(z' + \delta'), \quad (x', y', z', \delta') = 1,$$

where

$$x = x'd, \qquad y = y'd, \qquad z = z'd, \qquad \delta = \delta'd,$$

and conversely. Thus any solution of (1.3) with $(x, y, z, \delta) = d$, may be obtained from a solution $x'\,y'\,z'$ with a parameter δ', satisfying $(x', y', z', \delta') = 1$. Hence we may assume $(x, y, z, \delta) = 1$. Now suppose that (1.3) holds with δ even. If both x and y are even, we have $(x, y, z, \delta) > 1$. If both x and y are odd, z must be even. But then the left-hand side of (1.3) is $\equiv 2 \pmod 4$ and the right-hand side is $\equiv 0 \pmod 4$. Hence we may assume, without loss of generality, that x is odd and y is even if δ is even. We prove

THEOREM 1. *Let δ be a non-negative integer. Suppose that*

$$x(x + \delta) + y(y + \delta) = z(z + \delta), \quad (x, y, z, \delta) = 1 \tag{1.4}$$

for integers x, y, z. Define

$$\left.\begin{aligned} d_1 &= (x + z + \delta, y), & d_2 &= (x + z + \delta, y + \delta), \\ d_3 &= (y + z + \delta, x), & d_4 &= (y + z + \delta, x + \delta). \end{aligned}\right\} \tag{1.5}$$

Then

$$(d_1, d_4) = (d_2, d_3) = 1. \tag{1.6}$$

Moreover, if δ is odd,

$$\left.\begin{aligned} x = d_3\,((2 - \theta)\,d_1 - d_4), \qquad y = d_1\,((2 - \theta)\,d_3 - d_2), \\ z = d_1\,d_2 + d_3\,d_4 - (1 + \theta)\,d_2\,d_4, \end{aligned}\right\} \tag{1.7}$$

$$(1 + \theta)\,d_2\,d_4 - (2 - \theta)\,d_1\,d_3 = \delta, \qquad \theta = 0 \text{ or } 1. \tag{1.8}$$

If δ is even,

$$x = d_3 (d_1 - d_4), \quad y = d_1 (d_3 - \tfrac{1}{2} d_2), \quad z = \tfrac{1}{2} d_1 d_2 + d_3 d_4 - d_2 d_4, \tag{1.9}$$

$$d_2 d_4 - d_1 d_3 = \delta, \qquad d_1 \equiv d_2 \equiv 0 \pmod 2). \tag{1.10}$$

Conversely, given any natural numbers d_1, d_2, d_3, d_4 satisfying (1.6) *and integers x, y, z satisfying* (1.7), (1.8) *for δ odd*; (1.9), (1.10) *for δ even*; *the relations* (1.4), (1.5) *hold.*

Example. A case in point is $x = 3$, $y = 6$, $z = 7$, $\delta = 2$. Then $d_1 = 6$, $d_2 = 4$, $d_3 = 3$, $d_4 = 5$.

Next we give in Section 2 an infinite subclass of solutions of (1.4) which includes the solutions (1.2) as special cases. Section 2 concludes by proving, as an application of Theorem 1, that

$$T_x^{\,4}(1) + T_y^{\,4}(1) = 4T_z^{\,2}(1)$$

has infinitely many solutions in integers x, y, z.

Section 3 deals with the case $m = 3$. In connection with Fermat's conjecture, A. Ben-Menahem recently raised the question of the solvability of the diophantine equation

$$T_x^{\,3}(\delta) + T_y^{\,3}(\delta) = T_z^{\,3}(\delta). \tag{1.11}$$

M. Shimshoni with the assistance of Miss E. Sadie searched for solutions by means of a computer for $1 \leqslant \delta \leqslant 10$. For each of these values of δ, several solutions were found. The relatively large number of these solutions suggests the possibility that equations of the form (1.11) have infinitely many solutions (if $\delta = 0$, there is of course no solution).

Sierpiński (1962) proved that (1.11) has infinitely many solutions for $\delta = 1$. In other words, the sum (or difference) of two tetrahedral numbers is a tetrahedral number, that is

$$T_x^{\,3}(1) + T_y^{\,3}(1) = T_z^{\,3}(1) \tag{1.12}$$

for an infinity of positive integer triplets (x, y, z). See also Sierpiński (1964) and Wunderlich (1962).

Definition. Any solution of (1.11) in positive integers x, y, z is called *primitive* if either $(x, \delta) = 1$ or $(y, \delta) = 1$. Any other solution in positive integers is called *non-primitive.*

Multiplying (1.12) by δ^3 shows that (1.11) has infinitely many non-primitive solutions for all $\delta > 1$. In Section 3 we show that there is always also an infinity of non-trivial, i.e., primitive solutions:

THEOREM 2. *Equation* (1.11) *has infinitely many primitive solutions for all natural numbers* δ.

The first part of our proof works for all $\delta > 2$. It breaks down for $\delta = 2$, which thus requires special attention. This case was settled by means of a computer. Thus, not only the motivation of the theorem, but also part of the proof was obtained by the use of a computer, which proved, in fact, that for $\delta = 2$, (1.11) has infinitely many solutions.

For the case $m > 3$, we conjecture that (1.1) has only a finite number of solutions for any δ. However, for $\delta \neq 0$, unlike the (expected) behavior of Fermat's case, we have the solution

$$T_m{}^m(1) + T_m{}^m(1) = T_{m+1}{}^m(1)$$

for every m.

The final Section 4 lists all the solutions found by computer for $m = 3$, $1 \leqslant \delta \leqslant 10$, $0 < z \leqslant 10{,}000$; $m = 4$, $1 \leqslant \delta \leqslant 10$, $0 < z \leqslant 20{,}000$; $m = 5$, $1 \leqslant \delta \leqslant 10$, $0 < z \leqslant 6{,}300$; $m = 6$, $1 \leqslant \delta \leqslant 10$, $0 < z \leqslant 1{,}000$.

2. Sums of Generalized Triangular Numbers

We begin by proving Theorem 1. Suppose that (1.4) holds and that d_1, d_2, d_3, d_4 are given by (1.5). Let $d_{14} = (d_1, d_4)$. By (1.5), $d_{14}|\delta$, $d_{14}|y$, $d_{14}|x$, $d_{14}|z$. Hence $d_{14} = 1$ by the second part of (1.4). Similarly, $(d_2, d_3) = 1$, which proves (1.6).

Note the identities

$$y(y + \delta) = (z - x)(x + z + \delta), \tag{2.1}$$

$$2(x + z + \delta)(y + z + \delta) = (x + y + z + \delta)(x + y + z + 2\delta) \tag{2.2}$$

implied by (1.4). Let $d_{12} = (d_1, d_2)$. Then

$$\frac{d_1}{d_{12}} \Big| x + z + \delta, \qquad \frac{d_2}{d_{12}} \Big| x + z + \delta.$$

Since

$$\left(\frac{d_1}{d_{12}}, \frac{d_2}{d_{12}}\right) = 1,$$

we have $D|x + z + \delta$, where

$$D = \frac{d_1}{d_{12}} \frac{d_2}{d_{12}}.$$

Since $d_1|y$, $d_2|y+\delta$, we have by (2.1),

$$d_{12}^2 \left| (z-x)\frac{x+z+\delta}{D}\right. .$$

Let

$$d = (d_{12}, z-x). \tag{2.3}$$

Then $d|\delta$, $d|y$, $d|2x$, $d|2z$.

Case I: δ odd. Then d is odd. Hence $d|x$, $d|y$, $d|z$, $d|\delta$. Thus $d=1$. It follows that

$$d_{12}^2 \left| \frac{x+z+\delta}{D}\right. ,$$

and

$$\frac{y}{d_1}\,\frac{y+\delta}{d_2} = (z-x)\frac{x+z+\delta}{d_1\,d_2},$$

where $(x+z+\delta)/d_1 d_2$ is an integer. Moreover,

$$\left(\frac{y}{d_1}\,\frac{y+\delta}{d_2}, \frac{x+z+\delta}{d_1\,d_2}\right) = 1$$

by (1.5). Hence

$$x+z+\delta = d_1\,d_2. \tag{2.4}$$

Similarly,

$$y+z+\delta = d_3\,d_4. \tag{2.5}$$

Substituting into (2.2),

$$2d_1\,d_2\,d_3\,d_4 = (x+y+z+\delta)(x+y+z+2\delta).$$

By (1.5),

$$d_1|x+y+z+\delta, \qquad d_2|x+y+z+2\delta$$

$$d_3|x+y+z+\delta, \qquad d_4|x+y+z+2\delta.$$

Let

$$d_{13} = (d_1, d_3), \qquad d_{24} = (d_2, d_4).$$

Then

$$2\frac{d_1}{d_1/d_{13}}\,\frac{d_3}{d_3/d_{13}}\,\frac{d_2}{d_2/d_{24}}\,\frac{d_4}{d_4/d_{24}}=2d_{13}{}^2\,d_{24}{}^2$$

$$=\frac{x+y+z+\delta}{D_{13}}\,\frac{x+y+z+2\delta}{D_{24}},$$

where

$$D_{13}=\frac{d_1}{d_{13}}\,\frac{d_3}{d_{13}},\qquad D_{24}=\frac{d_2}{d_{24}}\,\frac{d_4}{d_{24}},$$

and where each of the two factors on the right are integers. Let $d = (d_{13},\ x+y+z+2\delta)$. Then $d|\delta$, $d|y$, $d|x$, $d|z$. Hence $d=1$. Similarly, $(d_{24}, x+y+z+\delta)=1$. It follows that

$$2=\frac{x+y+z+\delta}{d_1\,d_3}\,\frac{x+y+z+2\delta}{d_2\,d_4},$$

where each of the two factors on the right is an integer. Thus,

$$x+y+z+\delta=(2-\theta)\,d_1\,d_3,\qquad x+y+z+2\delta=(1+\theta)\,d_2\,d_4, \quad (2.6)$$

where $\theta = 0$ or 1. This implies (1.8); and (2.4), (2.5), (2.6), imply (1.7).

Case II. δ even. By (1.5), $d_1 \equiv d_2 \equiv 0 \pmod 2$. The number d given by (2.3) is now even, and

$$\frac{d}{2}\Big|\,x,\qquad \frac{d}{2}\Big|\,y,\qquad \frac{d}{2}\Big|\,z,\qquad \frac{d}{2}\Big|\,\delta.$$

Hence $d = 2$. By (2.1),

$$\frac{y}{d_1}\,\frac{y+\delta}{d_2}=\frac{z-x}{2}\cdot\frac{x+z+\delta}{D}\Big/\left(2\cdot\frac{d_{12}{}^2}{4}\right),$$

where $D = d_1\,d_2/d_{12}{}^2$, and

$$\frac{z-x}{2},\qquad \frac{x+z+\delta}{D}$$

are integers. Now

$$\left(\frac{z-x}{2},\frac{d_{12}{}^2}{4}\right)=1.$$

Hence

$$\frac{y}{d_1}\,\frac{y+\delta}{d_2}=\frac{z-x}{2}\cdot\frac{x+z+\delta}{D_{12}}\Big/2,$$

where $D_{12} = d_1 d_2/4$ and $(x + z + \delta)/D_{12}$ is an integer. If $2 \| z - x$,

$$\frac{y}{d_1} \frac{y+\delta}{d_2} = \frac{z-x}{2} \cdot \frac{x+z+\delta}{d_1 d_2/2}, \tag{2.7}$$

where the two factors on the right are integers. Moreover,

$$\left(\frac{y}{d_1} \frac{y+\delta}{d_2}, \frac{x+z+\delta}{d_1 d_2/2} \right) = 1$$

by (1.5). Hence

$$x + z + \delta = \frac{d_1 d_2}{2} \tag{2.8}$$

in this case. If $2 \| d_{12}$, then $2 \| d_1$, say. Now $2^k \| d_2$ for some $k \geqslant 1$. Then $2^k | x + z + \delta$. Also,

$$\frac{d_1}{2} \frac{d_2}{2} \Big| x + z + \delta.$$

But the highest power of 2 in $\dfrac{d_1}{2} \dfrac{d_2}{2}$ is only $k-1$. Hence in fact

$$2 \frac{d_1}{2} \frac{d_2}{2} \Big| x + z + \delta.$$

Thus (2.7) and hence (2.8) hold also in this case. Similar considerations show that

$$y + z + \delta = d_3 d_4. \tag{2.9}$$

From (2.2),

$$d_1 d_2 d_3 d_4 = (x + y + z + \delta)(x + y + z + 2\delta).$$

Similar to Case I we conclude

$$x + y + z + \delta = d_1 d_3, \qquad x + y + z + 2\delta = d_2 d_4. \tag{2.10}$$

Equations (2.8), (2.9), (2.10) imply (1.9), (1.10).

Conversely, suppose first that δ is odd, and that $x, y, z, d_1, d_2, d_3, d_4$ satisfy (1.6), (1.7), (1.8). Substituting x, y, z from (1.7) into (1.4) and using (1.8) gives an identity provided θ assumes one of the values 0 or 1. This proves the first part of (1.4). Now (1.7), (1.8) imply (2.4), (2.5), (2.6). Hence $(x+z+\delta,\ y) = (d_1 d_2, d_1((2-\theta)d_3 - d_2)) = d_1(d_2, (2-\theta)d_3 - d_2)$. If $\theta = 0$, d_2 is odd by (1.8), since δ is odd. Hence $(x+z+\delta, y) = d_1$ by (1.6). The remaining three relations of (1.5) are proved similarly. Let $d = (x, y, z, \delta)$. In particular, $d|d_1$, $d|d_4$. Hence $d=1$, proving the second part of (1.4).

Now suppose that δ is even, and that $x, y, z, d_1, d_2, d_3, d_4$ satisfy (1.6), (1.9), (1.10). Substituting x, y, z from (1.9) into (1.4) and using (1.10) proves the first part of (1.4) as before. Equations (1.9), (1.10) imply (2.8), (2.9), (2.10). Hence

$$(x+z+\delta, y) = \left[\frac{d_1 d_2}{2}, d_1\left(d_3 - \frac{d_2}{2}\right)\right] = d_1\left(\frac{d_2}{2}, d_3 - \frac{d_2}{2}\right) = d_1$$

by (1.6). The remaining relations of (1.5) are proved similarly. As before, we see that $(x, y, z, \delta) = 1$, completing the proof of Theorem 1.

Remarks. (i) For $\delta = 0$, the second part of Theorem 1 reduces to the Pythagorean triplets theorem. Indeed, the four parameters of (1.5) collapse into two. If we let

$$d_1 = d_2 = (x+z, y) = 2a, \qquad d_3 = d_4 = (y+z, x) = a+b$$

for integers a, b, we obtain from (1.9),

$$x = a^2 - b^2, \qquad y = 2ab, \qquad z = a^2 + b^2.$$

Since x is odd, a, b are of opposite parity. Also $(d_1, d_4) = 1$ implies $(a, b) = 1$.

(In the Pythagorean triplets theorem, the exclusion of trivial cases requires the condition $(x, y) = 1$, on which the proof hinges strongly. In Theorem 1, however, such a condition cannot be assumed, as is seen, e.g., by the nontrivial solution in the example given after the statement of Theorem 1, where $(x, y) = 3$.)

(ii) For odd δ, we shall construct two infinite subclasses of solutions of (1.4) (for $\theta = 1$ and 0, respectively). For the first class, let a_1, a_2, a_3, a_4 be positive integers satisfying

$$(a_2, a_3) = (2a_2 k + a_1, a_3 k + a_4, \delta) = 1$$

for all non-negative integers k, and

$$2a_2 a_4 - a_1 a_3 = \delta.$$

Define

$$d_1 = 2a_2 k + a_1, \quad d_2 = a_2, \quad d_3 = a_3, \quad d_4 = a_3 k + a_4, \quad k = 0, 1, 2, \ldots.$$

It is easily verified that (1.6), (1.8) are satisfied with $\theta = 1$, for all k. Hence if we define x, y, z by (1.7), i.e.,

$$x = a_3 (2a_2 - a_3) k + a_3 (a_1 - a_4)$$

$$y = (a_3 - a_2)(2a_2 k + a_1)$$

$$z = (2a_2^2 + a_3^2 - 2a_2 a_3) k + a_1 a_2 + a_3 a_4 - 2a_2 a_4, \quad k = 0, 1, 2, \ldots,$$

then (1.4), (1.5) are satisfied. In particular, the special case $\delta = 1$, $a_1 = 1$, $a_2 = 2$, $a_3 = 3$, $a_4 = 1$ gives the first of Khatri's results (1.2).

The second class is constructed similarly. Let a_1, a_2, a_3, a_4 be positive integers satisfying

$$(a_1, a_4) = (2a_1 k + a_2, a_4 k + a_3, \delta) = 1$$

for $k = 0, 1, 2, \ldots$, and

$$a_2 a_4 - 2a_1 a_3 = \delta.$$

Again (1.6), (1.8) are satisfied for all k, this time with $\theta = 0$. Also, (1.4), (1.5) are satisfied by

$$x = (2a_1 - a_4)(a_4 k + a_3)$$

$$y = 2a_1 (a_4 - a_1) k + a_1 (2a_3 - a_2)$$

$$z = (2a_1^2 + a_4^2 - 2a_1 a_4) k + a_1 a_2 + a_3 a_4 - a_2 a_4, \quad k = 0, 1, 2, \ldots.$$

For $\delta = 1$, the special cases $a_1 = 3$, $a_2 = 5$, $a_3 = 4$, $a_4 = 5$ and $a_1 = 4$, $a_2 = 5$, $a_3 = 3$, $a_4 = 5$ give the second two equations of (1.2) respectively.

Similar infinite subclasses can be constructed for δ even.

(iii) For δ odd, let S_0, S_1 be the subclasses of solutions of (1.4) corresponding to $\theta = 0$ and $\theta = 1$ respectively. Then S_0, S_1 are both infinite (follows from (ii)), and $S_0 \cap S_1 = \emptyset$ (since $\theta = 0$ implies $x + y + z \equiv 1 \pmod 2$, and $\theta = 1$ implies $x + y + z \equiv 0 \pmod 2$ by (2.6)).

We shall now prove that

$$T_x^4(1) + T_y^4(1) = 4T_z^2(1)$$

has infinitely many solutions, using the infinitely many solutions of the Pell equation

$$\beta^2 - 17\alpha^2 = 1. \tag{2.11}$$

Apply Theorem 1 with $\delta = \theta = 1$ and

$$d_1 = 3\beta_i - 5\alpha_i + 4, \qquad d_2 = 2\beta_i + 2\alpha_i + 3,$$

$$d_3 = 3\beta_i + 5\alpha_i + 4, \qquad d_4 = 2\beta_i - 2\alpha_i + 3,$$

where (β_i, α_i), $i = 0, 1, 2, \ldots$ are the infinitely many solutions of (2.11). We have

$$\begin{aligned} 2d_2\,d_4 - d_1\,d_3 &= 2(2\beta_i + 2\alpha_i + 3)(2\beta_i - 2\alpha_i + 3) - (3\beta_i - 5\alpha_i + 4)(3\beta_i + 5\alpha_i + 4) \\ &= 2 + 17\alpha_i^2 - \beta_i^2 = 1. \end{aligned}$$

Thus (1.8) and also (1.6) hold. Using (1.7), an easy computation shows that

$$x = \frac{(2\beta_i - 2\alpha_i + 3)(2\beta_i - 2\alpha_i + 4)}{2} - 1,$$

$$y = \frac{(2\beta_i + 2\alpha_i + 3)(2\beta_i + 2\alpha_i + 4)}{2} - 1,$$

$$z = 3\beta_i^2 + 5\alpha_i^2 + 10\beta_i + 7.$$

Using the identity

$$\frac{(u+1)(u+2)}{2}\left(\frac{(u+1)(u+2)}{2} - 1\right) = \frac{1}{4}u(u+1)(u+2)(u+3),$$

we obtain—on using (2.11)—the identity

$$\begin{aligned} &(2\beta_i - 2\alpha_i + 2)(2\beta_i - 2\alpha_i + 3)(2\beta_i - 2\alpha_i + 4)(2\beta_i - 2\alpha_i + 5) \\ &\quad + (2\beta_i + 2\alpha_i + 2)(2\beta_i + 2\alpha_i + 3)(2\beta_i + 2\alpha_i + 4)(2\beta_i + 2\alpha_i + 5) \\ &\quad = 4(3\beta_i^2 + 5\alpha_i^2 + 10\beta_i + 7)(3\beta_i^2 + 5\alpha_i^2 + 10\beta_i + 8), \end{aligned}$$

proving the assertion.

3. Sums of Generalized Tetrahedral Numbers

Proof of Theorem 2. Consider the Pell equation

$$u^2 - Dv^2 = N, \tag{3.1}$$

where D is a positive integer which is not a perfect square, and N is any integer satisfying $|N| < \sqrt{D}$. Let

$$\sqrt{D} = [a_0, a_1, a_2, \ldots] \tag{3.2}$$

be the expansion of $\sqrt{D}$ into a simple continued fraction. Denote by

$$\frac{p_n}{q_n} = [a_0, a_1, \ldots, a_n]$$

its nth *convergent*, and by

$$a_n' = [a_n, a_{n+1}, a_{n+2}, \ldots]$$

its nth *complete quotient*.

The following facts are known from the theory of continued fractions (see, e.g., Perron (1929), Shanks (1962)):

(i) The values p_n, q_n are monotonically increasing sequences of positive integers.

(ii) The expansion (3.2) is periodic. We denote the period length by m.

(iii) $a_n' = (s_n + \sqrt{D})/t_n$, where s_n, t_n are natural numbers ($s_0 = 0$, $t_0 = 1$). The sequences s_n, t_n are each purely periodic with period length m (in particular: $t_m = 1$).

(iv) The diophantine equation (3.1) is solvable in integers u, v if and only if there exists an integer n such that $N = (-1)^n t_n$. If the condition is satisfied, there are actually infinitely many solutions. An infinity of solutions is given by $(u, v) = (p_{n-1+km}, q_{n-1+km})$, where $k = 1, 2, 3, \ldots$ if m is even; $k = 2, 4, 6, \ldots$ if m is odd.

From (iii) and (iv) it follows in particular that (3.1) is always solvable if $N = 1$, since $t_{km} = 1$, $k = 0, 1, 2, \ldots$.

Let

$$x = A(B\delta^4 v - u), \quad y = A(B\delta^4 v + u), \quad z = 2AB\delta^2(\delta^2 - 3A^2N)\, v, \tag{3.3}$$

where A, B are positive integers, $(A, \delta) = 1$, N an integer, and u, v integer variables. Define

$$\begin{aligned} F(x, y, z) &= T_{x-\delta}{}^3(\delta) + T_{y-\delta}{}^3(\delta) - T_{z-\delta}{}^3(\delta) \\ &= x^3 + y^3 - z^3 - \delta^2 x - \delta^2 y + \delta^2 z \\ &= (x + y)(x^2 - xy + y^2 - \delta^2) + \delta^2 z - z^3. \end{aligned}$$

Substituting from (3.3) and simplifying, we obtain

$$F(x, y, z) = 6A^3B\delta^4 v(u^2 - Dv^2 - N), \tag{3.4}$$

where

$$D = B^2\delta^2((\delta(\delta^2 - 6A^2N))^2 - 36A^6N^3). \tag{3.5}$$

We now consider two cases.

I. $\delta > 2$.

Put $A = N = 1$, B arbitrary. Then (3.5) becomes

$$D = B^2\delta^2(\zeta^2 - 36), \qquad \zeta = \delta(\delta^2 - 6).$$

Suppose that D is a perfect square. Then already $\zeta^2 - 36 = \eta^2$ is square. But the only solutions of $(\zeta - \eta)(\zeta + \eta) = 36$ are $\zeta = 6$, $\eta = 0$; $\zeta = 10$, $\eta = 8$, both of which are incompatible with $\zeta = \delta(\delta^2 - 6)$. Hence (3.1) is a genuine Pell equation for this case. Letting $(u, v) = (p_n, q_n)$ be an infinite class of solutions of this Pell equation, we see from (3.4) that $F(x, y, z) = 0$. In other words, (x, y, z) comprises an infinity of distinct integer solutions of

$$T_{x-\delta}{}^3(\delta) + T_{y-\delta}{}^3(\delta) = T_{z-\delta}{}^3(\delta).$$

From (3.1) and (3.5), $(p_n, \delta) = 1$. Hence also $(x, \delta) = (y, \delta) = 1$. Clearly $y > 0$, $z > 0$ for $\delta \geqslant 2$. From (3.1),

$$p_n - \sqrt{D}\, q_n = \frac{1}{p_n + \sqrt{D} q_n} < 1; \qquad p_n < \sqrt{D}\, q_n + 1.$$

Since

$$D = B^2\delta^2(\zeta^2 - 36) < B^2\delta^2\zeta^2; \qquad \sqrt{D} < B\delta\zeta = B\delta^2(\delta^2 - 6),$$

we have

$$p_n < B\delta^4 q_n - 6B\delta^2 q_n + 1 < B\delta^4 q_n.$$

Thus also $x > 0$, completing the proof for this case.

II. $\delta = 2$.

In (3.5) we put $A = 1$, $B = B'/2$, where B' is an arbitrary odd integer, $N = -15$. Then

$$D = 156844\, B'^2.$$

It turns out that

$$u^2 - 156844\,v^2 = -15$$

is solvable, the least positive solution being

$$(u, v) = (p_{16}, q_{16}) = (7995740689, 20189462).$$

As in Case I, we see that $(x, 2) = (y, 2) = 1$.

Now $y > 0$, $z > 0$ but $x < 0$. Hence we demonstrated existence of an infinity of positive solutions (x, y, z) of the equation

$$T_y{}^3(\delta) = T_x{}^3(\delta) + T_z{}^3(\delta). \tag{3.6}$$

Remarks. (i) The values of D and N for the case $\delta = 2$ were found by a computer search as follows: If $\delta = z$ and if D is even, it follows from (3.5), (3.1), (3.3) that N must be odd for primitive solutions. A computer program for expanding $\sqrt{D}$ into a simple continued fraction and computing its associated sequence t_n was written. It was implemented for D of the form (3.5), $A = B = 1$, and small odd negative values for N (so that $D > 0$). Any N for which $t_n = |N|$ for odd n gives rise to a Pell equation (3.1), the solutions of which induce solutions of (3.6). The computer produced the values $N = -11, \ -15, \ -19, \ -37$. The smallest positive solution of the corresponding four Pell equations is that corresponding to $N = -15$ given above. Mr. J. Perl assisted with the programming.

(ii) Sierpiński (1962) solved the case $\delta = 1$ by constructing a suitable Pell equation with $D = 5$, $N = 1$. Other values of D and N can be given for $\delta = 1$ or for any specified value of δ. For example, if $\delta = 1$, then (3.5) with $A = 3$, $N = -1$, B arbitrary gives $D = 29269\,B^2$. Since 29269 is prime and $\equiv 1 \pmod 4$, the Pell equation $u^2 - Dv^2 = -1$ is solvable if B is any power of 29269 (see Perron (1929), Satz 22, p. 108). For $\delta > 2$, the value of A in (3.5) can be varied within certain bounds and B can be varied at will, giving rise to an infinity of Pell equations of the desired type for each δ.

4. Computer Results

The first few primitive solutions of (1.1) were found by computer for $m = 3, 4, 5, 6$ and $1 \leqslant \delta \leqslant 10$. For $m = 3, 4$, these solutions are listed in Tables I and II; our results for the case $m = 3$, $\delta = 1$ are contained in Wunderlich (1962). For $m = 5$, the only solution with $1 \leqslant \delta \leqslant 10$, $1 \leqslant z \leqslant 6300$ is given by $\delta = 1$, $x = y = 5$, $z = 6$. For $m = 6$, there are three solutions with $1 \leqslant \delta \leqslant 10$, $1 \leqslant z \leqslant 1000$; they all have $\delta = 1$, and the values of (x, y, z) are $(6, 6, 7)$, $(9, 10, 11)$, $(14, 14, 16)$.

TABLE I. Primitive solutions of $x(x+\delta)(x+2\delta)+y(y+\delta)(y+2\delta) = z(z+\delta)(z+2\delta)$ for $z = 1, \ldots, 10000$, $1 \leqslant \delta \leqslant 10$.

= 1

NO	X	Y	Z	NO	X	Y	Z	NO	X	Y	Z
1	3	3	4	2	8	14	15	3	20	54	55
4	30	55	58	5	39	70	74	6	61	102	109
7	84	90	110	8	34	118	119	9	48	138	140
10	119	154	175	11	187	201	245	12	100	290	294
13	327	336	418	14	149	429	435	15	252	424	452
16	248	450	474	17	362	415	492	18	219	515	528
19	136	532	535	20	424	448	550	21	314	527	562
22	434	495	588	23	399	588	644	24	324	663	688
25	272	688	702	26	304	695	714	27	349	713	740
28	532	643	747	29	424	705	753	30	378	790	818
31	608	754	868	32	230	903	908	33	489	869	918
34	775	950	1098	35	703	1064	1158	36	878	1044	1220
37	968	1001	1241	38	922	1286	1428	39	290	1430	1434
40	367	1436	1444	41	855	1343	1450	42	504	1629	1645
43	897	1621	1708	44	750	1690	1738	45	1351	1478	1786
46	798	1818	1868	47	438	2164	2170	48	1146	2072	2183
49	1139	2115	2220	50	1609	1941	2256	51	1105	2303	2385
52	853	2417	2452	53	1103	2514	2583	54	1484	2584	2738
55	1089	2773	2828	56	834	2958	2980	57	528	3138	3143
58	1775	2954	3154	59	1484	3094	3204	60	2478	2726	3286
61	2099	3211	3486	62	729	3595	3605	63	2200	3660	3908
64	742	4415	4422	65	2116	4580	4726	66	2948	4408	4810
67	3138	4630	5068	68	2912	4838	5167	69	868	6034	6040
70	2252	6390	6482	71	5338	5608	6900	72	3570	7154	7439
73	1271	7554	7566	74	6152	6586	8034	75	1160	8070	8078
76	5300	7284	8120	77	5630	7105	8129	78	6340	6788	8280
79	4115	8034	8379	80	4015	8910	9174	81	7104	7847	9442
82	7062	8094	9592								

= 2

NO	X	Y	Z	NO	X	Y	Z	NO	X	Y	Z
1	82	101	117	2	85	103	120	3	93	111	130
4	129	187	206	5	57	259	260	6	980	1183	1375
7	710	1921	1953	8	1105	2603	2668	9	2385	2490	3073
10	1584	3157	3285	11	1561	3427	3532	12	1545	3635	3726
13	1836	3955	4083	14	1295	4413	4450	15	1475	7938	7955
16	5873	7180	8305	17	5866	7663	8671				

= 3

NO	X	Y	Z	NO	X	Y	Z	NO	X	Y	Z
1	7	14	15	2	37	38	48	3	64	92	102
4	74	109	120	5	68	115	123	6	174	245	272
7	284	294	365	8	289	297	370	9	317	514	552
10	388	632	678	11	484	806	861	12	396	869	896
13	191	1099	1101	14	404	1669	1677	15	662	1653	1688
16	1080	1532	1694	17	652	1812	1840	18	1257	1780	1969
19	847	2846	2871	20	637	2948	2958	21	2051	2587	2961
22	1350	3149	3230	23	2062	3413	3648	24	2016	3712	3901
25	3272	3364	4182	26	3313	3404	4233	27	2549	4222	4512
28	3550	5030	5562	29	3335	5542	5919	30	3824	6789	7172
31	4726	7680	8236	32	4408	8060	8478	33	5839	8195	9084

TABLE I. *Continued*

= 4

NO	X	Y	Z	NO	X	Y	Z	NO	X	Y	Z
1	157	346	357	2	221	460	477	3	385	1326	1337
4	987	1172	1371	5	1477	2155	2366	6	1739	2533	2782
7	1793	3359	3522	8	2873	2919	3650	9	2233	3391	3688
10	1582	4841	4897	11	1041	5871	5882	12	3675	6405	6786
13	907	7093	7098	14	2438	7379	7467	15	3802	8207	8471
16	1027	8541	8546	17	3739	9317	9514				

= 5

NO	X	Y	Z	NO	X	Y	Z	NO	X	Y	Z
1	12	16	19	2	67	82	96	3	72	92	106
4	46	142	144	5	104	136	155	6	92	148	160
7	149	322	333	8	222	347	376	9	267	422	456
10	527	584	703	11	492	660	742	12	634	762	888
13	611	858	952	14	798	1067	1200	15	907	1208	1360
16	1002	1496	1634	17	1519	1682	2023	18	1618	1659	2066
19	1510	2022	2272	20	1332	2192	2346	21	1643	2414	2646
22	1642	2707	2896	23	2010	3172	3422	24	1269	3721	3770
25	2864	4504	4862	26	2168	5251	5372	27	3776	5020	5651
28	3994	5567	6183	29	4273	5685	6398	30	4891	5492	6564
31	3859	9256	9475								

= 6

NO	X	Y	Z	NO	X	Y	Z	NO	X	Y	Z
1	13	39	40	2	153	203	230	3	385	435	520
4	394	539	603	5	549	575	710	6	526	1757	1773
7	1084	2559	2623	8	355	3953	3954	9	1790	4867	4947
10	1437	6071	6098	11	2323	6990	7075	12	3415	7130	7383

= 7

NO	X	Y	Z	NO	X	Y	Z	NO	X	Y	Z
1	7	19	20	2	11	21	23	3	14	22	25
4	22	54	56	5	104	112	138	6	109	112	141
7	107	199	210	8	153	176	210	9	105	204	214
10	151	187	217	11	136	260	273	12	316	476	520
13	538	548	686	14	541	556	693	15	406	739	779
16	155	833	835	17	518	950	1000	18	436	1057	1082
19	193	1146	1148	20	910	962	1182	21	860	1178	1316
22	217	1360	1362	23	1030	1141	1373	24	301	1551	1555
25	385	1573	1581	26	913	1646	1736	27	833	1690	1756
28	522	1741	1757	29	598	1904	1924	30	731	1961	1995
31	1526	1690	2033	32	1430	1812	2072	33	1588	2006	2296
34	1635	2485	2703	35	801	2948	2968	36	1373	2982	3077
37	1953	2840	3121	38	1596	3078	3216	39	1958	2968	3230
40	2316	2774	3234	41	406	3418	3420	42	2403	3006	3451
43	2651	3353	3835	44	870	3857	3872	45	2924	3585	4144
46	3038	3873	4418	47	3406	3783	4543	48	1002	4611	4627
49	3566	3966	4760	50	2431	5355	5518	51	578	5768	5770
52	3296	5959	6279	53	3318	5990	6313	54	4012	5921	6482
55	2724	6692	6840	56	3799	7163	7504	57	2877	7564	7701
58	4216	7612	8022	59	592	8456	8457	60	3796	8260	8520
61	7000	7568	9193	62	3762	9129	9338	63	6313	8358	9420
64	5076	9408	9878								

TABLE I. *Continued*

= 8

NO	X	Y	Z	NO	X	Y	Z	NO	X	Y	Z
1	19	69	70	2	106	123	147	3	145	180	209
4	105	250	257	5	115	341	346	6	517	674	765
7	427	950	979	8	1131	1373	1594	9	662	1555	1595
10	1587	1797	2142	11	673	2831	2844	12	2127	2532	2959
13	2428	3045	3493	14	1241	3535	3586	15	2041	3814	4001
16	1491	3933	4004	17	1617	6559	6592	18	1938	6587	6643
19	4885	6363	7208	20	4905	7367	8032	21	5451	7453	8322
22	2559	8625	8700								

= 9

N	X	Y	Z	NO	X	Y	Z	NO	X	Y	Z
1	17	21	26	2	22	86	87	3	25	101	102
4	68	264	266	5	251	258	323	6	208	444	460
7	327	407	470	8	395	406	507	9	200	511	522
10	475	555	655	11	236	978	983	12	897	901	1135
13	950	1102	1302	14	658	1344	1396	15	743	1704	1751
16	875	1797	1865	17	1138	1760	1908	18	1720	1911	2296
19	1784	2068	2442	20	1749	2398	2677	21	1190	3166	3222
22	2387	3486	3827	23	1528	4530	4588	24	2051	4483	4623
25	548	5356	5358	26	4062	5026	5791	27	6032	7090	8322
28	2741	8772	8861	29	1121	9796	9801	30	6800	8610	9842

= 10

NO	X	Y	Z	NO	X	Y	Z	NO	X	Y	Z
1	21	84	85	2	458	1443	1459	3	1079	1653	1796
4	1802	2553	2825	5	2554	5083	5291	6	2295	5642	5767

TABLE II. Solutions of $x(x+\delta)(x+2\delta)(x+3\delta) + y(y+\delta)(y+2\delta)(y+3\delta) = z(z+\delta)(z+2\delta)(z+3\delta)$ for $z = 1, \ldots, 20000$, $1 \leqslant \delta \leqslant 10$.

δ	x	y	z	δ	x	y	z	δ	x	y	z
1	4	4	5	4	105	166	173	8	12	19	21
1	129	187	197	5	203	205	244	8	348	375	433
2	38	51	55	5	300	814	818	8	1964	3861	3925
2	4308	12901	12941	6	183	352	359	9	15	26	28
2	5289	16558	16601	7	19	44	45				

References

Khatri, M. N. (1955). Triangular numbers and Pythagorean triangles. *Scripta Math.* **21,** 94.

Perron, O. (1929). "Die Lehre von den Kettenbrüchen". Teubner, Leipzig. (Reprinted by Chelsea, New York).

Shanks, D. (1962). "Solved and Unsolved Problems in Number Theory", Vol. 1. Spartan, Washington, D.C.

Sierpiński, W. (1962). Sur une propriété des nombres tétraédraux. *Elem. Math.* **17,** 29–30.

Sierpiński, W. (1964). "Elementary Theory of Numbers". (Translated from Polish by A. Hulanicki.) Warsaw.

Wunderlich, M. (1962). Certain properties of pyramidal and figurate numbers. *Math. Comp.* **16,** 482–486.

On the Representation of an Integer as the Sum of Four Integer Cubes

L. J. MORDELL

St. Johns College, Cambridge, England

It has been proved that all integers $n \not\equiv \pm 4 \pmod 9$ can be expressed as a sum of four integer cubes. The proof depends upon various identities, e.g.,

$$(x+2)^3 + (6x-1)^3 + (8x-2)^3 + (-9x+2)^3 = 18x + 7.$$

Numerical evidence suggests that numbers $\equiv 4 \pmod 9$ can also be so represented. This would be so if four polynomials P, Q, R, S in x with integer coefficients and degrees $\leqslant 4$ could be found such that

$$P^3 + Q^3 + R^3 + S^3 = 9x + 4. \tag{1}$$

It is known that no such representation is possible with linear polynomials.

Schinzel (1968) has recently proved the more general result that such a representation with quartic polynomials not all constant cannot hold for

$$P^3 + Q^3 + R^3 + S^3 = Lx + M, \tag{2}$$

where L and M are integer constants and $M \equiv 4 \pmod 9$. Write (2) as

$$\sum (ax^4 + bx^3 + cx^2 + dx + e)^3 = 3^\alpha Lx + M, \quad \alpha \geqslant 1, \tag{3}$$

where all summations used refer to four sets of integer constants a_r, b_r, c_r, d_r, $r = 1, 2, 3, 4$. His proof is a 3-adic one. He shows that if a representation (3) is taken such that the product of the leading coefficients in P, Q, R, S is a minimum, then

$$a \equiv 0 \pmod{81},\ b \equiv 0 \pmod{27},\ c \equiv 0 \pmod 9,\ d \equiv 0 \pmod 3.$$

A contradiction now arises if x is replaced by $\frac{1}{3}x$ in (3).

The proof is rather complicated since it requires an expansion of (3) in powers of x. I have found a simpler proof, a 3^λ—adic one, where $\lambda = \frac{1}{4}, \frac{1}{3}$.

This has the great advantage that if for an integer n, $n \equiv 0 \pmod{3^{1/4}}$, then $n \equiv 0 \pmod 3$. This leads to simplicity in the expansion and presentation. A full proof will appear in Mordell (1970) but I give an outline here. Both proofs depend upon two lemmas.

LEMMA 1. *The only integer solution of*

$$\Sigma e^3 \equiv 4 \pmod 9,$$

is given by $e \equiv 1 \pmod 3$.

LEMMA 2. *The only integer solution of*

$$\Sigma a^3 \equiv 0 \pmod 9, \quad \Sigma a^2 \equiv 0 \pmod 3$$

is given by $a \equiv 0 \pmod 3$.

Now from (3), we have at once

$$\Sigma a^3 \equiv \Sigma b^3 \equiv \Sigma c^3 \equiv \Sigma d^3 \equiv 0, \text{ or } \Sigma a \equiv \Sigma b \equiv \Sigma c \equiv \Sigma d \equiv 0 \pmod 3. \tag{4}$$

This shows that $\alpha \geqslant 2$. Then from $x = \pm 1$, we deduce

$$(a + c) \equiv 0, \quad (b + d) \equiv 0 \pmod 3,$$

and so identically

$$\Sigma\big(a(x^4 - x^2) + b(x^3 - x) + e\big)^3 \equiv 4 \pmod 9.$$

This gives

$$\Sigma(ax + b)^2 \equiv 0 \pmod 3,$$

and so

$$\Sigma a^2 \equiv \sum b^2 \equiv 0 \pmod 3.$$

Since $\Sigma a^3 = 0$, we have $a = 3a'$.
Now (3) becomes

$$\Sigma(bx^3 + cx^2 + dx + e)^3 \equiv 4 \pmod 9.$$

Then $\Sigma b^3 \equiv 0 \pmod 9$ and since $\Sigma b^2 \equiv 0 \pmod 3$, $b = 3b'$, and $c = 3c'$, $d = 3d'$. Write (3) as

$$\Sigma(3a' x^4 + 3b' x^3 + 3c' x^2 + 3d' x + e)^3 = 9L' x + M. \tag{5}$$

My 3^λ—adic method can be applied since the coefficient of x^4 is $\equiv 0 \pmod 3$. Replace x by $x/3^{1/4}$ and then

$$\Sigma(a' x^4 + 3^{1/4}b' x^3 + 3^{2/4}c' x^2 + 3^{3/4}d' x + e)^3 = 3^{7/4}L' x + M. \tag{6}$$

In taking this as a congruence mod $3^{5/4}$, it is easy to pick out the necessary terms, and so

$$\Sigma(a' x^4 + e)^3 + \Sigma(3^{1/4}b' x^3)^3 \equiv 4 \pmod{3^{5/4}}. \tag{7}$$

The coefficients of x^{12} and x^8 in (7) give

$$\Sigma a'^3 \equiv 0 \pmod{3^{5/4}} \equiv 0 \pmod 9, \quad \Sigma a'^2 \equiv 0 \pmod{3^{1/4}} \equiv 0 \pmod 3.$$

Hence $a' = 3a''$.
The process can be continued. Now (6) becomes

$$\Sigma(3^{1/4}b' x^3 + 3^{2/4}c' x^2 + 3^{3/4}d' x + e)^3 \equiv 3^{7/4}L' x + M \pmod 9,$$

or

$$\Sigma\big((3^{1/4}b' x^3 + e)^3 + 3(3^{1/4}b' x^3 + e)^2(3^{2/4}c' x^2 + 3^{3/4}d' x) + (3^{2/4}c' x^2)^3\big)$$
$$\equiv 3^{7/4}L' x + M \pmod 9.$$

Then from the coefficients of x^9, x^6 and x^2,

$$\Sigma 3^{3/4}b'^3 \equiv 0 \pmod 9, \quad \Sigma b'^3 \equiv 0 \pmod 9,$$
$$\Sigma 3^{6/4}b'^2 + 3^{6/4}\Sigma c'^3 \equiv 0 \pmod 9, \quad 3^{6/4}\Sigma c' \equiv 0 \pmod 9.$$

Then

$$\Sigma b'^2 \equiv 0 \pmod 3 \text{ and } b' = 3b''.$$

On continuing the process, we find

$$a'' = 3a''', \; c' = 3c'' \text{ and finally } b'' = 3b''', \; a''' = 3a''''.$$

This gives the contradiction.

The argument does not apply to polynomials of the fifth degree, and so one may conjecture that a representation holds with such polynomials.

Note added in proof: Dr. J. H. E. Cohen and I have since proved that no representation exists with polynomials of the fifth and sixth degrees. To be published in *J. London Math. Soc.*

References

Mordell, L. J. (1970). On sums of four cubes of polynomials. *Acta Arith.* **XVI,** 365–369.

Schinzel, A. (1968). On sums of four cubes of polynomials. *J. London Math. Soc.* **43,** 143–146.

Arithmetic Properties of Linear Recurrences

R. R. LAXTON

University of Nottingham, England

1. Introduction

The only fact known about the set of prime divisors of a general second order linear recurrence is that it is infinite (see Ward, 1954; we are excluding those recurrences which are periodic). The Riemann hypothesis for number fields implies that the divisor set has positive density. Furthermore, it indicates that 'most' linear recurrences (with a given companion polynomial $f(x)$) have a fixed density c_f of prime divisors; these results are discussed in section 2.

Other results connected with the description of the set of prime divisors of a linear recurrence are considered in section 3.

Section 2 provides the group theoretic background necessary to present our heuristic arguments.

The author thanks Dr. C. Laughlin of the University of Nottingham for computing the results contained in this article.

2. The Density of Prime Divisors of Linear Recurrence

Let $f(x) = x^2 - Px + Q = (x - \theta_1)(x - \theta_2) \in Z[x]$, $Q \neq 0$ and $(P, Q) = 1$; we shall assume that θ_1/θ_2 is not a root of unity. A sequence $W = [w_0, w_1] = \{\ldots, w_{-1}, w_0, w_1, \ldots, w_n, \ldots\}$ of rational numbers with $w_i \in Z$ for all i sufficiently large is said to be a rational integral linear recurrence with companion polynomial (c.p.) $f(x)$ if $w_{n+2} = Pw_{n+1} - Qw_n$ for all $n \in Z$. We may assume that $w_0, w_1 \in Z$ and if we write $A = w_1 - w_0\theta_2$, $B = w_1 - w_0\theta_1$, then

$$w_n = (A\theta_1{}^n - B\theta_2{}^n)/(\theta_1 - \theta_2)$$

for all $n \in Z$. $\Delta(W) = |A \,.\, B|$ is called the invariant of W and we consider only those W for which $\Delta(W) \neq 0$; the collection of all such sequences with c.p. $f(x)$ will be denoted by $F(f)$. An integer m is called a *divisor* of W—denoted by $m|W$—if m divides some term w_n of W. If $V = [v_0, v_1] \in F(f)$ is given by $v_n = (C\theta_1{}^n - D\theta_2{}^n)/(\theta_1 - \theta_2)$, then, as demonstrated in Laxton (1969), the sequence T given by $t_n = (AC\theta_1{}^n - BD\theta_2{}^n)/(\theta_1 - \theta_2)$ is in $F(f)$; T is called the product of W and V.

In Laxton (1969) we defined an equivalence relation on $F(f)$; $W = \{w_n\}$ and $V = \{v_n\}$ are said to be equivalent if there exist non-zero integers k, l and an integer t such that $kw_n = lv_{n+t}$ for all $n \in Z$. The set of equivalence classes (W) form an abelian group $G(f)$ with the product operation defined above. An integer is said to be a divisor of $(W) \in G(f)$ if it is a divisor of every recurrence in (W). The following sub-groups of $G(f)$ were defined in Laxton (1969), section 3; $G(f, p^n)$ consisting of all $(W) \in G(f)$ which are divisible by the prime power p^n, $H(f, p)$ consisting of all (W) which have finite order modulo $G(f, p)$ and $K(f, p)$ consisting of all (W) which contain a recurrence V for which $(\Delta(V), p) = 1$.

Let $I = [0, 1] = \{ \dots, i_0, i_1, \dots, i_n, \dots \}$, where $i_0 = 0$, $i_1 = 1$, be the Lucas sequence associated with $f(x)$. It is given by the formula $i_n = (\theta_1{}^n - \theta_2{}^n)/(\theta_1 - \theta_2)$. The class (I) is the identity element of $G(f)$. Every prime divides (I) (see section 3).

In this section we are only concerned with the density of prime divisors of a linear recurrence W—or what amounts to the same thing—of a class (W) of $G(f)$. Consequently we may neglect a finite number of primes and in what follows we shall assume that the primes p do not divide $Q(P^2 - 4Q)$.

Theorem 3.7 of Laxton (1969) asserts that $K(f, p) = H(f, p)$ for all such primes, and therefore we know that $(W) \in H(f, p)$ for all but a finite number of primes. The *rank* $e(p)$ of the prime p in $G(f)$ is the least positive integer n for which $i_n \equiv 0 (\text{mod}\, p)$. Let $m(p) = p + 1$ if $Q(\theta_1) \neq Q$ and (p) is a prime ideal in $Q(\theta_1)$ and be $p - 1$ otherwise. For each p we have $m(p)/e(p) \in Z$ and it was shown, in the above mentioned theorem, that each factor group $K(f, p)/G(f, p)$ has order $m(p)/e(p)$. Consequently we would expect, on heuristic grounds, *a* linear recurrence $W \in F(f)$ to have density

$$c_f = \lim_{x \to \infty} \frac{1}{\pi(x)} \sum_{p < x} \frac{e(p)}{m(p)} \tag{1}$$

of prime divisors, provided this limit exists.

We have worked out the theory and made computations for the cases when $f(x) = (x - a)(x - 1)$, $a \in Z$, and we shall present the results here. $e(p)$ is now just the exponent of a modulo p and $m(p) = p - 1$ so that (1) becomes

$$c_a = \lim_{x \to \infty} \frac{1}{\pi(x)} \sum_{p < x} \frac{e(p)}{p - 1} \tag{2}$$

On the assumption that the Riemann hypothesis of the Dedekind Zeta functions over certain Galois fields is true, P. D. T. A. Elliott, P. J. Stephens and the author have shown that the above limits exist (the method is based in part on that of Hooley (1967).) and have determined their values. When a is a prime p these are

$$c_p = \left(1 + d_p \cdot \frac{p}{p^3 - p - 1}\right) C, \tag{3}$$

where

$$C = \prod_q \left(1 - \frac{q}{q^3 - 1}\right)$$

and the correction factors $d_2 = \frac{1}{64}$, $d_p = \frac{1}{20}$ if $p \equiv 3 (\bmod 4)$ and $d_p = \frac{2}{5}$ if $p \equiv 1 (\bmod 4)$.

The following is a table for c_p consisting of the conjectured value calculated from (3) and computed value using the 550 primes between 233 and 4409 (the decimals are correct to the nearest 0.005).

TABLE I.

c_p	Conjectured Value	Computed Value
c_2	0·570	0·580
c_3	0·580	0·585
c_5	0·585	0·610
c_7	0·575	0·565
c_{11}	0·575	0·570
c_{13}	0·575	0·575
c_{17}	0·575	0·565
c_{19}	0·575	0·590

Similar methods yield the density $\delta(W)$ of prime divisors of a linear recurrence $W \in F(f)$. We need only consider those W given by $w_n = (a^n - B)/(a - 1)$, where $\Delta(W) = B$, and when a and B are distinct primes we have

$$\delta(W) = \lim_{x \to \infty} \frac{1}{\pi(x)} \sum_{\substack{p < x \\ p \mid W}} 1 = \tag{4}$$

$$\left(1 + e_1 \frac{a}{a^3 - a - 1} + e_2 \frac{B}{B^3 - B - 1} + e_3 \frac{aB}{(a^3 - a - 1)(B^3 - B - 1)}\right) C,$$

where the constants e_1, e_2 and e_3 have been determined and only depend on the values of a and B modulo 4. In particular, the value of e_1 is $-\frac{1}{64}$ if $a = 2$, $\frac{1}{20}$ if $a \equiv 3 \pmod 4$ and $\frac{2}{5}$ if $a \equiv 1 \pmod 4$ so that $\delta(W) \to c_a$ as $\Delta(W) \to \infty$.

The following is a table of densities of linear recurrences for $a = 2$ and 3; we exhibit the conjectured values using (4) and computed values using the same range of primes as before and the same rounding off.

TABLE II.

B	$2^n - B$		$3^n - B$	
	Conjectured Value	Computed Value	Conjectured Value	Computed Value
3	0·575	0·580	—	—
5	0·565	0·580	0·585	0·575
7	0·570	0·580	0·580	0·595
11	0·570	0·590	0·580	0·585
59	0·570	0·575	0·580	0·580
61	0·570	0·575	0·580	0·590
67	0·570	0·590	0·580	0·600
71	0·570	0·590	0·580	0·585
73	0·570	0·575	0·580	0·595
79	0·570	0·590	0·580	0·585
83	0·570	0·580	0·580	0·590
89	0·570	0·565	0·580	0·570
97	0·570	0·570	0·580	0·600
101	0·570	0·570	0·580	0·585
	Mean = ·580. Computed c_2 = ·580.		Mean = ·590. Computed c_3 = ·585.	

It is clear that a much greater range of primes would be needed to experimentally test the equations (3) and (4) for the correction factors d_p, e_1, e_2 and e_3.

Formulae for c_f and $\delta(W)$ have been obtained when neither a nor B are necessarily prime. The formulae were more complicated but of the same form. (These are contained in P. Stephen's Ph.D. thesis to be submitted to the University of Nottingham).

Stephens has shown, *without any hypothesis*, that provided N is large enough,

$$\frac{1}{N} \sum_{a \leqslant N} \sum_{p \leqslant x} \frac{e_p(a)}{p-1} = C\,li\,x + 0(x/(\log x)^E),\ E \geqslant 2,$$

and obtained similar average results for densities of prime divisors (averaged over both a and $\Delta(W)$). These results will be appearing elsewhere. The arguments used by Stephens (1969) to obtain average results for Artin's conjecture give an idea of the methods used.

On the basis of these calculations and computations we conjecture for *all* $f(x) = x^2 - Px + Q = (x - \theta_1)(x - \theta_2)$, where $Q \neq 0$, $(P, Q) = 1$ and θ_1/θ_2 is not a root of unity, that there exists a constant c_f, $0 < c_f < 1$, such that given any $\in > 0$, $c_f - \in < \delta(W) < c_f + \in$ for all $(W) \in G(f)$ with $\Delta(W)$ greater than some integer $N(\in)$, except possibly for those (W) which are of finite order or those of the form $(V)^n$, where $n \in Z$, $|n| > 1$ and $(V) \in G(f)$.

3. On Sets of Prime Divisors

We call a set P of primes a *divisor set* for $G(f)$ if there is an element in $G(f)$ whose set of prime divisors is, with but a finite number of exceptions, P (one could modify this to allow a set of exceptions to have zero density). The results of Section 2 suggest that 'most' divisor sets for $G(f)$ have a density approaching c_f. Of course, a set with density $> c_f$ may well be a divisor set, e.g., the set of all primes is a divisor set of (I). On the other hand, the evidence available suggests that sets of density greatly less than c_f are unlikely to be divisor sets for $G(f)$. We know the following facts about divisor sets.

(i) Such a set must be infinite.

(ii) A divisor set for $G(f)$ must contain all but a finite number of those primes p with $e(p) = p - 1$.

(iii) A set of primes with density one is the divisor set of the identity element (I) only. (This follows from a generalization of a result of Schinzel (1960).)

(iv) The set of primes p with $e(p) \equiv 0 \pmod 2$ is a divisor set of $(E) \in G(f)$, where $E = [2, P]$ and $f(x) = x^2 - Px + Q$. On the other hand, the set of all primes with $e(p) \equiv 1 \pmod 2$ cannot be a divisor set of any $G(f)$. (On average the density of the first set is $\frac{2}{3}$ and that of the second is $\frac{1}{3}$).

We may also consider the following related problem. Recall that every prime divides the identity element $(I) \in G(f)$. Does there exist an infinite set P of primes for which the only element of $G(f)$ which is divisible by an infinite number of primes of P is the (I)? We note that if such a set P exists the contribution of the $\{e(p)/m(p), p \in P\}$ to the sum (1) is zero.

We can prove the following. Let the roots θ_1, θ_2 of $f(x)$ be real and P be the collection of all primes of the form $(\theta_1{}^n - \theta_2{}^n)/(\theta_1 - \theta_2)$, a term of the

Lucas sequence $I \in F(f)$. Now if $(W) \in G(f)$ but $(W) \neq (I)$, then at most a finite number of primes in P divide (W). The method of proof is to compare the size of terms of W with those of I. Thus, for example, if there exists an infinite number of primes of the form $2^n - 1$, then a set P, with the property described above, exists for $G(f)$, where $f(x) = x^2 - 3x + 2$. (Actually a weaker condition than this will give us our desired set. If there exists an infinite set of primes p such that $p | 2^n - 1$ and $p > \sqrt{2^n}$, then the answer to our question is again in the affirmative.) On the other hand, if $f(x) = (x - a^2)(x - 1)$, $a \in Z$, no term of the Lucas sequence $I \in F(f)$ is prime (except if $a = 2$ and $n = 1$). It is interesting to observe that in this case the answer to our question is in the negative since an odd prime divides one and only one of the three elements (A), (B), $(E) \in G(f)$ (see Laxton (1969), section 4).

References

Hooley, C. (1967). On Artin's conjecture, *J. fur die reine und angewandte Math.*, **225,** 209–220.

Laxton, R. R. (1969). On group of linear recurrences I. *Duke Math. J.* **36,** 721–736.

Schinzel, A. (1960). On the congruence $a^x \equiv b (\mathrm{mod}\, p)$. Academie Polonaise des Sciences, Serie des sci. math. astr. et phys. Vol. VIII, No. 5, 307–309.

Stephens, P. J. (1969). An average result for Artin's conjecture. *Mathematika.* **16,** 178–188.

Ward, M. (1954). Prime divisors of second order recurring sequences. *Duke Math. J.* **21,** 607–614.

Some Relationships Satisfied by Additive and Multiplicative Recurrent Congruential Sequences, with Implications for Pseudorandom Number Generation

I. J. GOOD AND R. A. GASKINS

Virginia Polytechnic Institute, Blacksburg, Virginia, U.S.A. and Hampden-Sydney College, Virginia, U.S.A.

Abstract

The theory of the periods of linear recurrent sequences is considered from as elementary a point of view as possible. Multiplicative recurrent sequences are also considered and some remarkable properties are reported that were found empirically. An example of a result that has been proved is

$$148176G^2|\Delta_G{}^2\, 5^{u_n}$$

where u_n denotes the nth Fibonacci number and $G = 2^\alpha 3^\beta 7^\gamma (\alpha \geqslant 4, \beta \geqslant 1)$. The work is intended to be of interest in the theory of numbers and of relevance to the generation of pseudorandom numbers.

Consider a sequence $u_1, u_2 \ldots$ that satisfies the linear recurrence relation

$$u_n = c_m u_{n-1} + c_{m-1} u_{n-2} + \ldots + c_1 u_{n-m} \quad (n = m+1, m+2, \ldots), \quad (1)$$

where $c_1, c_2, \ldots, c_m$ are constant integers with $c_1 \neq 0$ and $m > 1$. Usually $u_1, u_2, \ldots, u_m$ have given values. The Fibonacci sequence is defined by the parameters $m = 2, c_1 = c_2 = 1, u_1 = u_2 = 1$.

The periodic properties of linear recurrent sequences reduced modulo an integer N have been discussed several times. For this and related matters see, for example, Ward (1933), Rees (1946), Duparc, Lekkerkerker, and Peremans (1953), Zierler (1959), Heimer (1964), Tausworthe (1965), Golomb (1967), and further references in these works. The subject is of interest in the theory of numbers and in the theory of the generation of pseudorandom numbers. Among other things we shall mention and prove some of the results and we shall also consider the "multiplicative recurrent" congruential sequence (v_n) exemplified by

$$v_n \equiv 3^{u_n} \pmod{2^{s+2}}, \quad (2)$$

or, with slightly greater generality,

$$v_n \equiv v_{n-1}{}^{c_m} \dots v_{n-m}{}^{c_1} \pmod{2^{s+2}} \qquad \text{(all } v\text{'s odd)}, \tag{3}$$

the case where the c's are all either 0 or 1 being of most interest. (The number 3 could be replaced by other odd numbers). It can be proved (compare LeVeque, 1956, p. 54) that

$$3^{2^s} \equiv 1 \pmod{2^{s+2}} \qquad (s = 1, 2, 3, \dots) \tag{4}$$

so that if we take the v_n's modulo 2^{s+2} we should take the u_n's modulo 2^s. We shall also consider moduli other than powers of 2.

On most modern computers a multiplication time is only slightly longer than an addition time, so that multiplicative methods do not suffer in comparison with additive methods. (Very fast methods are useful when random-number generation forms part of the inner loop of a calculation). Nevertheless we shall see that the sequence (v_n) has some very interesting properties that might make it unsuitable for the generation of pseudorandom numbers.

The first suggestion in unclassified literature for the use of a multiplicative method was published by Lehmer (1951), the modulus being a Mersenne prime, but moduli that are powers of 2 are about equally convenient and have also been used: see, for example, Taussky and Todd (1956). The more general "mixed congruential method", with $x_n \equiv ax_{n-1} + b \pmod{N}$ has also been considered and it appears to be a good method for some values of the parameters, but not for all: see, for example Allard, Dobell, and Hull (1963). For some values of the parameters it shows too much correlation between successive numbers. It may be that the slightly more elaborate hybrid method, with N a power of 2,

$$w_n \equiv (i + nj)w_{n-1} \pmod{2^t} \qquad (t \geqslant 3,\ w_1 \text{ odd}) \tag{5}$$

will turn out to be satisfactory with i odd and j twice an odd number. We see at once that all the terms w_n are odd. It might also be worth considering adding a constant to the right side of (5).

We have not tested the statistical properties of the sequence (w_n) but we shall now prove that it has the cycle 2^{t-1}, when $t \geqslant 3$, where by "cycle" we mean the smallest period.

The proof depends on the following analogue of Wilson's theorem:

$$1.3.5.7 \dots (2^t - 1) \equiv 1 \pmod{2^t} \qquad \text{for } t \geqslant 3. \tag{6}$$

This theorem is included in a result stated by Gauss (1801), Art. 78, according to Dickson (1919–1950), but since Gauss's book is one of the most famous ever published on any subject it is unobtainable in most large libraries. It might therefore be of value to include a proof here.

Consider the group of odd residues modulo 2^t. In this group the elements -1 and $2^{t-1}+1$ are both of order 2 and are distinct if $t > 2$. But is is known that the product of all the elements of an Abelian group having more than one element of order 2 is the identity of the group. (The product of the elements of order greater than 2 is the identity since each of these elements can be paired off with its inverse; and the product of the elements of order 2 is also the identity as can easily be seen by an inductive argument). This proves (6) for $t \geqslant 3$.

A necessary condition for the sequence $w_1, w_2, \ldots$ to reveal a period is that for some n and n' we have both $w_n \equiv w_{n'} \pmod{2^t}$ and $w_{n+1} \equiv w_{n'+1}$, that is, $(i+nj)w_n \equiv (i+n'j)w_n \pmod{2^t}$ and hence that $i+nj \equiv i+n'j \pmod{2^t}$ since w_n is odd, i.e. that $n \equiv n' \pmod{2^{t-1}}$ since j is twice an odd number. So 2^{t-1} must divide any period. Therefore the cycle is 2^{t-1} if

$$w_{2^{t-1}+1} \equiv w_1 \pmod{2^t},$$

that is, if

$$(i+j)(i+2j)\ldots(i+2^{t-1}j) \equiv 1 \pmod{2^t}. \tag{7}$$

Now for any pair of integers r and r', we have $(i+rj)-(i+r'j) = (r-r')j$ and this is a multiple of 2^t if and only if $r - r' \equiv 0 \pmod{2^{t-1}}$. Therefore the 2^{t-1} factors of the left side of (7) are distinct modulo 2^t, so that the left side is congruent to the left side of (6). Therefore by the analogue of Wilson's theorem it follows that the cycle of (w_n) is 2^{t-1}. We see also that the sequence "closes" with term number $2^{t-1}+1$, that is, the sequence "bites its own tail".

We now turn our attention to the sequences (u_n) and (v_n), and we denote the modulus generically by N. In the first place it is almost obvious that if $N = N_1 N_2$, where N_1, N_2 are coprime, then the cycle of the sequence (u_n) (or of the sequence (v_n)) is the L.C.M. of the cycles of the sequences taken modulo N_1 and N_2 separately. Hence it is enough to consider the cycles when N is a prime power p^s. We shall be especially concerned with the case $p = 2$, partly because this case is the easiest one to consider on a binary computer and is of most interest in connection with the generation of pseudorandom numbers.

When the sequence (u_n) is reduced modulo N let us denote the length of its cycle by $P(N, f, u_1, u_2, \ldots, u_m)$ where f is the monic polynomial defined by

$$f(x) = x^m - c_m x^{m-1} - \ldots - c_1 \qquad (m > 1). \tag{8}$$

The coefficients $c_1, \ldots, c_m$ need be defined only modulo N.

Let the maximum cycle of the sequence (u_n), over all choices of the first

m terms, be denoted by $P(N, f)$, otherwise expressed,

$$P(N, f) = \max_{u_1, \dots, u_m} P(N, f, u_1, \dots, u_m).$$

We shall prove later, without claiming originality, that if f is irreducible modulo p (that is, if it does not factorize modulo p into factors each of degree at least 1), then

$$P(p^s, f) | (p^m - 1) p^{s-1} \qquad (s = 1, 2, \dots). \tag{9}$$

Note that c_1 must be prime to p if f is irreducible (mod p) so that the sequence can be uniquely generated backwards. It follows that the sequence "bites its own tail" when it first completes a cycle.

Now there are irreducible polynomials, which we shall call *primeval* (mod p), for which

$$P(p, f) = p^m - 1. \tag{10}$$

[See note following Equation (15) below. It might seem natural to call a primeval polynomial "primitive", but this has been preempted. Of course f is primeval if $p^m - 1$ is prime.] But since it is easily seen that

$$P(p, f) | P(p^s, f)$$

it follows that when f is primeval (mod p) we have

$$P(p^s, f) = (p^m - 1) p^t \tag{11}$$

for some integer t for which $0 \leqslant t \leqslant s - 1$. We conjecture that t is always equal to $s - 1$ when f is primeval (mod p). We shall be mainly concerned with the case $p = 2$. The periods of (u_n) and (v_n) are equal, when (u_n) is taken modulo 2^s and (v_n) modulo 2^{s+2}.

The above conjecture is consistent with all 32 of the results shown in the table presented by Heimer (1964). We have also checked that when $p = 2$ we have $t = s - 1$ for

(i) the multiplicative sequences for $f(x) = x^3 - x - 1$, with $v_1 = 1$, $v_2 = 3$, $v_3 = 3$ or 5 or 7; $1 \leqslant s \leqslant 8$ and $s = 22$;

(ii) the multiplicative sequence for $f(x) = x^5 - x^2 - 1$, with $v_1 = v_2 = v_3 = v_4 = 1$, $v_5 = 3$; $1 \leqslant s \leqslant 8$ and $s = 22$;

(iii) the additive sequences with $u_1 = c_1, \dots, u_m = c_m$ for $f(x) = x^2 - x - 1$, $x^3 - x^2 - 1$, and $x^5 - x^2 - 1$; $19 \leqslant s \leqslant 32$.

Moreover we found the following remarkable relationship to be satisfied in all cases examined, and a proof will be given later:

$$v_{n+rG} - v_n \equiv r(v_{n+G} - v_n) \pmod{2^{2\nu}} \tag{12}$$

where

$$r = 0, 1, 2, \ldots; \quad n = 1, 2, 3, \ldots; \quad \nu = 2, 3, 4, \ldots,$$

and

$$G = (2^m - 1)2^{\nu-2} \tag{13}$$

The relationship (12) can be expressed in the form:

$$v_{n+G} - v_n, \quad \text{that is,} \quad 3^{u_{n+G}} - 3^{u_n} \pmod{2^{2\nu}},$$

regarded as a function of n, has period G. In other symbols,

$$\Delta_G{}^2 v_n \equiv 0 \pmod{2^{2\nu}}, \tag{12A}$$

where Δ_G is the operator of differencing at distance G.

Now the fact that $v_n \pmod{2^{\nu+1}}$ has period G can be expressed in the form $\Delta_G v_n \equiv 0 \pmod{2^{\nu+1}}$ so it is natural to examine other powers of the operator Δ_G. We found that

$$\Delta_G{}^k v_n \equiv 0 \pmod{2^{2\nu+k-2}} \quad (k \geqslant 2) \tag{12B}$$

for the case $f(x) = x^2 - x - 1, v_1 = v_2 = 3$, with $\nu = 2, k = 2, 3, 4, 5$ and $v = 3, k = 2, 3, 4$. These were the only cases examined, but they suggest that (12B) might always be true when f is irreducible (mod 2). But we regard this conjecture as quite weak since it is difficult to see how the proof of (12A), which we give later, could be extended to values of $k > 2$.

The theorem given by (12) or (12A) seems to be new even for the "multiplicative Fibonacci" and "multiplicative Lucas" sequences as they might be called. The number 3 can be replaced by any other odd number that is as "primitive as can be" modulo $2^{2\nu}$. The theorem was verified for a variety of special cases, for the polynomials $f(x) = x^2 - x - 1, x^3 - x^2 - 1$, and $x^5 - x^2 - 1$, before we found a proof.

For $f(x) = x^3 - x^2 - 1$; $v_1 = v_2 = 1$, $v_3 = 3$; we found that

$$v_{rG+2} + v_{rG} \equiv v_2 + v_0 \pmod{2^{2\nu}} \tag{14}$$

when $\nu = 16, 1 \leqslant r \leqslant 100$. This can be largely explained in terms of (12) by the following argument.

If it so happens that

$$v_{G+\mu} - v_\mu \equiv -(v_{G+\mu'} - v_{\mu'}) \pmod{2^{2\nu}} \tag{14A}$$

Then by (12A) we have

$$v_{rG+\mu} - v_{(r-1)G+\mu} \equiv -(v_{rG+\mu'} - v_{(r-1)G+\mu'}),$$

that is,

$$v_{rG+\mu} + v_{rG+\mu'} \equiv v_{(r-1)G+\mu} + v_{(r-1)G+\mu'}$$

so that (14A), with $\mu = 2$ and $\mu' = 0$, when combined with (12), implies (14) by induction.

Thus, if a relation resembling (14) occurs "by chance" in some sequence (v_n) when $r = 1$, it will continue to happen for all values of r in that sequence. Note too that since G is a period of (v_n) when this sequence is taken modulo 2^{v+1}, the left side of (14A) can take only 2^{v-1} distinct values; so it is not a very great "coincidence" that (14) is true.

Odd Primes

We have checked that the cycle of the Fibonacci sequence (u_n) modulo p^s is $(p^m - 1)p^{s-1}$ for $m = 2$ (of course), $p = 3, 1 \leqslant s \leqslant 4$. Also, writing $v_n = 2^{u_n}$, we have found that $\Delta_{24}{}^2 v_n \equiv 0 \pmod{3^5}$, and that 24 is the smallest interval for this phenomenon. This suggests that the appropriate form of (12B) for odd primes p might be

$$\Delta_G{}^k a^{u_q} \equiv 0 \quad (\text{mod } p^{2v+k-1}), \tag{12C}$$

where $G = (p^m - 1)p^{v-1}$, a is a primitive root of $p, k \geqslant 0$, and $v \geqslant 2$. It would not be surprising if odd primes behave slightly differently from the even prime since this is a feature of primitive roots. The conjecture (12C) has not been tested beyond the evidence already mentioned.

Further Explanations

A polynomial d(x) (with integer coefficients) that can be expressed in the form

$$g(x) = qQ(x) + F(x)\, f(x)$$

is said to be congruent to 0 to the moduli q and $f(x)$, where q is a positive integer and $Q(x)$, $F(x)$, and $f(x)$ are polynomials. Also two polynomials $g(x)$ and $h(x)$ are said to be congruent to the moduli q and $f(x)$ if their difference $g(x) - h(x)$ is congruent to 0. This congruence is written in the familiar notation

$$g(x) \equiv h(x) \quad (\text{modd } q, f(x)).$$

Congruence to a single modulus q is the special case in which $f(x) = 0$.

An irreducible polynomial (mod p) is a polynomial with integer coefficients, of degree at least 1, which cannot be expressed (mod p) as the product of two polynomials each of degree at least 1. This definition is familiar for any integer p but we shall always assume p to be prime. It is known that every irreducible polynomial $f(x)$ (mod p) of degree not exceeding m must divide

$$x^{p^m-1} - 1 \qquad (\text{mod } p);$$

that is

$$x^{p^m-1} \equiv 1 \qquad (\text{modd } p, f(x)). \tag{15}$$

For a proof of this fact see, for example, Adamson (1964), p. 126. It was also used by Rees (1946), who mentions further that there are irreducible polynomials f (mod p) for which the smallest integer q for which $f \mid x^q - 1$ is $q = p^m - 1$. These are our primeval polynomials. In fact there are $\phi(p^m - 1)/m$ such monic polynomials, where ϕ is Euler's function (see, for example, Golomb, 1967, p. 49).

We now generalize (15) to recurrence relations modulo p^s.

Theorem. If $f(x)$ is irreducible (mod p), *then, for every positive integer s we have*

$$x^{(p^m-1)p^{s-1}} \equiv 1 \qquad (\text{modd } p^s, f(x)). \tag{16}$$

Proof by Induction: The statement is true when $s = 1$ since this is the theorem just stated. Suppose then that it is true for any particular value of s. Then

$$x^{(p^m-1)p^{s-1}} \equiv 1 + p^s Q(x) \qquad (\text{modd } p^{s+1}, f(x)),$$

where $Q(x)$ is some polynomial. Therefore, by raising both sides to the pth power we see that

$$x^{(p^m-1)p^s} \equiv 1 + \binom{p}{1} p^s Q(x) + \ldots + \binom{p}{p} p^{sp} (Q(x))^p \qquad (\text{modd } p^{s+1}, f(x)).$$

$$\equiv 1 \qquad (\text{modd } p^{s+1}, f(x)),$$

as required. [We mention parenthetically that the polynomial

$$x^{(p^m-1)p^{s-1}} - 1$$

has a unique decomposition into factors modulo p^s only when $s = 1$. For example,

$$\begin{aligned} x^6 - 1 &\equiv (x+1)(x-1)(x^2 - x + 1)(x^2 + x + 1) \\ &\equiv (x+1)(x-1)(x^2 - x - 1)(x^2 - x - 1) \qquad (\text{mod } 4).] \end{aligned}$$

The application of the theorem to linear recurrence relations will now be explained.

Suppose we have the linear recurrence relation (1). It is a familiar elementary fact that the generating function $U(x) = u_1 x + u_2 x^2 + \dots$ is of the form

$$\frac{xg(x)}{1 - c_m x - \dots - c_1 x^m} \tag{17}$$

where $g(x)$ is a polynomial of degree less than m. From this it can be proved that, when $f(x)$ is irreducible modulo N, the sequence $u_1, u_2, \dots$ has as its cycle the smallest integer P for which

$$x^P \equiv 1 \quad (\text{modd}\, N, f(x)), \tag{18}$$

provided that the H.C.F. of $u_1, \dots, u_m$ and N is 1. This is proved, for example, for the case $N = 2$, by Golomb (1967), pp. 32–33, and the proof goes through virtually unchanged for arbitrary values of N.

Now by (16) we know that the congruence (18) will be true with $P = (p^m - 1)p^{s-1}$ when $N = p^s$ and this proves (9). Moreover the cycle must be a multiple of $p^m - 1$ if f is primeval. Therefore we have the

THEOREM. *A necessary and sufficient condition for a primeval linearly recurrent sequence* (mod p^s) *to have cycle length* $(p^m - 1)p^{s-1}$ *is that* $(p^m - 1)p^{s-2}$ *is not a period.*

The condition of this theorem is easy to check on a binary computer (without using end-round carry) when $p = 2$ and s does not exceed the word length, since the appropriate powers of x can be obtained by successive squaring of x^{2m-1}. For example, for the Fibonacci sequence, which is primeval, we can obtain the coefficients from the relationships:

$$a_0 = 2, b_0 = 1, \quad a_{r+1} = 2a_r b_r + a_r^2, \quad b_{r+1} = a_r^2 + b_r^2 \quad (\text{mod } 2^s)$$

where

$$x^{3.2^r} \equiv a_r x + b_r \quad (\text{modd } 2^s, x^2 - x - 1).$$

In this manner we found that the primeval sequences with $u_1 = c_1, \dots, u_m = c_m$ for $f(x) = x^2 - x - 1$, $x^3 - x^2 - 1$, and $x^5 - x^2 - 1$ all have cycle $(2^m - 1)2^{s-1}$ (mod 2^s) for $19 \leqslant s \leqslant 32$. In view of what has already been said it would be sufficient here for at least one of the numbers $u_1, u_2 \dots, u_m$ to be odd.

When $f(x)$ is primeval (mod p) and $N = p^s$, suppose that the cycle length is $(p^m - 1)p^t$ $(t \leqslant s - 1)$. Of the p^{ms} possible "beginnings" $(u_1, \dots, u_m)$ there are $p^{m(s-1)}$ for which all the u's are divisible by p. These generate shorter cycles but all the other beginnings do not. Hence there are

$$\frac{p^{ms} - p^{m(s-1)}}{(p^m - 1)p^t} = p^{m(s-1)-t}$$

independent cycles of length $(p^m - 1)p^t$ corresponding to $f(x)$. (As we said before, perhaps $t = s - 1$ always.).

Proof of (12). Let $p = 2$, let $f(x)$ be of degree m and irreducible (mod 2), and let $G = (2^m - 1)\,2^{\nu-2}$ $(\nu \geqslant 2)$. We have

$$x^G - 1 \equiv 0 \quad (\text{modd } 2^{\nu-1}, f(x))$$

so that

$$(x^G - 1)^k \equiv 0 \quad (\text{modd } 2^{(\nu-1)k}, f(x)),$$

and it follows readily that

$$\Delta_G{}^k\, u_n \equiv 0 \quad (\text{mod } 2^{(\nu-1)k}) \;\; (k \geqslant 1). \tag{19}$$

For $k = 2$ this gives

$$u_{2G+n} - u_{G+n} \equiv u_{G+n} - u_n \quad (\text{mod } 2^{2\nu-2}). \tag{20}$$

We denote $u_{G+n} - u_n$ by λ for the time being. Then, by (4),

$$3^\lambda - 1 \equiv 0 \quad (\text{mod } 2^{\nu+1}),$$

so that

$$(3^\lambda - 1)^2 \equiv 0 \quad (\text{mod } 2^{2\nu+2}),$$

and therefore

$$3^\lambda - 1 \equiv 1 - 3^{-\lambda} \quad (\text{mod } 2^{2\nu+2}).$$

(Since $(2, 3) = 1$ the use of negative powers of 3 is meaningful and legitimate.) But

$$3^{u_{2G+n} - u_{G+n}} \equiv 3^\lambda \quad (\text{mod } 2^{2\nu}),$$

by (4) and (20). Therefore

$$3^{u_{2G+n} - u_{G+n}} - 1 \equiv 1 - 3^{u_n - u_{G+n}} \quad (\text{mod } 2^{2\nu}).$$

Therefore

$$\Delta_G{}^2\, 3^{u_n} \equiv 0 \quad (\text{mod } 2^{2\nu}),$$

as asserted in (12A).

A Result for More Complicated Moduli

The above method of proof can be extended to give interesting results for more complicated moduli. The following example will give an adequate impression of this extension and of the kinds of results that can be obtained.

Let

$$G = 2^{\alpha} 3^{\beta} 7^{\gamma} \qquad (\alpha \geqslant 4 \ \beta \geqslant 1, \gamma \geqslant 0). \tag{21}$$

We have not included 5 as a factor of G because we decided to take $f(x)$ as the "Fibonacci polynomial" $x^2 - x - 1$ which is reducible modulo 5 but irreducible moduli 2, 3, and 7. Accordingly we write u_n for the nth Fibonacci number, although we have previously used this symbol mainly in a more general sense, and we let

$$y_n = 5^{u_n}. \tag{22}$$

Here we have chosen the number 5 because it is the smallest number prime to 2, 3, and 7, and is also a primitive root of these three primes.

One period, perhaps the cycle, of the sequence (y_n) modulo $2^A 3^B 7^C$ is the L.C.M. of 2^{A-2}, 2.3^{B-1}, and 6.7^{C-1} and is therefore $2^{A-2} 3^{B-1} 7^{C-1}$, so that

$$5^{2^{A-2} 3^{B-1} 7^{C-1}} \equiv 1 \quad (\mathrm{mod}\ 2^A 3^B 7^C) \quad (A \geqslant 3, B \geqslant 2, C \geqslant 1). \tag{23}$$

This device *gives more information than Euler's theorem* since $\phi(2^A 3^B 7^C)$ is 24 times the period mentioned here.

Now

$$\begin{aligned} G &= (2^2 - 1)\, 2^{\alpha} . 3^{\beta-1} 7^{\gamma} \\ &= (3^2 - 1)\, 3^{\beta} . 2^{\alpha-3} 7^{\gamma} \\ &= (7^2 - 1)\, 7^{\gamma} . 2^{\alpha-4} 3^{\beta-1}, \end{aligned}$$

so that

$$x^G \equiv 1 \text{ modulo } 2^{\alpha+1} 3^{\beta-1} 7^{\gamma},\ 2^{\alpha-3} 3^{\beta+1} 7^{\gamma}, \text{ and } 2^{\alpha-4} 3^{\beta-1} 7^{\gamma+1},$$

and therefore modulo the L.C.M. of these numbers; i.e.

$$\begin{aligned} x^G &\equiv 1\ (\mathrm{modd}\ 2^{\alpha+1} 3^{\beta+1} 7^{\gamma+1}, x^2 - x - 1) \\ &\equiv 1\ (\mathrm{modd}\ 42G, x^2 - x - 1). \end{aligned} \tag{24}$$

Therefore, for any integer $k > 0$,

$$(x^G - 1)^k \equiv 0 \quad (\mathrm{modd}\ 2^{(\alpha+1)k} 3^{(\alpha+1)k} 7^{(\gamma+1)k}, f),$$

and

$$\Delta_G{}^k u_n \equiv 0 \quad (\mathrm{mod}\ 2^{(\alpha+1)k} 3^{(\beta+1)k} 7^{(\gamma+1)k}). \tag{25}$$

For $k = 2$ this gives

$$u_{2G+n} - u_{G+n} \equiv u_{G+n} - u_n \pmod{2^{2\alpha+2}\, 3^{2\beta+2}\, 7^{2\gamma+2}}. \tag{26}$$

Denote $u_{G+n} - u_n$ by λ. We have, by (25) with $k = 1$,

$$\lambda \equiv 0 \pmod{2^{\alpha+1}\, 3^{\beta+1}\, 7^{\gamma+1}}. \tag{27}$$

Therefore by (23) we have

$$5^{\lambda} - 1 \equiv 0 \pmod{2^{\alpha+3}\, 3^{\beta+2}\, 7^{\gamma+2}},$$
$$\Delta_G y_n \equiv 0 \pmod{2^{\alpha+3}\, 3^{\beta+2}\, 7^{\gamma+2}}, \tag{28}$$

and

$$5^{\lambda} - 1 \equiv 1 - 5^{-\lambda} \pmod{2^{2\alpha+6}\, 3^{2\beta+4}\, 7^{2\gamma+4}}.$$

We can replace the first λ here by $u_{2G+n} - u_{G+n}$ if, by (23) and (26), we use the modulus $2^{2\alpha+4}\, 3^{2\beta+3}\, 7^{2\gamma+3}$. Hence

Hence

$$\Delta_G{}^2\, 5^{u_n} \equiv 0 \pmod{2^{2\alpha+4}\, 3^{2\beta+3}\, 7^{2\gamma+3}},$$

that is,

$$\Delta_G{}^2\, 5^{u_n} \equiv 0 \pmod{148{,}176G^2} \quad (G = 2^{\alpha}\, 3^{\beta}\, 7^{\gamma},\ \alpha \geqslant 4,\ \beta \geqslant 1), \tag{29}$$

when (u_n) is the Fibonacci sequence. The simplest example of this is

$$\Delta_{48}{}^2\, 5^{u_n} \equiv 0 \pmod{2^{12}\, 3^5\, 7^3 = 341{,}397{,}504}. \tag{30}$$

We checked this for $n = 1$ on a computer. The case $n = 0$ is

$$5^{u_{96}} + 1 \equiv 2\,.\,5^{u_{48}} \pmod{2^{12}\, 3^5\, 7^3}.$$

Modulo $2^{12}\, 3^5\, 7^3$, this gives a solution of the congruence

$$z^l + 1 \equiv 2y^l,$$

where $l = u_{48} = 4{,}807{,}526{,}976$, since $u_{48} | u_{96}$. (Note the resemblance to Fermat's equation to a large modulus.) It is also pleasing to write the example in the form

$$5^{u_{48}u'_{48}} + 1 \equiv 2\,.\,5^{u_{48}} \pmod{2^{12}\, 3^5\, 7^3}, \tag{31}$$

where u'_{48} denotes the 48th Lucas number. This congruence is not obvious since u_{48}, being $2^6\,.\,3^2\,.\,7\,.\,23\,.\,47\,.\,1103$, is not a multiple of L.C.M. $(\phi(2^{12}), \phi(3^5), \phi(7^3)) = 2^{11}\, 3^4\, 7^2$.

References

Adamson, I. T. (1964). "Introduction to Field Theory" Oliver and Boyd, Edinburgh, London, and New York.

Allard, J. L., Dobell, A. R., and Hull, T. E. (1963). Mixed congruential random number generators for decimal machines. *J. Assoc. Computing Machinery*, **10,** 131–141.

Dickson, L. W. (1919/50). "History of the Theory of Numbers," Vol. 1. Carnegie Institution Washington; and Chelsea, New York.

Duparc, H. J. A., Lekkerkerker, C. G., and Peremens, W. (1953). Reduced sequences of integers and pseudo-random numbers, Math. Centrum Amsterdam, Rapport ZW 1953–002, 15 pp. (Reviewed by D. H. Lehmer in *Math. Rev.* **14,** 770.).

Gauss, K. F. (1801). "Disquisitiones arithmeticae".

Golomb, S. W. (1967). "Shift Register Sequences". Holden–Day; San Francisco, Cambridge, London, Amsterdam.

Heimer, R. L. (1964). Further comments on the periodicity of the digits of the Fibonacci sequence. *The Fibonacci Quarterly* **2,** 211–214.

Lehmer, D. H. (1951). *In* "Second Harvard Symposium on Large Scale Automatic Computing Machinery". Harvard, 145.

LeVeque, W. J. (1956). "Topics in Number Theory, I". Addison–Wesley, Reading (Mass.), Palo Alto and London.

Rees, D. (1946). Note on a paper by I. J. Good. *J. London Math. Soc.* **21,** 169–172.

Taussky, O., and Todd, J. (1956). Generation and testing of pseudo-random numbers. *In* Symposium on Monte Carlo Methods, 15–28. John Wiley and Chapman and Hall, New York and London.

Tausworthe, R. C. (1965). Random numbers generated by linear recurrence modulo two. *Math. Comp.* **19,** 201–209.

Ward, M. (1933). The arithmetical theory of linear recurring sequences. *Trans. Amer. Math. Soc.* **35,** 600–628.

Zierler, N. (1959). Linear recurring sequences. *J. Soc. Indust. Appl. Math.* **7,** 31–48.

Investigation of *T*-Numbers and *E*-Sequences

David G. Cantor*

University of California, Los Angeles

A *PV-number* θ is a real algebraic integer >1, whose algebraic conjugates have absolute value <1. A *T-number* is a real algebraic integer >1, whose algebraic conjugates have absolute value $\leqslant 1$, with at least one conjugate having absolute value $=1$. If x is a real number we shall denote by $N(x)$ the "nearest" integer to x, i.e., $N(x)$ is the unique integer satisfying $x - \frac{1}{2} < N(x) \leqslant x + \frac{1}{2}$; and by $\|x\|$ we shall denote $|N(x) - x|$. It is known, Salem (1945), that the set of PV-number forms a closed, non-discrete subset of the real numbers and that the closure of the set of T-numbers includes the set of PV-numbers. However, nothing else is known concerning the closure of the set of T-numbers. The purpose of this investigation is to obtain numerical evidence concerning the distribution of T-numbers.

Let θ be a PV or T-number which is a zero of the monic polynomial $\sum_{i=0}^{n} c_i X^i$. It is known, Salem (1945), that there exist infinitely many λ in the field $\mathbf{Q}(\theta)$ such that if $a_n = N(\lambda\theta^n)$ then

$$|a_{n+1} - a_n^2/a_{n-1}| < \tfrac{1}{2} \tag{1}$$

for all large n. (If θ is PV any $\lambda \in \mathbf{Q}(\theta)$ such that $\lambda\theta^n$ is an algebraic integer for large n will do. In fact, a_n is the trace of $\lambda\theta^n$ and $\Sigma_i\, c_i\, a_{n-i} = 0$ for all large n.) Sequences $\{a_n\}$ satisfying (1) or, what is the same thing, $a_{n+1} = N(a_n^2/a_{n-1})$ are called *E-sequences*. Pisot (1938) and Flor (1960) have shown that if $a_1 \geqslant a_0 + 2\sqrt{a_0}$, then $\phi = \lim_{n\to\infty} a_{n+1}/a_n$ exists and is > 1. Furthermore $|\phi - a_1/a_0| < \sqrt{3/32a_0}$. Such numbers $\phi > 1$ are called *E-numbers*. It follows that the set of *E-numbers* contains both the set of *PV*-numbers and the set of T-numbers. It is not known if the PV- and T-numbers comprise the E-numbers.

* The author wishes to thank the Sloan Foundation and NSF Grant GP 13164 for support during the preparation of this paper.

We calculate E-sequences $\{a_n\}$ whose limiting ratios ϕ are close to 1, and check whether these E-sequences satisfy linear recurrence relationships with constant coefficients. (In actuality we can only check a finite initial segment of such an E-sequence.) If the E-sequence satisfies the recurrence $\sum_{i=0}^{r} c_i a_{n+i} = 0$, then ϕ is a root of $\sum_{i=0}^{r} c_i X^i = 0$ and is a PV- or T-number.

The sequence $\{a_n\}$ satisfies a linear recurrence relation if and only if the corresponding generating function $a = \sum_{n=0}^{\infty} a_n X^n$ is rational. To determine this, we work in the field of formal Laurent series of the shape $b = \sum_{n=n_0}^{\infty} b_n X^n$, where n_0 is an integer (possibly negative). We put $\lambda(b) = \sum_{n=n_0}^{0} b_n X^n$, more generally $\lambda_k(b) = \sum_{n=n_0}^{k} b_n X^n$. If $b_{n_0} \neq 0$, we put $\operatorname{ord}(b) = n_0$, and $\operatorname{ord}(0) = \infty$. We apply the analogue of the standard continued fraction algorithm for real numbers to a. The polynomials in $1/X$ play the role of the integers. This algorithm will terminate if and only if a is rational. (For a related algorithm see Berlekamp (1968)). Since this algorithm does not seem to be in the literature, we describe it here. (A special case for so-called J-fractions appears in Wall (1948).) Put $A_0 = a$ and $c_0 = \lambda(A_0)$. Inductively, for $n = 1, 2, 3, \ldots$ put $A_n = 1/(A_{n-1} - c_{n-1})$ and $c_n = \lambda(A_n)$. Then

$$a = c_0 + \cfrac{1}{c_1 + \cfrac{1}{c_2 + \cdots}},$$

where the c_i's are polynomials in $1/X$. Put $p_{-2} = 0, p_{-1} = 1, q_{-2} = 1, q_{-1} = 0$ and inductively for $n = 0, 1, 2. \ldots$ let $p_n = c_n p_{n-1} + p_{n-2}$ and $q_n = c_n q_{n-1} + q_{n-2}$. Then p_n/q_n are approximants to a in the sense that $\operatorname{ord}(q_n a - p_n) = -\operatorname{ord}(q_{n+1})$ if a_{n+1} is defined, and otherwise $q_n a - p_n = 0$. If p, q are polynomials in $1/X$ satisfying $\operatorname{ord}(qa - p) > -\operatorname{ord}(q)$ then the fraction p/q is in fact among the fractions p_n/q_n. This means that any linear recurrence of degree d, which is satisfied by the sequence $\{a_n\}$ for d or more consecutive times, will give rise to one of the rational approximants p_n/q_n. The algorithm, as described, is unwieldy, for it involves working with the entire power series a, or at least with the entire initial segment of interest. The modification we describe now brings in coefficients of a when they are needed and no sooner.

Put $h_n = q_n a - p_n$ and $h_n^{(m)} = q_n \lambda_m(a) - p_n$. The identity $A_n = -h_{n-2}/h_{n-1}$ is easy to verify. It follows that $c_n = -\lambda(h_{n-2}/h_{n-1})$ and in fact $c_n = -\lambda(h_{n-2}^{(m)})$ for all sufficiently large m. The algorithm follows:

I. (Initialize).
Put $h_{-2} = a_0, h_{-1} = -1, q_{-2} = 1, q_{-1} = 0, n = 0, m = 0.$

II. (Increase m if necessary).

If $h_{n-1} = 0$ or if $\operatorname{ord}(q_{n-2}) - \operatorname{ord}(h_{n-1}) + m < 0$ or if $\operatorname{ord}(q_{n-1}) + \operatorname{ord}(h_{n-2}) - 2\operatorname{ord}(h_{n-1}) + m < 0$ then

A. Put
$$h'_{n-2} = h_{n-2} + q_{n-2}\, a_{m+1}\, X^{m+1}$$
$$h'_{n-1} = h_{n-1} + q_{n-1}\, a_{m+1}\, X^{m+1}$$
$$m' = m + 1.$$

B. Replace m, h_{n-1}, h_{n-2} by m', h'_{n-1}, h'_{n-2}, respectively.

C. Repeat Step II.

III. (Calculate c_n and h_n).

When the conditions of step II are no longer met, put $c_n = -\lambda(h_{n-2}/h_{n-1})$, $q_n = c_n q_{n-1} + q_{n-2}$, and $h_n = c_n h_{n-1} + h_{n-2}$. Increase n by 1 and go to step II.

The algorithm terminates when m (or n) is large enough. Note that the quantity called h_n in the algorithm is in fact what was denoted by $h_n^{(m)}$ earlier. When using this algorithm one finds that the coefficients of the power series (which must be kept exactly) are rational numbers whose numerators and denominators are enormous integers with varying, unpredictable numbers of digits. For this reason a multiple-precision arithmetic package with automatic storage allocation was written.†

The results of these calculations are still preliminary in nature and will be described later. However, it is noteworthy that the E-sequences we have calculated (mostly with $a_0 \leqslant 20$) seem to satisfy recurrences of low degree if they satisfy any recurrence we can find. Furthermore the coefficients of these recurrences are always small integers. Paul Galyean has made use of this empirical fact by performing the continued fraction algorithm (mod p) where p is a large prime, in our case $2^{31} - 1$, thus speeding calculations considerably. These results are then checked using multiple precision arithmetic. In no case has a false recurrence been found.

Pisot (1938) showed that all E-sequences with $a_0 = 2$ and $a_0 = 3$ satisfy linear recurrence relations of low degrees and actually obtained their coefficients. It is not surprising that he did not continue for $a_0 = 4$, for our calculations show that the E-sequence with $a_0 = 4$, $a_1 = 13$ satisfies no recurrence relation of degree $\leqslant 100$. A somewhat surprising result was that certain initial segments of E-sequences satisfied recurrences, not of the PV or T type, for many terms. For example the sequence with $a_0 = 8$, $a_1 = 10$ satisfied the recurrence $a_n = a_{n-1} + a_{n-6}$ for $6 \leqslant n \leqslant 37$. Since the roots of $x^6 = x^5 + 1$ are not PV or T numbers the above recurrence cannot hold for all n.

† This package, written in 360 assembly language for use with PL/I, is suitable for large 360's and is available from the author.

References

Berlekamp, E. (1968). "Algebraic Coding Theory". 178–192. McGraw Hill, New York.

Flor, P. (1960). Über eine Klasse von Folgen natürlicher Zahlen. *Math. Ann.* **140,** 299–307.

Pisot, C. (1938). La Répartition Modulo un et les nombres algébriques, *Annali dr. r. Scuola Normal Sup. Pisa*, Ser. 2 **7,** 205–248.

Salem, R. (1945). Power Series with Integral Coefficients, *Duke Math. J.* **12,** 153–172.

Wall, H. W. (1948). "Analytic Theory of Continued Fractions". Van Nostrand, New York (reprinted (1967) by Chelsea, New York.).

Use of Computers in Cyclotomy

Joseph B. Muskat

Department of Mathematics, Bar-Ilan University, Ramat Gan, Israel

1. Cyclotomy of order e over $GF(p)$. For any prime $p \equiv 1 \pmod{e}$, define the cyclotomic number $(h, k)_e$, $0 \leqslant h, k \leqslant e-1$, as the number of solutions of

$$\text{ind}_g\, n \equiv h \pmod{e},\ \text{ind}_g\,(n+1) \equiv k \pmod{e},\ 1 \leqslant n \leqslant p-2, \qquad (1)$$

where g is a fixed primitive root of p. A good general reference is Dickson (1935a).

The Jacobi sums of order e

$$R_e(u, v) = \sum_{z=2}^{p-1} \beta^{v\, \text{ind}_g\, z + u\, \text{ind}_g (1-z)}, \qquad \beta = \exp(2\pi i/e), \qquad (2)$$

are closely connected with the cyclotomic numbers (Whiteman (1960) (2.7)):

$$e^2(h, k)_e = \sum_{u,v=0}^{e-1} (-1)^{uf} R_e(u, v)\, \beta^{-hu-kv}, \qquad f = (p-1)/e. \qquad (3)$$

If the exponents of β in (2) which are congruent $(\text{mod}\, e)$ are grouped, one may write

$$R_e(u, v) = \sum_{j=0}^{e-1} C(j, u, v)\beta^j. \qquad (4)$$

The integers $C(j, u, v)$ will be called *Jacobi coefficients.*

The basic problems of cyclotomy are

(a) determining relationships between pairs of Jacobi sums of orders e and divisors of e,

(b) from these relationships deriving relationships between Jacobi coefficients,

(c) deriving from

$$|R_e(u, v)|^2 = p, \qquad u, v, u + v \not\equiv 0 \pmod e$$

diagonal quadratic forms which represent p or a fixed multiple of p,

(d) finding formulas for the cyclotomic numbers. These formulas have the form

$$\sum_{y=1}^{t} Q_{h,k,y} J_y = (h, k)_e, \tag{5}$$

where the J_y include p, a constant, and certain linear combinations of Jacobi coefficients, the latter being replaced by coordinates of the above-mentioned diagonal quadratic forms where possible.

There are several ways in which using the computer can facilitate finding formulas for cyclotomic numbers. If e is a prime, one may derive from the analysis of Dickson (1935b) the number of variables, say t, in the formulas for the cyclotomic numbers of order e. Compute the cyclotomic numbers for t primes $p \equiv 1 \pmod e$. From these, calculate the needed Jacobi coefficients. Insert the values of the cyclotomic numbers and the values of the J's into (5) to obtain for each $(h, k)_e$ a set of t simultaneous linear equations in the t unknowns $Q_{h,k,y}$, $y = 1, \ldots, t$. Solve these equations to determine the formulas.

This method was applied for primes $e = 7, 11, 13$. The computer was used to calculate the cyclotomic numbers and the Jacobi coefficients as well as to solve the simultaneous linear equations. For $e = 13$ the matrix (J_y) was 30×30 and there were 35 vectors of constants.

Exact solutions were needed. It was found that when the right side of (5) is multiplied by $4e^2$, the new Q's turn out to be integers. Rounding the computed solutions gave exact results.

If e is composite, however, the famous identity of Davenport and Hasse (1935) (0.9)

$$\prod_{q=0}^{d-1} \tau(qc + r) = \beta^{-rd \operatorname{ind}_g d}\, \tau(rd) \prod_{q=1}^{d-1} \tau(qc), \qquad e = cd, \tag{6}$$

where $\tau(s)$ denotes the Gaussian sum

$$\sum_{z=1}^{p-1} \beta^{s \operatorname{ind}_g z} \exp(2\pi i z/p),$$

expressed in terms of Jacobi sums provides additional relationships between Jacobi coefficients. The implication is that if the distinct prime divisors of e are $d_1, \ldots, d_m$, we can group the primes $p \equiv 1 \pmod e$ into equivalence classes based on $\text{ind}_g\, d_j$ (mod e/d_j), $j = 1, \ldots, m$. (The class into which a prime falls usually depends upon the choice of g.) Then for the primes in each class there is a set of cyclotomic number formulas in terms of fewer variables than are needed in a set of formulas which hold for all primes $p \equiv 1 \pmod{e}$.

When the number of variables is known, formulas can be determined by solving simultaneous linear equations, as above. This technique was applied for $e = 9$ and $e = 20$. In the latter case this was done only for those classes having enough (i.e., 18) primes for which the necessary data had been computed. The resulting formulas were very valuable in checking out the rather lengthy computer program which generated the formulas for the cyclotomic numbers directly (i.e., without reference to computed values). (Some early efforts to determine formulas from computed values failed because too few variables were used (Lehmer, 1954; Bruck, 1956).

In order to derive the formulas directly, one needs the solution to problem *b*) above. This, in turn, requires solving problem *a*). Specifically, one needs to know when one Jacobi sum is a 2*e*th root of unity times an automorphic image of another, and what is the root of unity (Zee, 1968).

One may calculate the Jacobi sums for several primes and look for relationships, then try to prove the observed patterns using elementary relationships between Jacobi sums and the Davenport–Hasse identity. A drawback is that one may overlook some of the relationships. A preferred approach is to have the computer check, using the formula of Kummer (Dickson, 1935c; Theorem 4), whether two Jacobi sums have the same prime ideal factorization. If so, the mathematician may refer to computations to help formulate the arithmetic aspect of the relationship, i.e., to what root of unity is their ratio equal. This information, in turn, will give guidance for the proof. For example it is apparent that in order to establish (Muskat, 1968)

$$R_{30}(2, 2) = \beta^{-2\text{ ind }2-6\text{ ind }3+5\text{ ind }5}\, R_{30}(16, 9)$$

the Davenport–Hasse identity (6) must be applied with $d = 2$, 3, and 5.

For certain values of e, however, applying the elementary relationships and the Davenport–Hasse identity does not give all the relationships between pairs of Jacobi sums. Other information is needed in determining which square root is to be taken. Such *sign ambiguities* have been encountered for $e = 12$, 15, 24 Muskat, 1968, 20 (Muskat and Whiteman, 1970), 21, 28 and 39 (Zee and Muskat, 1970).

The following procedure was helpful in studying sign ambiguities, First calculate the Jacobi sums involved for several primes to verify that sometimes one square root is taken and sometimes the other, and that the choice is independent of the indices of divisors of e and the parity of f. (This is necessary, as a $\pm$ left by Dickson (1935c; 44) in studying $e = 18$ was resolved recently by Baumert and Fredricksen (1967; 4.11) using (6).) Then make lists of primes for which each square root is selected and include possibly relevent binary quadratic decompositions of each prime. Stare at the numbers in the quadratic decompositions until a distinction between the lists is discovered. Then try to prove this distinction (perhaps by means of congruence properties of sums of cyclotomic numbers). Example:

$$R_{21}(1,4)^2 = \beta^{14 \text{ ind } 7} R_{21}(3,6)^2, \qquad p = A^2 + 7B^2.$$

$R_{21}(1,4) = \beta^{7 \text{ ind } 7} R_{21}(3,6)$ if and only if $3 \mid B$; otherwise $3 \mid A$.

For $e = 12$, 20, and 28 the situation is somewhat more complicated. There arise identities of the form

$$R_e(u,v)^2 = m^2 \beta^{2w} R_e(y,z)^2, \qquad m^2 = \pm 1.$$

The sign of m^2 can be determined with the aid of (6). In addition, it can be associated with a congruence condition on X and Y in

$$p = X^2 + 4Y^2. \tag{7}$$

If $m^2 = +1$, the sign of m can be associated with further congruence conditions, but not by means of (6). If $m^2 = -1$, however, the sign of m depends upon the choice of the primitive root g.

The program which computed prime ideal decompositions of Jacobi sums also determined the automorphisms leaving a decomposition invariant. A Jacobi sum with decomposition invariant under "many" automorphisms had its arithmetic properties investigated (by hand). The most interesting discovery was $R_{39}(1,16)$, with decomposition invariant under $\phi(39/2 = 12$ automorphisms. This suggested that $R_{39}(1,16)$ might be associated with a binary quadratic decomposition of p. The following was found: (Zee, 1968; Chapter 8)

$$R_{39}(1,16) = m\beta^{13 \text{ ind } 13} R_{39}(2,32), \qquad m = \pm 1.$$

Since $p \equiv 1 \pmod 3$, $4p = L^2 + 27M^2$.

$m = 1$ if and only if $M \equiv 0, \pm 4 \pmod{13}$ if and only if

$$\beta^{13 \text{ ind } 13} R_{39}(1,16) = E + F\sqrt{-39},$$

so that $p = E^2 + 39F^2$.

$m = -1$ if and only if $L \equiv 0, \pm 4 \pmod{13}$ if and only if

$$\beta^{13 \text{ ind } 13} R_{39}(1,16) = G\sqrt{-3} + H\sqrt{13},$$

so that $p = 3G^2 + 13H^2$.

The prime ideal decompositions of Jacobi sums were computed for $e \leqslant 100$ and selected larger values of e. Examination of these tables revealed no other instance of the conditions associated with the known occurrences of sign ambiguities; perhaps there are no more than the seven sign ambiguities noted above (but see Zee and Muskat (1970) for related phenomena).

Computers have also played a role in various applications of cyclotomy, such as the existence of residue difference sets (Baumert and Fredricksen, 1967) and power residuacity (Muskat, 1964).

2. Cylotomy of order *e* over $GF(p^2)$. To compute the cyclotomic numbers of order e over $GF(p)$, a table of ind n, $1 \leqslant n \leqslant p-1$, is needed. The size of the computer memory limits the size of p that can be handled. This limit can be extended by storing only the first half of the table and, with some sacrifice of speed, by packing several entries into one memory location. The limits imposed by the memory size of the IBM 7090 caused the author no hardships.

The cyclotomic numbers of order e over $GF(p^2)$ are defined as in (1) except that n ranges over all values of $GF(p^2)$ but 0 and -1. Constructing an index table has both time and space requirements which are $O(p^2)$. The former is essentially an inconvenience if one's computer is not saturated, but the latter is a severe restriction, The space (but not the time) requirement can be reduced to $O(p)$ as follows:

Elements of $GF(p^2)$ can be represented in the form $a + bx$, $a, b \in GF(p)$, x^2 a quadratic nonresidue of p. If $a \neq 0$, $\text{ind}(a + bx) = \text{ind}\, a + \text{ind}(1 + a^{-1}bx)$. If $a = 0$, $\text{ind}(a + bx) = \text{ind}\, b + \text{ind}\, x$. Thus three tables are needed: ind a, $1 \leqslant a \leqslant p-1$; a^{-1}, $1 \leqslant a \leqslant p-1$; $\text{ind}(1 + bx)$, $0 \leqslant b \leqslant p-1$.

The time required to compute each of these tables is $O(p)$. Thus adapting this approach to the calculation of the Eisenstein sum

$$\sum_{b=0}^{p-1} \exp(2\pi i \,\text{ind}\,(1 + bx)/e)$$

yielded an algorithm with both space and time requirements which are $O(p)$. Giudici (1967) examined values of Eisenstein sums in studying the evaluation of the character sums defined by Brewer (1966).

3. A conjecture. Reliability and speed are characteristics which make the digital computer a useful tool for research in number theory. A drawback is the time lag between conceiving an idea and getting numerical evidence from a computer program written to gain insight into a particular situation. The author is enthusiastic about performing interactive computing for preliminary explorations, based upon his experiences with PTSS (Pitt Time Sharing System) over the past few years. Programming is simpler, for one need not anticipate all eventualities at first. Having access to any desired intermediate results hastens error detection and removal. One can often receive useful results after a single session at a terminal.

The time lag can also be shortened somewhat if flexible programs had been prepared previously for related problems or if good subroutine packages are available. For example, while looking something up in Hasse (1958) recently, the author took another look at the expression for ind 2 (mod 32) and compared it with results cited in Whiteman (1954). A generalization was suggested. Since testing it required only minor modifications of existing programs, significant supporting evidence was available within a couple of days.

Let $D_e(j) = C(j, 1, \frac{1}{2}e) - C(j + \frac{1}{2}e, 1, \frac{1}{2}e)$, e even. Then

$$R_e(1, \tfrac{1}{2}e) = \sum_{j=0}^{\frac{1}{2}e-1} D_e(j)\beta^j.$$

The conjecture takes two forms. Let $n > 3$.

$$\begin{aligned} \tfrac{1}{2}D_4(1) + \sum_{t=3}^{n-1} 2^{t-2} \sum_{j=0}^{2^{t-3}-1} D_{2^t}(2j+1) & \\ \equiv (2^{n-2}+1)\,\text{ind}\,2 \quad (\text{mod } 2^n); & \end{aligned} \tag{8}$$

$$\begin{aligned} \tfrac{1}{2}D_4(1) + \sum_{t=3}^{n-2} 2^{t-2} \sum_{j=0}^{2^{t-3}-1} D_{2^t}(2j+1) & \\ - 2^{n-3} \sum_{j=0}^{2^{n-4}-1} D_{2^{n-1}}(2j+1) \equiv \text{ind}\,2 \quad (\text{mod } 2^n). & \end{aligned} \tag{9}$$

To perceive more clearly the relationship of the conjecture to the formulations in Whiteman (1954) and Hasse (1958), note that the terms in the summations in (8) and (9) occur in the quadratic partitions (Giudici, Muskat and Robinson (to be published))

$$p = D_e(0)^2 + 2\sum_{y=1}^{\frac{1}{4}e-1} D_e(y)^2, \qquad e = 2^m, \quad m > 2,$$

while $\frac{1}{2}D_4(1) = \pm Y$ in (7).

The conjecture has now been verified for

$$\begin{aligned} n &= 6, 7, \quad p \equiv 1 \ (\text{mod } 2^n), \quad p < 20{,}000, \\ n &= 8, \quad p \equiv 1 \ (\text{mod } 2^n), \quad p < 19{,}000. \end{aligned}$$

The author recognizes that proof of the conjecture will provide a satisfactory theorem "for those who regard the Jacobi sums as known" (Bruck, 1956). Those who look upon the quadratic partitions as basic will prefer another formulation.

Acknowledgement

At the time of writing, the author was a member of the Computer Center of the University of Pittsburgh, U.S.A. He thanks them for liberal access to its various computing systems and thanks the National Science Foundation and the National Institute of Health for their grants.

References

Baumert, L. D. and Fredricksen, H. (1967). The Cyclotomic Numbers of Order Eighteen with Applications to Difference Sets. *Math. Comp.* **21,** 204–219.

Brewer, B. W. (1966). On Primes of the Form $u^2 + 5v^2$. *Proc. Amer. Math. Soc.* **17,** 502–509.

Bruck, R. H. (1956). Computational Aspects of Certain Combinatorial Problems. *In* "Proceedings of Symposia in Applied Mathematics", Vol. VI, "Numerical Analysis", 31–43. Amer. Math. Soc., Providence, Rhode Island.

Davenport, H. and Hasse, H. (1935). Die Nullstellen der Kongruenzzetafunktionen in gewissen zyklischen Fällen. *J. reine angew. Math.* **172,** 151–182.

Dickson, L. E. (1935a). Cyclotomy, Higher Congruences and Waring's Problem. *Amer. J. Math.* **57,** 391–424.

Dickson, L. E. (1935b). Cyclotomy and Trinomial Congruences. *Trans. Amer. Math. Soc.* **37,** 363–380.

Dickson, L. E. (1935c). Cyclotomy when e is Composite. *Trans. Amer. Math. Soc.* **38,** 187–200.

Giudici, R. H. E. (1967). On the Evaluation of Character Sums. Dissertation, University of Pittsburgh.

Giudici, R. H. E., Muskat, J. B. and Robinson, S. F. On the Evaluation of Brewer Sums. To be published.

Hasse, H. (1958). Der 2^n – *te* Potenzcharakter von 2 im Körper der 2^n - ten Einheitswurzeln. *Rend. Circ. Mat. di Palermo*, Serie II, **7,** 185–243.

Lehmer, E. (1954). On Cyclotomic Numbers of Order Sixteen. *Canadian J. Math.* **6,** 449–454.

Muskat, J. B. (1964). Criteria for Solvability of Certain Congruences. *Canadian J. Math.* **16,** 343–352.

Muskat, J. B. (1968). On Jacobi Sums of Certain Composite Orders. *Trans. Amer. Math. Soc.* **134,** 483–502.

Muskat, J. B. and Whiteman, A. L. (1970). The Cyclotomic Numbers of Order Twenty. *Acta Arithmetica* **17,** 185–216.

Whiteman, A. L. (1954). The Sixteenth Power Residue Character of 2. *Canadian J. Math.* **6,** 364–373.

Whiteman, A. L. (1960). "The Cyclotomic Numbers of Order Ten". *In* "Proceedings of Symposia in Applied Mathematics, Vol. X, "Combinatorial Analysis", 95–111. Amer. Math. Soc., Providence, Rhode Island.

Zee, Y. C. (1968). Analysis of Jacobi Sums of Various Orders. Dissertation, University of Pittsburgh.

Zee, Y. C. and Muskat, J. B. (1971). On Sign Ambiguities of Jacobi Sums. To be published.

Calculation of the First Factor of the Cyclotomic Class Number

ROBERT SPIRA

Department of Mathematics, Michigan State University,
Ann Arbor, Michigan, U.S.A.

Recall that the cyclotomic field is the simple algebraic extension of the rationals by $(e^{2\pi i/m})$, which is of degree $\phi(m)$, and is generated by any root of $Q_m(x)$, the mth cyclotomic polynomial. These fields are of basic scientific interest. In what follows, to avoid certain trivial exceptions, we take $m \not\equiv 2 \pmod 4$ and $m > 1$. From general algebraic number theory, we know that the integral ideals of this field may be partitioned into a finite number of classes, and this number is called the class number $H(m)$. We write, for this cyclotomic field,

$$H(m) = H_1(m) H_2(m)$$

in the classical manner, Metsankyla (1967), where $2H_1(m)$ is an integer.

The quantity $H_1(m)$ is of interest because of the Fermat problem. Indeed, in the case $m = p$, an odd prime, then $H_1(m)$ is an integer, and $H(p)$ is divisible by p if and only if $H_1(p)$ is divisible by p. A prime p which divides $H(p)$ is called *irregular*, and the least such prime is 37. The Fermat theorem is known for regular p.

A formula for $H_1(m)$ is

$$H_1(m) = (2m)^{1-\phi(m)/2}\, e(m) \prod_{k|m} \prod_{\chi \in P_k} (2k)^{-1} \sum_{l=1}^{k} (-\chi(l)l)$$

where

$$e(m) = \begin{cases} 1 & \textit{if } m \textit{ odd}, \\ \tfrac{1}{2} & \textit{otherwise} \end{cases}$$

$$P_k = \{\chi \mid \chi(-1) = -1,\ \chi \textit{ a primitive character} \bmod k\}.$$

The sums in the above formula are similar to sums for certain quadratic field class numbers. Also, it is clear that the product over P_k need not be repeatedly calculated.

The calculation of $H_1(m)$ can be outlined as follows:

1. Factor m,
2. Calculate $\phi(m)$, $e(m)$,
3. Find all divisors k of m,
4. For each k find all primitive characters mod k with $\chi(-1) = -1$,
5. Calculate the sums,
6. Multiply the sums together,

where of course the actual calculation will be in a different form to avoid repetition.

The steps above represent the several portions of the program. Steps 1 and 2 are obvious. The divisors in step 3 are found by trial.

The generation of the primitive characters is more interesting. For each k we find a basis of $M(k)$, the multiplicative group of residues prime to k. In case k is odd and

$$k = p_1^{\alpha_1} \dots p_R^{\alpha_R},$$

we set $g(p_j)$ = the least positive number which is a primitive root mod p_j^{α} for all α. Then we solve

$$B_j \equiv g_j \,(\mathrm{mod}\, p_j^{\alpha_j}), \qquad B_j \equiv 1 \,(\mathrm{mod}\, p_i^{\alpha_i}) \qquad i \neq j.$$

For k divisible by 2, slightly different procedures are necessary. Now any character χ is determined by its values at $B_1, \dots, B_R$. We determine $\beta_1, \dots, \beta_R$ by

$$\chi(B_j) = e^{2\pi i \beta_j / h_j}, \qquad 0 \leqslant \beta_j < h_j,$$

where

$$h_j = \phi(p_j^{\alpha_j}).$$

Then we number the character χ by the Cantor numbering system:

$$N = \beta_1 + \beta_2 h_1 + \beta_3 h_1 h_2 + \dots + \beta_R h_1 h_2 \dots h_{R-1}, \qquad 0 \leqslant N \leqslant \phi(k) - 1.$$

Given N, we can also find χ_N. We can generate the N's which belong to primitive χ's.

Thus, step 4 is really

4a generate N's belonging to primitive χ's,

4b generate for each such N, χ_N,

4c check to see that $\chi_N(-1) = -1$.

Now $\chi_N(j) = e^{2\pi i \tau_j/\phi(k)}$, and the generation of χ_N gives the integers τ_j, $1 \leqslant \tau_j \leqslant \phi(k)$, or $\tau_j = 0$ if $\chi_N(j) = 0$. Previously, (Metsankyla, 1969), the sum

$$\sum_{l=1}^{k} \chi(l)l$$

was calculated in double precision floating point. However, there was no error analysis, though the results are probably correct. The process presently being used rests on the unique representation of an algebraic number as a residue of an irreducible polynomial. Set

$$e^{2\pi i \tau_j/\phi(k)} = a_0 + a_1\xi + a_2\xi_2^2 + \ldots + a_{\phi(m)-1}\,\xi^{\phi(m)-1}$$

where $a_0, \ldots, a_{\phi(m)-1}$ are integers and

$$\xi = e^{2\pi i/\phi(m)}, \qquad Q_{\phi(m)}(\xi) = 0.$$

Thus,

$$\chi_N(l) \leftrightarrow (a_0, a_1, \ldots, a_{\phi(m)-1})$$

and there are subroutines to add and multiply such vectors. To carry out the multiplications the cyclotomic polynomial $Q_{\phi(m)}(\chi)$ is used. Its coefficients are automatically generated.

Finally, when all these vectors are added and multiplied together, one gets the vector

$$(H_1(m), 0, 0, \ldots, 0),$$

The programming was done in ASA FORTRAN, since the underlying machines are always changing.

As it turns out, the quantities $2H_1(m)$ are integers which in many cases exceed the precision capability of present day machines. To handle this, a multiple precision routine was constructed, written also in FORTRAN.

This last routine is a simulation of an imaginary machine and is extremely flexible. It was designed along the lines of the IBM 704, with AC and MQ registers, and has a repertoire of about thirty commands including fixed point and floating point arithmetic.

The word structure is given by three parameters, N, β, k, and a simulator word consists of $N + 1$ machine words, the first a sign word, and the next N words each having k β-digits. The base β can be odd and it is usually chosen small for reasonable divide times. Each word of k β-digits is represented in the hardware machine as a single integer η, $0 \leqslant \eta < \beta^k$. It is planned to calculate all $H_1(m)$ with $m \leqslant 2000$.

References

Metsankyla, T. (1967). Über den ersten Faktor der Klassenzahl des Kreiskörpers, *Ann. Acad. Sci. Fenn.*, Ser: AI **416,** 48pp.

Metsankyla, T. (1969). Calculation of the first factor of the class number of the cyclotomic field. *Math. Comp.* **23,** 533–537.

A Numerical Study of Units in Composite Real Quartic and Octic Fields

HARVEY COHN*

University of Arizona, Tucson, Arizona, U.S.A.

Introduction

Composite-quadratic fields $Q(m_1^{\frac{1}{2}}, m_2^{\frac{1}{2}}, \ldots)$ are an attractive area of research on class number because of many formulas involving fundamental units quite simply. Such fields can be imbedded in a *maximal composite-quadratic* field where the corresponding m_i are positive or negative primes chosen so that each quadratic field has only one prime divisor in its discriminant, (i.e., $m_i = 2$ or $(-1)^{(p-1)/2} p$ for p an odd prime > 0). Thus the composite quadratic field can be imbedded in a *real* maximal composite-quadratic field precisely when every (square-free) m_i is the sum of two squares (or has only positive prime divisors $\not\equiv -1 \pmod 4$)). In what follows we restrict m_i accordingly, dealing with quartic and octic fields.

We follow the general pattern of work of Kuroda (1943), Kubota (1953), Wada (1966), in which no restriction to real fields was made. Their work was not largely computational; the restriction to real fields makes large scale computing more feasible. Class numbers of composite fields emerge in the process.

We presuppose knowledge of Ince's tables (1934) of fundamental unit and class number as restricted to real quadratic fields $Q(m^{\frac{1}{2}})$ for square-free $m < 2025$. Let us call $\varepsilon(m)$ the fundamental unit (for m square-free) and normalize

$$\varepsilon(m) > 1 \tag{1}$$

Our computation is primarily a table of units of type

$$\eta = \eta(m_1, m_2) = (\varepsilon(m_1)\,\varepsilon(m_2)\,\varepsilon(m_3))^{\frac{1}{2}} \tag{2}$$

* Research supported by the U.S. National Science Foundation Grant G–6423. Computer support on the CDC3600 at the Argonne National Laboratory, (Argonne, Ill. U.S.A.), made possible through the cooperation of Dr. Wallace J. Givens, Director Applied Mathematics Division, and Mr. Burton S. Garbow, Program Analyst.

where the norms satisfy

$$N\varepsilon(m_i) = -1 \tag{3}$$

and $m_1 m_2 m_3$ is a perfect square, or in usual terms,

$$m_1 m_2 m_3 =_2 1. \tag{4}$$

The computation is subject to the limitation of Ince's table, $m_i < 2025$. Such units (2) are indeed a biquadratic generalization of the quadratic units (3) satisfying the Pellian equation of norm -1.

Main Theorem

The following theorem governs the results achieved here. It is an extension of work of Kubota (1953) and Wada (1966). It will be proved in a more general form in a later paper:

Let k be a totally real unique factorization field of maximal unit-signature rank, (i.e., any totally positive unit is a perfect square). Let μ_1, μ_2 be totally positive integers in k (not necessarily distinct). Let η_1, η_2, η_3 be units as follows:

$$\eta_1 \in k(\mu_1^{\frac{1}{2}}), \qquad \eta_2 \in k(\mu_2^{\frac{1}{2}}), \qquad \eta_3 \in k((\mu_1 \mu_2)^{\frac{1}{2}}), \tag{5}$$

let the product $\eta_1 \eta_2 \eta_3 \gg 0$ (totally positive) and let

$$(N_k \eta_i N_k \eta_j)^{\frac{1}{2}} \in k, \qquad (i, j = 1, 2, 3), \tag{6}$$

where N_k is the norm relative to k in each case. Then there exists a totally positive $\mu \in k$, which serves the purpose that

$$(\mu\eta_1 \eta_2 \eta_3)^{\frac{1}{2}} \in k(\mu_1^{\frac{1}{2}}, \mu_2^{\frac{1}{2}}) = k_4. \tag{7}$$

This μ divides the relative discriminant of k_4 over k.

As a corollary, when $k = Q$ and μ_i are integers m_i of the type considered here, then $\mu = d$ where d is a square-free positive divisor of $m_1 m_2$. Thus in (2)

$$d^{\frac{1}{2}}\eta \in Q(m_1^{\frac{1}{2}}, m_2^{\frac{1}{2}}). \tag{8}$$

Actually the determination of d is not trivial when $m_1 m_2 m_3$ has more than 3 prime factors. (Compare the "Redei decomposition", Redei (1934) or the "principal divisors of the discriminant" which occur in quadratic and cubic field theory, Barrucand and Cohn (1970).

If we can take $d = 1$, then we call η a "Kuroda unit" and call $Q(m_1^{\frac{1}{2}}, m_2^{\frac{1}{2}})$ a "Kuroda field". Here η is *quartic*, and

$$\eta = \eta(m_1, m_2) \in Q(m_1^{\frac{1}{2}}, m_2^{\frac{1}{2}}). \tag{9}$$

By the main theorem, η is a Kuroda unit when m_1 and m_2 are (positive, rational) primes. (This was observed by Scholz (1934) as a consequence of class-field theory). If η is not a Kuroda unit then η is *octic* and

$$\eta = \eta(m_1, m_2) \in Q(m_1^{\frac{1}{2}}, m_2^{\frac{1}{2}}, d^{\frac{1}{2}}). \tag{10}$$

Computation of Table I

The method of computation is primarily double precision decimal (CDC 3600). First of all, the sub-table of $\varepsilon(m)$ with $N\varepsilon(m) = -1$, $m < 2025$ was fed into the computer (in integral form for decimal storage).

TABLE I. Coordinates of $d^{\frac{1}{2}}\eta = (d\varepsilon(M(J))\,\varepsilon(M(K))\,\varepsilon(M(L)))^{\frac{1}{2}}$. The NORM $= -d^2$ is truncated from an earlier print-out

N	T	U	M(J)	V	M(K)	W	M(L)	SGN	NORM
1	6	2	2	2	5	2	10	4	—1.0
2	10	6	2	2	13	2	26	2	—1.0
3	46	38	2	10	29	6	58	1	—1.0
4	52	34	2	8	37	6	74	4	—1.0
5	52	38	2	8	41	6	82	2	—1.0
6	350	278	2	54	53	34	106	7	—1.0
7	46	32	2	6	61	4	122	1	—1.0
8	68	46	2	8	65	6	130	2	—1.0
9	50	40	2	6	85	4	170	7	—25.0
10	540	398	2	56	101	38	202	7	—1.0
11	562	398	2	54	109	38	218	1	—1.0
12	340	234	2	32	113	22	226	1	—1.0
13	4868	3442	2	416	137	294	274	1	—1 0
14	100	70	2	8	145	6	290	2	—25.0
15	10910	7820	2	906	149	632	298	7	—1.0
16	674	478	2	54	157	38	314	1	—1.0
17	74	56	2	6	173	4	346	7	—1 0
18	350	242	2	26	181	18	362	4	—1.0
19	1040	730	2	76	185	54	370	4	—25.0
20	234660	161098	2	16232	197	11822	394	4	—1.0
21	86	64	2	6	229	4	458	1	—1.0
22	1172	814	2	72	265	50	530	1	—1.0
23	7376	5242	2	452	269	318	538	1	—1.0
24	46886	33164	2	2818	277	1992	554	7	—1.0
25	596326	401674	2	33186	293	24634	586	4	—1.0
26	173592	124798	2	9812	313	7054	626	7	—1.0
27	5350714	3748476	2	297742	317	212504	634	2	—1.0
28	21296	15058	2	1140	349	806	698	1	—1.0
29	110	80	2	6	365	4	730	1	—25.0
30	523044	369818	2	27080	373	19150	746	4	—1.0

N	T	U	M(J)	V	M(K)	W	M(L)	SGN	NORM
31	822108	581506	2	41696	389	29474	778	7	—1.0
32	22430	15864	2	1126	397	796	794	1	—1.0
33	12450144	8778118	2	615620	409	434050	818	4	—1.0
34	7838	5622	2	382	421	274	842	2	—1.0
35	56491224	39948286	2	2642548	457	1868702	914	7	—1.0
36	858646	608534	2	40082	461	28278	922	1	—1.0
37	61672	43030	2	2812	481	1962	962	4	—169.0
38	18500	13330	2	856	485	594	970	7	—25.0
39	1190	850	2	54	493	38	986	7	—289.0
40	623254	441102	2	27650	509	19534	1018	1	—1.0
41	1000484	707086	2	43832	521	30978	1042	1	—1.0
42	19198	14072	2	862	533	588	1066	1	—1.0
43	2917150	2062736	2	125418	541	88684	1082	2	—1.0
44	2643677360	1862811810	2	111623692	557	79207418	1114	2	—1.0
45	806	572	2	34	565	24	1130	7	—1.0
46	98733604	69815118	2	4139128	569	2926802	1138	1	—1.0
47	4110	2872	2	166	613	116	1226	4	—1.0
48	15606	10676	2	602	629	440	1258	2	—289.0
49	1580849254	1118386886	2	61894274	653	43744038	1306	1	—1.0
50	507874	359116	2	19754	661	13968	1322	4	—1.0
51	57023980	40969342	2	2226792	677	1549702	1354	7	—1.0
52	366	262	2	14	685	10	1370	7	—1.0
53	325297048	230011662	2	12285868	701	8687726	1402	2	—1.0
54	2617972	1851166	2	98320	709	69522	1418	4	—1.0
55	143658	98516	2	5146	733	3752	1466	4	—1.0
56	25905524	18317966	2	941552	757	665778	1514	4	—1.0
57	552	386	2	20	761	14	1522	4	—1.0
58	5426	3814	2	194	773	138	1546	2	—1.0
59	128300	89390	2	4512	785	3238	1570	2	—25.0
60	203465426	144196908	2	7223402	797	5096200	1594	1	—1.0
61	2122705300	1500978734	2	74630344	809	52771602	1618	1	—1.0
62	1717625818	1214607016	2	59948622	821	42387908	1642	7	—1.0
63	2556416	1807642	2	88788	829	62782	1658	1	—1.0
64	7999050	5656012	2	273874	853	193664	1706	4	—1.0
65	127520	89990	2	4356	857	3074	1714	4	—1.0
66	35019420	24762610	2	1190696	865	841954	1730	2	—25.0
67	693623584	490465098	2	23421964	877	16561858	1754	2	—1.0
68	4598	3254	2	150	941	106	1882	7	—1.0
69	68822	48672	2	2234	949	1580	1898	2	—169.0
70	16851390552	11915732534	2	545870084	953	385988438	1906	7	—1.0
71	78550	54080	2	2462	965	1788	1930	4	—25.0
72	1508	1066	2	48	985	34	1970	2	—1.0
73	8308812	5875246	2	263144	997	186070	1994	7	—1.0
74	7	5	5	3	13	1	65	7	—1.0
75	11	5	5	3	17	1	85	1	—1.0
76	46	18	5	8	26	4	130	4	—1.0
77	13	7	5	3	29	1	145	1	—1.0
78	54	22	5	8	37	4	185	2	—1.0
79	407	153	5	47	53	25	265	4	—1.0
80	106	46	5	14	58	6	290	1	—1.0
81	265	111	5	31	73	13	365	4	—1.0
82	424	192	5	50	74	22	370	1	—4.0
83	179	85	5	19	89	9	445	7	—1.0
84	906	394	5	92	97	40	485	1	—1.0
85	762	350	5	74	106	34	530	2	—1 0
86	883	393	5	83	113	37	565	4	—1 0
87	3112	1472	5	298	122	126	610	1	—4.0
88	2071	925	5	177	137	79	685	1	—1.0
89	137	63	5	11	157	5	785	7	—1.0
90	115673	56863	5	9667	173	3933	865	7	—1.0
91	13045	6071	5	939	193	437	965	7	—1.0

TABLE I (*continued*)

N	T	U	M(J)	V	M(K)	W	M(L)	SGN	NORM
92	188	88	5	14	197	6	985	7	—1.0
93	5146	2302	5	362	202	162	1010	2	—1.0
94	324	148	5	22	218	10	1090	7	—4.0
95	202	94	5	14	226	6	1130	1	—1.0
96	237	115	5	17	229	7	1145	1	—1.0
97	11647	5205	5	763	233	341	1165	4	—1.0
98	1183	509	5	71	257	33	1285	4	—1.0
99	828	364	5	50	274	22	1370	4	—4.0
100	29661	13271	5	1783	277	797	1385	1	—1.0
101	6974873	2901139	5	378983	293	182229	1465	4	—1.0
102	31936	14328	5	1850	298	830	1490	2	—4.0
103	637595	284785	5	36039	313	16097	1565	4	—1.0
104	262068	117364	5	14810	314	6614	1570	7	—4 0
105	10645625	4827443	5	606279	317	267397	1585	1	—1.0
106	361699	166705	5	19703	337	9081	1685	2	—1.0
107	9400	4176	5	502	346	226	1730	2	—4.0
108	15542	6950	5	832	349	372	1745	1	—1.0
109	4434	1954	5	236	353	104	1765	1	—1.0
110	164986	76222	5	8958	362	3878	1810	1	—1.0
111	89740	40128	5	4646	373	2078	1865	4	—1.0
112	1215222	542882	5	61222	394	27350	1970	1	—1.0
113	4613721	2062579	5	231473	397	103555	1985	2	—1.0
114	46	16	10	14	13	4	130	1	—1.0
115	52	16	10	12	17	4	170	2	—4.0
116	32	10	10	6	26	4	65	2	—1.0
117	34	10	10	6	29	2	290	4	—1.0
118	308	100	10	52	37	16	370	1	—4.0
119	46	14	10	6	53	2	530	2	—1.0
120	244	76	10	32	58	20	145	4	—4.0
121	8348	2618	10	1060	61	338	610	4	—4.0
122	1196	376	10	140	73	44	730	1	—4.0
123	380	120	10	44	74	28	185	2	—4 0
124	301108	95216	10	30572	97	9668	970	2	—4.0
125	572	178	10	56	101	18	1010	2	—1.0
126	24484	7724	10	2378	106	1504	265	4	—1.0
127	460	146	10	44	109	14	1090	2	—4.0
128	2892	914	10	272	113	86	1130	4	—1.0
129	1264	398	10	108	137	34	1370	1	—1.0
130	540	170	10	44	149	14	1490	2	—4.0
131	251132	79294	10	20012	157	6338	1570	2	—4.0
132	4908	1514	10	364	173	118	1730	2	—4.0
133	1917122	606398	10	142534	181	45062	1810	1	—1.0
134	26410372	8351692	10	1901060	193	601168	1930	4	—4.0
135	444	142	10	32	197	10	1970	1	—1.0
136	238	76	10	16	226	10	565	1	—1.0
137	5132	1622	10	310	274	196	685	4	—4.0
138	780	248	10	44	314	28	785	2	—4.0
139	1262668	399988	10	68000	346	42932	865	7	—4.0
140	1992764	630418	10	100394	394	63520	985	7	—1.0
141	3856	1210	10	182	442	116	1105	2	—1.0
142	2572	812	10	120	458	76	1145	4	—4.0
143	945724	299064	10	40180	554	25412	1385	1	—4.0
144	408041632708	129034093948	10	16856044480	586	10660698580	1465	4	—4.0
145	554	174	10	22	626	14	1565	4	—1.0
146	805632005044	254763209186	10	31995734326	634	20235879176	1585	4	—1.0
147	31916	10092	10	1208	698	764	1745	4	—4.0
148	8144300	2575452	10	298184	746	188588	1865	4	—4.0
149	37704956	11923418	10	1338106	794	846288	1985	7	—1.0
150	8992	2372	13	1406	37	410	481	2	—1.0

N	T	U	M(J)	V	M(K)	W	M(L)	SGN	NORM
151	71	19	13	11	41	3	533	4	—1.0
152	7304	2032	13	962	58	266	754	7	—4.0
153	15187	4211	13	1777	73	493	949	2	—1.0
154	188	52	13	22	74	6	962	1	—4.0
155	28144	8072	13	3214	82	862	1066	7	—4.0
156	3025	785	13	307	85	91	1105	4	—25.0
157	472	132	13	50	89	14	1157	2	—1.0
158	54083	14999	13	5491	97	1523	1261	4	—1.0
159	290	78	13	28	101	8	1313	4	—1.0
160	47590	13198	13	4622	106	1282	1378	4	—1.0
161	1679671	466939	13	161257	109	44621	1417	7	—1.0
162	1354	386	13	126	122	34	1586	7	—1.0
163	1557	433	13	133	137	37	1781	2	—1.0
164	133	37	13	11	149	3	1937	1	—1.0
165	84	20	17	16	26	4	442	2	—4.0
166	67	17	17	13	29	3	493	1	—1.0
167	50	12	17	8	37	2	629	2	—1.0
168	370	90	17	58	41	14	697	1	—1.0
169	60	14	17	8	53	2	901	4	—1.0
170	1004	244	17	132	58	32	986	7	—4.0
171	225	55	17	29	61	7	1037	7	—1.0
172	6050	1490	17	762	65	182	1105	1	—25.0
173	1086074	263442	17	127130	73	30830	1241	1	—1.0
174	17876	4348	17	2084	74	504	1258	7	—4.0
175	4451842	1079754	17	452026	97	109630	1649	1	—1.0
176	303	73	17	29	109	7	1853	4	—1.0
177	1341536942	325422122	17	126221202	113	30608282	1921	7	—1.0
178	2032	414	26	392	29	74	754	1	—4.0
179	124	24	26	20	37	4	962	2	—4.0
180	66540	13090	26	10424	41	2038	1066	1	—1.0
181	2450	494	26	346	53	66	1378	7	—1.0
182	3186	628	26	410	61	80	1586	1	—1.0
183	14116	2768	26	1652	73	324	1898	4	—4.0
184	57812	11364	26	6736	74	2636	481	1	—4.0
185	92	18	26	10	82	4	533	2	—4.0
186	3856	762	26	298	170	116	1105	7	—1.0
187	8840	1734	26	622	202	244	1313	7	—1.0
188	5766164	1131284	26	390688	218	153180	1417	1	—4.0
189	2450	480	26	148	274	58	1781	4	—1.0
190	38116	7492	26	2208	298	868	1937	2	—4.0
191	2424	436	29	386	37	74	1073	4	—1.0
192	2931	541	29	455	41	85	1189	4	—1.0
193	225565	41475	29	28597	61	5363	1769	2	—1.0
194	39308	5162	58	4570	74	1200	1073	7	—1.0
195	9586	1258	58	1058	82	278	1189	2	—1.0
196	1473596	193404	58	133352	122	35036	1769	4	—4.0
197	420	52	65	32	170	20	442	2	—100.0
198	64874	8054	65	4774	185	2958	481	7	—1.0
199	6700	828	65	392	290	244	754	2	—4.0
200	3173	391	65	165	365	103	949	2	—1.0
201	808	100	65	42	370	26	962	4	—1.0
202	340	42	65	16	445	10	1157	2	—25.0
203	16690	2076	65	760	485	470	1261	1	—25.0
204	7944	988	65	346	530	214	1378	1	—1.0
205	345676	42876	65	13996	610	8680	1586	2	—4.0
206	3585	445	65	137	685	85	1781	7	—25.0
207	2004	248	65	74	730	46	1898	2	—1.0
208	296	32	85	26	130	14	442	1	—4.0
209	155	17	85	13	145	7	493	7	—1.0
210	177	19	85	185	185	7	629	4	—1.0
211	2572	280	85	158	265	86	901	7	—1.0

TABLE I (*continued*)

N	T	U	M(J)	V	M(K)	W	M(L)	SGN	NORM
212	440	48	85	26	290	14	986	1	—4.0
213	244305	26195	85	12641	365	6935	1241	2	—25.0
214	36958	4010	85	1922	370	1042	1258	1	—1.0
215	294732	32128	85	13450	485	7258	1649	7	—1.0
216	1418092685	153704503	85	59617235	565	32354965	1921	4	—25.0
217	33500	2936	130	2780	145	1220	754	4	—100.0
218	3100	272	130	228	185	100	962	7	—100.0
219	344264	30194	130	21148	265	9274	1378	7	—1.0
220	10020	878	130	524	365	230	1898	4	—100.0
221	847880	74362	130	44078	370	38660	481	4	—25.0
222	31730	2782	130	1174	730	1030	949	4	—25.0
223	3438700	301594	130	110410	970	96836	1261	2	—4.0
224	19132	1678	130	602	1010	528	1313	2	—1.0
225	11109220	974602	130	336578	1090	295120	1417	1	—25.0
226	4220	370	130	114	1370	100	1781	4	—100.0
227	4400	386	130	114	1490	100	1937	7	—25.0
228	628	52	145	48	170	20	986	2	—4.0
229	17230	1430	145	1266	185	526	1073	2	—1.0
230	40292	3092	170	2964	185	1136	1258	1	—4.0
231	444	34	170	26	290	20	493	2	—4.0
232	654	50	170	34	370	26	629	4	—1.0
233	600	46	170	26	530	20	901	2	—25.0
234	24550	1882	170	994	610	762	1037	4	—1.0
235	977220	75056	170	36220	730	27740	1241	7	—100.0
236	60994824	4678090	170	1958426	970	1502044	1649	7	—1.0
237	860	66	170	26	1090	20	1853	2	—100.0
238	2549337628	195508088	170	75831556	1130	58165260	1921	4	—4.0
239	63548	3732	290	3304	370	1940	1073	1	—4.0
240	77964820	4578250	290	3156702	610	1853680	1769	2	25.0
241	15543268	739318	442	708712	481	438230	1258	2	—169.0
242	56950	2710	442	2566	493	2074	754	7	—289.0
243	1054	50	442	42	629	34	962	4	—289.0
244	995488	47342	442	37700	697	30490	1066	4	—169.0
245	20788	988	442	692	901	560	1378	4	—4.0
246	47710	2270	442	1482	1037	1198	1586	1	—169.0
247	3632412	172778	442	103112	1241	83378	1898	2	—1.0
248	15887	715	493	633	629	485	1073	2	—1.0
249	113135	5095	493	4285	697	3281	1189	2	—289.0
250	8690961	391465	493	269915	1037	206635	1769	1	—289 0
251	479820	17474	754	15470	962	14648	1073	2	—1.0
252	306375290	11157530	754	9383722	1066	8885110	1189	2	—169.0
253	1850151680	67378550	754	46457490	1586	43988932	1769	4	—169 0
254	2556200	81406	986	78036	1073	72070	1258	7	—1.0

The main loop was an enumeration of triples (m_1, m_2, m_3) satisfying (2). and (3). (These values were printed as $M(J), M(K), M(L)$). Then, as the integer $d(\geqslant 1)$ went through square-free divisors of $m_1 m_2 m_3$, the machine tried to write

$$(d\varepsilon(m_1)\,\varepsilon(m_2)\,\varepsilon(m_3))^{\frac{1}{2}} = (T + U(m_1)^{\frac{1}{2}} + V(m_2)^{\frac{1}{2}} + W(m_3)^{\frac{1}{2}})/4 \quad (11)$$

in integers T, U, V, W (which must be > 0 since the left-hand quantity dominates its conjugates). We accomplish (11) by writing a total of four conjugates and solving for T, U, V, W but the choice of signs on the left is not known beforehand. Now, in our ignorance of the signs we have 8 a priori possibilities leading to a 4 × 12 matrix system:

$$\begin{pmatrix} \frac{1}{4} & m_1^{\frac{1}{2}}/4 & m_2^{\frac{1}{2}}/4 & m_3^{\frac{1}{2}}/4 \\ \frac{1}{4} & m_1^{\frac{1}{2}}/4 & -m_2^{\frac{1}{2}}/4 & -m_3^{\frac{1}{2}}/4 \\ \frac{1}{4} & -m_1^{\frac{1}{2}}/4 & m_2^{\frac{1}{2}}/4 & -m_3^{\frac{1}{2}}/4 \\ \frac{1}{4} & -m_1^{\frac{1}{2}}/4 & -m_2^{\frac{1}{2}}/4 & m_3^{\frac{1}{2}}/4 \end{pmatrix} \begin{pmatrix} T \\ U \\ V \\ W \end{pmatrix} = \begin{pmatrix} (d\varepsilon(m_1)\,\varepsilon(m_2)\,\varepsilon(m_3))^{\frac{1}{2}} \\ \pm\,(d\varepsilon(m_1)/\varepsilon(m_2)\,\varepsilon(m_3))^{\frac{1}{2}} \\ \pm\,(d\varepsilon(m_2)/\varepsilon(m_1)\,\varepsilon(m_3))^{\frac{1}{2}} \\ \pm\,(d\varepsilon(m_3)/\varepsilon(m_1)\,\varepsilon(m_2))^{\frac{1}{2}} \end{pmatrix} \tag{12}$$

Thus we really have 8 columns on the right corresponding to the $(\pm, \pm, \pm)$ signs. They are coded as binary digits

$$\left.\begin{aligned} &\mathrm{SGN}(+,+,+) = 0, \quad \mathrm{SGN}(+,+,-) = 1, \quad \mathrm{SGN}(+,-,+) = 2, \\ &\qquad\qquad\qquad\qquad\qquad\qquad \mathrm{SGN}(+,-,-) = 3 \\ &\mathrm{SGN}(-,+,+) = 4, \quad \mathrm{SGN}(-,+,-) = 5, \quad \mathrm{SGN}(-,-,+) = 6, \\ &\qquad\qquad\qquad\qquad\qquad\qquad \mathrm{SGN}(-,-,-) = 7 \end{aligned}\right\} \tag{13}$$

Thus for each d (until the right one is revealed) the machine checks to see if any one of the 8 columns T, U, V, W are integral. When this happens the machine prints the values as well as the check value of the (absolute) norm relative to Q:

$$\mathrm{NORM} = N(d^{\frac{1}{2}}\eta). \tag{14}$$

Kubota (1953) showed that the norm of η relative to the conjugates of $Q(m_1^{\frac{1}{2}}, m_2^{\frac{1}{2}})$ is -1, so actually, only SGN $= 1, 2, 4, 7$ occurs, and

$$\mathrm{NORM} = -d^2. \tag{15}$$

As a further consequence, we can normalize our Kuroda unit by choosing integers a_1, a_2, a_3 (mod 2) so that the unit

$$\varepsilon(m_1, m_2) = \varepsilon(m_1)^{a_1}\,\varepsilon(m_2)^{a_2}\,\varepsilon(m_3)^{a_3}\,\eta(m_1, m_2) \tag{16}$$

has only one negative conjugate corresponding to $-m_1^{\frac{1}{2}}, -m_2^{\frac{1}{2}}, +m_3^{\frac{1}{2}}$. (Thus $\eta(m_1, m_2)$ is already normalized if SGN $= 1$).

This table of the $d^{\frac{1}{2}}\eta$ required several trials to decide on the best "epsilon" for integer discrimination. In earlier runs the errors in T, U, V, W were printed out so that an "epsilon" could be chosen which would make exactly one column T, U, V, W look like an integer. A suitable "epsilon" came to 10^{-13} by inspection.

Class Number

Let us consider H the class number of the field $Q(p_1^{\frac{1}{2}}, p_2^{\frac{1}{2}}, \ldots, p_s^{\frac{1}{2}})$ where p_i are different positive primes $\not\equiv -1 \pmod 4$. Let $g(m)$ be the number of classes per genus of $Q(m^{\frac{1}{2}})$. Then (see Wada, 1966),

$$H = UG/2^E. \tag{17}$$

Here $E = 2^s - 1 - s$, so $E = 0$ for $s = 1$, $E = 1$ for $s = 2$, $E = 4$ for $s = 3$, etc. Also $G = \prod g(m_j)$ where m_j runs over all $2^s - 1$ divisors of $p_1 \ldots p_s$ except 1. Finally U is the index of a certain subgroup in the groups of all units of $Q(p_1^{\frac{1}{2}}, p_2^{\frac{1}{2}}, \ldots, p_s^{\frac{1}{2}})$, namely the subgroup generated by units in the subfields $Q(m_j^{\frac{1}{2}})$.

When $s = 2$, we find $U = 2$ so

$$H = G \tag{18}$$

regardless of whether or not $Q(p_1^{\frac{1}{2}}, p_2^{\frac{1}{2}})$ is a Kuroda field. In one case $\{\varepsilon(p_1), \varepsilon(p_2), (\varepsilon(p_1)\,\varepsilon(p_2)\,\varepsilon(p_1 p_2))^{\frac{1}{2}}\}$ is a fundamental system of units and in the other case $\{\varepsilon(p_1), \varepsilon(p_2), \varepsilon(p_1 p_2)^{\frac{1}{2}}\}$ is such a system. Actually $H = G$ in principle when there is a unit of index 2 introduced for every m_j not equal to a prime (as just illustrated for $m_j = p_1 p_2$).

When $s = 3$ however, the situation becomes more interesting and more complicated. For example, let p_1, p_2, p_3 be primes such that $Q(p_1^{\frac{1}{2}}, p_2^{\frac{1}{2}})$, $Q(p_1^{\frac{1}{2}}, p_3^{\frac{1}{2}})$, $Q(p_1^{\frac{1}{2}}, (p_2 p_3)^{\frac{1}{2}})$ have Kuroda units. It is then not hard to see that, in the normalized form (16), the units

$$\eta_1 = \varepsilon(p_1, p_2), \qquad \eta_2 = \varepsilon(p_1, p_3), \qquad \eta_3 = \varepsilon(p_1, p_2 p_3) \tag{19}$$

satisfy all the conditions of the Main Theorem where

$$k = Q(p_1^{\frac{1}{2}}), \qquad \mu_1 = p_2, \qquad \mu_2 = p_3. \tag{20}$$

The only difficulty is that p_2 and p_3 might cease to be primes in $Q(p_1^{\frac{1}{2}})$! Thus we only have a weaker extension of Scholz result as cited earlier.

Octic Form of Main Theorem

If the field $Q(p_1^{\frac{1}{2}}, (p_2 p_3)^{\frac{1}{2}})$ is a Kuroda field and the (Kronecker) characters satisfy

$$(p_3/p_1) = (p_2/p_1) = -1 \tag{21}$$

then there is a so-called "Kuroda unit" with the property (see (19))

$$\xi = (\eta_1 \eta_2 \eta_3)^{\frac{1}{2}} \in Q(p_1^{\frac{1}{2}}, p_2^{\frac{1}{2}}, p_3^{\frac{1}{2}}). \tag{22}$$

TABLE II. Classification of octic field $Q(p_1^{\frac{1}{2}}, p_2^{\frac{1}{2}}, p_3^{\frac{1}{2}})$ according to the unit structure.

Type	When is $N\varepsilon(p_i p_{i+1}) = -1$?	Value of $N\varepsilon(p_1 p_2 p_3)$	When is $\eta(p_i, p_{i+1} p_{i+2})$ a Kuroda unit?	Fundamental units are $\{\varepsilon(p_1), \varepsilon(p_2), \varepsilon(p_3), \eta(p_1, p_2)$, and ...	U
A	all i	-1	all i	$\eta(p_2, p_3), \eta(p_1, p_3), \xi(p_1, p_2, p_3)\}$	32
B	,,	,,	$i = 2, 3$	$\eta(p_2, p_3), \eta(p_1, p_3), \eta(p_2, p_1 p_3)\}$	16
C	,,	,,	never	,,	16
D	$i = 1, 3$	,,	$i = 2, 3$	$\eta(p_1, p_3), \varepsilon(p_2 p_3)^{\frac{1}{2}}, \eta(p_2, p_1 p_3)\}$	16
E	,,	,,	never	,,	16
F	,,	$+1$	—	$\eta(p_1, p_3), \varepsilon(p_2 p_3)^{\frac{1}{2}}, \varepsilon(p_1 p_2 p_3)^{\frac{1}{2}}\}$	16
G	$i = 1$	-1	—	$\varepsilon(p_1 p_3)^{\frac{1}{2}}, \varepsilon(p_2 p_3)^{\frac{1}{2}}, \eta(p_3, p_1 p_2)\}$	16

In this case, $U = 32$ and

$$H = 2G. \quad (23)$$

In all other cases, (i.e., where this Kuroda unit is lacking), we conjecture on the basis of cases which actually occur that

$$H = G. \quad (24)$$

There are 33 cases where $p_1p_2p_3 < 2025$ and they are tabulated in Table III according to seven types classified in Table II. (Other types are possible in addition to those which actually occurred).

Although it is not strictly necessary for the calculation of U, a separate program was undertaken to calculate the value of ξ for type A, as follows:

$$\xi^2 = \varepsilon(p)^a\, \varepsilon(q)^b\, \varepsilon(r)^c\, \eta(p,q)\, \eta(p,r)\, \eta(p,qr) \quad (25)$$

where a, b, c are selected (0 or 1) so as to make ξ^2 totally positive. (This is equivalent to the normalization (16)). The basis coordinates are written as

$$\xi = [T + Up^{\frac{1}{2}} + Vq^{\frac{1}{2}} + Wr^{\frac{1}{2}} + X(pq)^{\frac{1}{2}} + Y(qr)^{\frac{1}{2}} + Z(pr)^{\frac{1}{2}} + S(pqr)^{\frac{1}{2}}]/8 \quad (26)$$

The results are shown in Table IV.

TABLE III. Octic fields for $pqr < 2025$ according to type and class number. (Here p, q, r are a rearrangement of p_1, p_2, p_3).

p	q	r	Type	G	H
2	5	13	A	1	2
,,	,,	17	E	1	1
,,	,,	29	B	2	2
,,	,,	37	C	1	1
,,	,,	41	F	2	2
,,	,,	53	A	1	2
,,	,,	61	E	1	1
,,	,,	73	E	3	3
,,	,,	89	F	1	1
,,	,,	97	E	1	1
,,	,,	101	D	2	2
,,	,,	109	E	3	3
,,	,,	113	A	4	8
,,	,,	137	B	2	2
,,	,,	149	E	1	1
,,	,,	157	C	1	1
,,	,,	173	C	3	3
,,	,,	181	D	2	2
,,	,,	193	E	1	1
,,	,,	197	A	1	2
2	13	17	G	2	2
,,	,,	29	E	1	1
,,	,,	37	C	1	1
,,	,,	41	B	2	2
,,	,,	53	D	2	2
,,	,,	61	D	2	2
,,	,,	73	E	1	1
,,	17	29	E	1	1
,,	,,	37	E	1	1
,,	,,	41	F	1	1
,,	,,	53	F	1	1
5	13	17	E	1	1
,,	,,	29	F	2	2

TABLE IV. Octic units ξ for $Q(p^{\frac{1}{2}}, q^{\frac{1}{2}}, r^{\frac{1}{2}})$ in coordinate form for Type A, and $\mu^{\frac{1}{2}}\xi$ for Type B. (Again p, q, r are a rearrangement of p_1, p_2, p_3)

Type	A	A	A	A	B	B	B
p	2	2	2	2	5	2	2
q	5	5	5	5	2	5	13
r	13	53	113	197	29	137	41
a	0	1	1	1	1	1	0
b	1	1	0	1	0	0	0
c	0	1	1	1	0	1	0
T	68	7680	70478	432916	276	1173958	9534
U	46	5434	49898	306118	120	830074	6824
V	32	3440	31614	193532	200	526454	2642
W	20	1056	6630	30844	52	100298	1478
X	22	2430	22302	136842	92	372186	1856
Y	8	472	2974	13788	36	44978	418
Z	14	746	4694	21810	24	70918	1052
S	6	334	2098	9570	16	31798	292
μ	—	—	—	—	$(11+5^{\frac{1}{2}})/2$	$(13+4\cdot 2^{\frac{1}{2}})$	$(7+2\cdot 2^{\frac{1}{2}})$

It is quite clear how the numerical procedure of Table I is generalized. There are $2^7 = 128$ different sign possibilities for the conjugates of ξ. Thus an 8×136 matrix, like (12), has to be solved for one column of eight (hopefully) *integral* coefficients out of 128 columns. The integer discrimination constant 10^{-5} was a satisfactory "epsilon".

Concluding Remarks

In Type A, by (23), $2|H$.

Furthermore $2|H$ for Type B by the octic form of the Main Theorem. For as long as $Q(p^{\frac{1}{2}}, (qr)^{\frac{1}{2}})$ is a Kuroda Field, $\xi \in Q(p^{\frac{1}{2}}, q^{\frac{1}{2}}, r^{\frac{1}{2}})$ unless q or r fails to be a prime in $Q(p^{\frac{1}{2}})$. Thus $(r/p) = 1$, and $2|g(pr)$ hence $2|G$ and H. (Thus for example in Table IV we also show $\mu^{\frac{1}{2}}\xi$ in coordinate form, analogously with (26), where μ is the totally positive irrational divisor of r in $Q(p^{\frac{1}{2}})$).

For Type C, this discussion fails and we can have $H = 1$.

In a similar manner there is a remarkable (but unexplained) difference between Type D and Type E. For Type D, again, it seems $2|G$ and H but not for Type E. Actually, $(p_3/p_2) = 1$, but this no longer implies $2|g(p_2 p_3)$ since $N\varepsilon(p_2 p_3) = +1$. Nevertheless for Type D we obtain (from Table III) pairs $(p_2, p_3) = (5,101), (5,181), (13,53), (13,61)$ where the *ambiguous class without ambiguous ideal is a perfect square*, or $2|g(p_2 p_3)$. This is *not* true for Type E. Here, Table III shows $(p_2, p_3) = (2,17)$, $(5,61)$, $(2,73)$, $(2,97)$, $(5,109)$, $(5,149)$, $(2,193)$, $(13,29)$, $(13,17)$, where $g(p_2 p_3)$ is odd.

As a final observation from the data, $Q((p_1 p_2)^{\frac{1}{2}}, (p_1 p_3)^{\frac{1}{2}})$ seems to be a Kuroda field only for Type A (not B).

While it is tempting to speculate on composite fields of degree $16(s = 4)$ the "perfect" situations for η and ξ (see (2) and (22)) might not exist since no quartic field $Q(p_1^{\frac{1}{2}}, p_2^{\frac{1}{2}})$ can have an unfactored rational prime. A further barrier to speculation is the limit of Ince's Table.

Prof. H. Hasse has kindly called attention to a comprehensive study by Varmon (1924) extending the theory to relative composite fields of prime-degree radicals over the cyclotomic field.

Bibliography

Barrucand, P., and Cohn, H. (1970). A rational genus, class number divisibility, and unit theory for pure cubic fields. *J. Number Theory.* **2,** 7–21.

Ince, E. L. (1934). "Cycles of reduced Ideals in Quadratic Fields". British Association Math Tables IV, London.

Kubota, T. (1953). Uber die Beziehung der Klassenzahlen der Unterkörper des bizyklischen biquadratischen Zahlkörpers. *Nagoya Math. J.* **6,** 119–127.

Kuroda, S. (1943). Uber die Dirichletschen Körper. *J. Fac. Sci. Univ. Tokyo,* **4,** 382–406.

Redei, L. (1934). Arithmetischer Beweis des Satzes uber die Anzahl der durch 4 teilbaren Invarianten der absoluten Klassengruppe in quadratischen Zahlkörper. *J. fur Math.* **171,** 55–60.

Scholz, A. (1934). Uber die Lösbarkeit der Gleichung $t^2 - Du^2 = -4$. *Math. Zeitschrift* **39,** 93–111.

Wada, H. (1966). On the class number and the unit group of certain algebraic number fields, *J. Fac. Sci. Univ. Tokyo,* Sec. I, **13,** 201–209.

Varmon, J. (1924). Uber abelsche Körper, deren alle Gruppeninvarianten aus einer Primzahl bestehen und über abelsche Körper als Kreiskörper, *Diss.* Lund.

Class Numbers and Units of Complex Quartic Fields

RICHARD B. LAKEIN*

State University of New York at Buffalo, New York, U.S.A.

1. Introduction

After the quadratic fields, the algebraic number fields most accessible to computation are the class of complex quartic fields K which are quadratic over the Gauss field. We discuss here various methods for calculating the fundamental unit and class number of such fields K, and we give the results of a computation of 1000 cases with "prime discriminant." Our results show striking similarities to the real quadratic case and suggest the extension of an old conjecture.

First some notation. Denote the Gauss field by $F = Q(\sqrt{-1})$, and let $K = F(\sqrt{\mu})$, where μ is a square-free Gaussian integer. Denote by δ the relative discriminant of K/F; so $\delta = \mu$, $(1+i)^2\mu$, or 4μ, and $K = F(\sqrt{\delta})$. (Cf. Hilbert (1894).) If $\delta = \pi \equiv \pm 1 \bmod 4$, π a Gaussian prime, we shall say simply, K has prime discriminant. Since K is totally imaginary of degree 4, K has one fundamental unit denoted by E_1 or occasionally by E_K.

2. Continued Fractions Over *F*.

Since the ring O_F of Gaussian integers is euclidean, as well as a discrete lattice in the complex plane, it is natural to consider continued fraction algorithms for complex numbers α over O_F. In fact A. Hurwitz (1887) showed that the most natural regular continued fraction is convergent, and is periodic if and only if α is a quadratic irrationality over F. (Call it for the moment algorithm (A)—here the partial quotient for z is the Gaussian integer at the center of the unit square in which z lies.) Hence we may use the periodic continued fraction for $\sqrt{\mu}$, $\mu \in O_F$, to find a unit ε_A of $K = F(\sqrt{\mu})$. Unfortunately there is no certainty that ε_A will be the fundamental unit E_1 of K. In

* Research supported by a grant from the Joint Awards Council of the Research Foundation of State University of New York.

the absence of any other theoretical information, in order to be sure of finding E_1 one would be forced to search for units smaller than $\varepsilon_A = \frac{1}{2}(t + u\sqrt{\mu})$ among all the integers $\frac{1}{2}(r + s\sqrt{\mu})$ of K with $|s| \leqslant |u|$. (Cf. Fjellstedt 1953; pp. 445, 448.)

3. J. Hurwitz's Algorithm

There is another algorithm, due to J. Hurwitz (1902), which is known to give a partial solution to the problem of calculating the fundamental unit of K. The algorithm, denoted by (J), is a semiregular continued fraction

$$\alpha = a_0 - \frac{1}{a_1 -}\,\frac{1}{a_2 -}\cdots = \langle a_0, a_1, a_2, \ldots\rangle.$$

The partial quotients a_n are *even* Gaussian integers $r + si$, $r \equiv s \bmod 2$; and the complex plane is partitioned into squares of diagonal 2, centered at the even lattice points, by the lines $y = \pm x + m$, m an odd rational integer. Eventually A. Stein (1927) worked out the proofs that (J) is about as good as possible, in that it yields a unit ε_J which at worst is the cube of E_1.

Start with a "normalized basis number" ω_0 for K/F, essentially $\omega_0 = \sqrt{\mu}$, $\frac{1}{2}(1 + \sqrt{\mu})$, or $(1 + \sqrt{\mu})/(1 + i)$ cf. Stein (1927), p. 73. Let a_0 be the nearest even Gaussian integer to ω_0, and set $\omega_0 = a_0 - (1/\omega_1)$. Then ω_1 has a purely periodic continued fraction of the type (J). Furthermore, suppose that the development proceeds thus:

$$\omega_1 = \langle a_1, a_2, \ldots, a_N, \rho\omega_1\rangle,$$

with $\rho\omega_1$ the first reappearance of ω_1 with possible a unit factor $\rho = \pm 1$, $\pm i$. Then $\langle a_1, a_2, \ldots, a_N\rangle$ is called the "partial period" for ω_1. Compute recursively the denominators q_n of the convergents to ω_1. Then

$$\varepsilon_J = \rho q_N \omega_1 - q_{N-1} = E_{1+i} \tag{1}$$

is the fundamental unit of the order O_{1+i} of K, where

$$O_{1+i} = [1, (+\, i)\,\omega_0] = \{a + b\,(1 + i)\,\omega_0 \,|\, a, b \in O_F\}.$$

But E_{1+i} must be either E_1, $E_1^{\,2}$, cr $E_1^{\,3}$, and so a (theoretically) minor amount of testing leads to E_1.

In particular for odd discriminants $\delta = \mu \equiv \pm 1 \bmod 4$, there are two cases. For $\delta \equiv \pm 5 \bmod 4 + 4i$ either $E_{1+i} = E_1$ or $E_{1+i} = E_1^{\,3}$ can occur. If however $\delta \equiv \pm 1 \bmod 4 + 4i$, it must be the case that $\varepsilon_J = E_{1+i} = E_1$. For reasons that will become clear in the next section, the (minor) testing in the former case may be completely infeasible. Consequently I restricted my attention to the latter case, and more especially to such *prime* discriminants.

4. Computation of Fundamental Units

I calculated on a CDC 6400 computer, using algorithm (J) and formula (1), the fundamental unit

$$E_K = E_{1+i} = E_1 = \tfrac{1}{2}(t + u\sqrt{\delta}) \tag{2}$$

for the first 1000 prime relative discriminants $\delta = \pi \equiv \pm 1 \bmod 4 + 4i$. Here π is a primitive (non-rational) Gaussian prime [see remark in last section] with positive real and imaginary parts. The largest prime π has norm 83089. (We should note that actually each case stands for the *pair* of complex-conjugate fields $K = F(\sqrt{\pi})$ and $\bar{K} = F(\sqrt{\bar{\pi}})$, since clearly $E_{\bar{K}} = \bar{E}_K$.) The length N of the partial period is always even, and tends to increase as $|\delta|$ increases, eventually (in the 1000th case) becoming as large as $N = 264$, with $|t| > 10^{69}$.

Here I should be a little more precise. What I actually computed was the sequence of partial quotients $a_1, a_2, \ldots, a_N$ of the partial period for ω_1. The a_n's are all quite small, of the order of $\sqrt{|\delta|}$. The real and imaginary parts of the sequence of partial denominators q_n were then computed recursively, using the machine's double-precision hardware (96 bits). Finally the real and imaginary parts t_1, t_2, u_1, u_2 of the coefficients t and u of E_1 were calculated according to formula (1), again in double precision.

Thus while E_1 is known in principle, still in nearly 400 cases (including 300 of the last 500)—essentially all cases with $N > 104$—we have only a 96-bit = 29-digit approximation for at least one of t_1, t_2, u_1, u_2. The exact computation would require the use of a multiprecise arithmetic package. Hence we see that in the excluded case $\delta \equiv \pm 5 \bmod 4 + 4i$, the loss of precision may make it impossible, at present, to determine whether $\varepsilon_J = E_1$ or $E_1{}^3$. Even when this loss of precision does not occur, the actual amount of testing required is very likely infeasible when t and u are fairly large.

Even single (48-bit) precision is however quite enough to compute the *regulator* of K: $R_K = 2 \log |E_K|$, with sufficient accuracy to tempt us to use the analytic class number formula for calculating the class number of K.

5. The Analytic Class Number Formula

The class number of $K = F(\sqrt{\delta})$ can be calculated like that of a real quadratic field. Given a suitable definition of reduced binary quadratic form over F, one would list all the reduced forms with discriminant δ and find a complete set of inequivalent "periods" (cf. Mathews, 1913).

But consider the class number formula. The Dedekind zeta function for K factors: $\zeta_K(s) = \zeta_F(s)\, L(s, \chi)$, where $L(s, \chi)$ is the L-series for K/F:

$$L(s, \chi) = \Sigma\chi(\alpha)/N\alpha^s = \Pi(1 - \chi(\pi)/N\pi^s)^{-1}. \tag{3}$$

(Summation over nonzero integral ideals (α) of F; product over prime ideals (π).) Here χ is the character for the quadratic extension K/F—like the Kronecker symbol; essentially a Jacobi symbol (δ/α) in F. The limit formula gives

$$h_K = |\delta|\, L(1, \chi)/\pi R_K. \qquad (\pi = 3.14159...) \tag{4}$$

Now the fields $K = F(\sqrt{\delta})$ with prime relative discriminant δ, like the quadratic fields with prime discriminant, have an odd class number, which I expected to be quite small (hence lots of equivalent reduced forms). Furthermore, I had previously written a routine to calculate $\chi(\alpha) = (\delta/\alpha)$, patterned after the rational Jacobi symbol, while on the other hand it was not immediately clear how to generate efficiently all the reduced forms for a given δ. Therefore, the natural choice, at least for a first computation, was to use the Gaussian–Jacobi-symbol routine to estimate the Euler product for $L(1, \chi)$ and so to approximate the class number.†

6. Estimation of $L(1, \chi)$

Since I expected the class number, as given in (4), to be quite small, I arbitarily cut off the Euler product in (3) after the 46 factors corresponding to Gaussian primes with norm < 200. This estimate for $L(1, \chi)$ was used in (4) to approximate the (small, odd, integral) class number h_K for the same 1000 prime discriminants as in Section 4. For $h \leqslant 7$ the estimates fell in the intervals (0.944, 1.063), (2.876, 3,146), (4.862, 5.248), (6.845, 7.329).

There is no useful theoretical bound on the size of the error introduced by truncating the Euler product. Consequently, in the 44 cases where the first estimate for h was $\geqslant 9$, increasingly more factors of $L(1, \chi)$ were included. Thus: 302 factors for $h = 9$, up to 548 factors for $h = 17$, with a maximum error of .126 in the estimate for h. Finally in the 9 cases with $h > 20$, the first 2000 factors in (3) were used, up to $N\pi = 17477$. This gave an error of at most .078.

Clearly an arithmetic method, such as the one of reduced forms, is preferable: both to certify my analytically approximated results, and for any

† I am indebted to Dr. Shanks for this suggestion.

more extended computation of this case; as well as for the case $\delta \equiv \pm 5 \bmod 4 + 4i$, where it may not be feasible to find E_1.

7. Tabulation of Class Numbers for the First 1000 Prime Discriminants

I calculated (i.e., closely approximated) the class number h_K for the same 1000 prime discriminants $\delta = \pi \equiv \pm 1 \bmod 4 + 4i$. (Just as in the computation of units, we actually have 1000 pairs of complex-conjugate fields K and $\bar{K}$, with $h_{\bar{K}} = h_K$.) The largest prime π has norm 83089.

Here is my tabulation of the number of cases with class number $h = 1, 3, 5$, etc., among these 1000 quartic fields. For comparison I list also the corresponding distribution for the first 1000 real quadratic prime discriminants, copied from Shanks (1969).

TABLE I. Distribution of 1000 class numbers.

	$h = 1$	3	5	7	9	11	13	15	17	19	21	23	25	27	29	33	43
Quartic	808	103	29	16	7	9	7	6	6	—	1	1	1	3	1	2	—
Quadratic	816	101	35	22	9	6	5	2	1	—	1	—	—	1	—	—	1

The complete table, listing δ, t_1, t_2, u_1, u_2, N, R_K, $L(1, \chi)$ and h_K, has been deposited in the Unpublished Mathematical Tables file of Mathematics of Computation.

8. Remarks

First we explain why the case where π is a rational prime $p = 3 \bmod 4$ has been excluded. In this case K contains a real quadratic subfield $Q(\sqrt{p})$ with fundamental unit ε_p, and E_K is found simply as $E_K = \sqrt{(i\varepsilon_p)}$. (Cf. Kuroda, 1943; p. 399.) For $N(1+\varepsilon_p)$ and $-N(1-\varepsilon_p) = 2a^2$ and $2pb^2$, or vice versa, and we have $\sqrt{(i\varepsilon_p)} = \frac{1}{2}(1+i)(a + b\sqrt{p})$. Furthermore h_K equals the product of the class numbers of its quadratic subfields. (Cf. Hilbert, 1894; p. 50–52. The result is due to Dirichlet.)

It has been oberved that the real quadratic fields with prime discriminant exhibit a distribution of class numbers which is as striking as it is mysterious: 80% with $h = 1$, 10% with $h = 3$, etc. (See Shanks, 1969) On the basis of our results it seems certain that a similar—perhaps the identical—situation prevails for the quartic fields considered here. It is not too much to expect similar results for another class of quartic fields: the quadratic extensions of $Q(\sqrt{-3})$ and of the other three euclidean complex quadratic fields.

It is conceivable to conjecture even further: quadratic extensions of *any* complex quadratic field. This general case however is not suitable for efficient computation since there is no reasonable way to find the fundamental unit.

References

Fjellstedt, L. (1953). On a class of Diophantine equations of the second degree in imaginary quadratic fields. *Arkiv för Mat.* **2,** 435–461.

Hilbert, D. (1894). Über den Dirichletschen biquadratischen Zahlkörper. *Math. Ann.* **45,** 309–340 (Werke I, 24–52).

Hurwitz, A. (1887). Über die Entwicklung complexer Grössen in Kettenbrüche. *Acta. Math.* **11,** 187–200 (Werke II, 72–83).

Hurwitz, J. (1902). Über die Reduction der binären quadratischen Formen mit complexen Coefficienten und Variablen. *Acta. Math.* **25,** 231–290.

Kuıoda, S. (1943). Über den Dirichletschen Körper, *J. Fac. Sci. Imp. Univ. Tokyo* Section I, **4,** 383–406.

Matthews, G. B. (1913). A theory of binary quadratic arithmetical forms with complex integral coefficients. *Proc. Lond. Math. Soc.* (2) **11,** 329–350.

Shanks, D. (1969). Review of table. Class number of primes of the form $4n + 1$. *Math. Comp.* **23,** 213–214.

Stein, A. (1927). Die Gewinnung der Einheiten in gewissen relativ-quadratischen Zahlkörpern durch das J. Hurwitzsche Kettenbruchverfahren. *J. reine angew. Math.* **156,** 69–92.

The Diophantine Equation $x^3 - y^2 = k$*

MARSHALL HALL, JR.

Department of Mathematics, California Institute of Technology, Pasadena, California, U.S.A.

Dedicated to Professor L. J. Mordell

1. Introduction

The solution in integers of the equation

$$x^3 - y^2 = k \tag{1.1}$$

is not only interesting in itself but is related to the representation of integers by binary cubic forms. If

$$f(x, y) = ax^3 + bx^2y + cxy^2 + dy^3 = m \tag{1.2}$$

then f has as covariants the Hermitian form H, the Jacobian form J, and discriminant D where

$$\begin{aligned} H &= (3ac - b^2)x^2 + (9ad - bc)xy + (3bd - c^2)y^2 \\ J &= (9abc - 2b^3 - 27a^2d)x^3 + (18ac^2 - 27abd - 3b^2c)x^2y \\ &\quad + (-18b^2d + 27acd + 3bc^2)xy^2 + (2c^3 - 9bcd + 27ad^2)y^3 \\ D &= 27a^2d^2 - b^2c^2 + 4ac^3 + 4b^3d - 18abcd. \end{aligned} \tag{1.3}$$

These are related by the syzygy

$$4H^3 + J^2 = 27Df^2. \tag{1.4}$$

If (x, y) is a solution of (1.2) then putting $X = -4H$, $Y = 4J$ and $k = -432Dm^2$ we have from (1.4)

$$X^3 - Y^2 = k. \tag{1.5}$$

* This research was supported in part by ONR contract N00014–67–A–0094–0010.

Knowing X and Y in (1.5) gives us the values for H and J, and we may solve $H = r$, $J = s$ as simultaneous equations to determine the x, y of (1.2). Thus to solve (1.2) it is sufficient to solve (1.5). Conversely an equation (1.1) can be shown to be equivalent to a finite number of equations (1.2). There are other relations between (1.1) and (1.2). For example, Alan Baker uses the fact that the discriminant of $X^3 - 3xXY^2 - 2yY^3$ is $-108(x^3 - y^2) = -108k$.

From the work of Mordell (1913) and the Thue–Siegel–Roth theorems it follows that for a given $k \neq 0$, the equation (1.1) has only a finite number of solutions in integers (x, y). In order to find these solutions we need to obtain an upper bound on $|x|$, (or $|y|$) in terms of $|k|$. This has recently been done in a major contribution by Baker (1967–68) in which he has shown

$$\max(|x|, |y|) < \exp\{10^{10}|k|^{10^4}\}. \tag{1.6}$$

Equivalently we have a lower bound on $|k|$ in terms of $|x|$

$$|k| > 110^{-10}(\log x)^{10^{-4}}. \tag{1.7}$$

This paper is concerned with lower bounds for $|k|$ in terms of x, y and the numerical analysis of finding sequences for which $|k|$ is small in comparison with x and y. The identity which appeared in the four author paper (1965), shows that for t an odd integer we have

$$\begin{aligned} &x^3 - y^2 = k \\ &x = 6561t^{10} - 1458t^7 + 135t^4 - 4t \\ &y = 531{,}441t^{15} - 177{,}147t^{12} + 26244t^9 - 11944t^6 + \tfrac{135}{2}t^3 - \tfrac{1}{2} \\ &k = \tfrac{1}{4}(-81t^6 + 14t^3 - 1). \end{aligned} \tag{1.8}$$

This proves that there are infinitely many cases in which $0 < |k| < \frac{1}{9}x^{3/5}$. It has been proved by H. Davenport (1965) that if $f^3(t) - g^2(t) = h(t)$ where $f(t)$, $g(t)$ are polynomials over the complex field of degrees $2r$, $3r$ respectively then either $h(t)$ is identically zero or is of degree at least $r+1$. If f, g are integer valued polynomials in t and $h(t)$ is of degree $r+i$, then we have infinitely many cases of $z^3 - y^2 = k$ with $0 < |k| < Cx^e$ where $e = (r+i)/2r$ and C is a constant depending on the leading coefficients of $f(t)$ and $h(t)$. It is plausible that there are enough cases with $h(t)$ of degree $r+i$ and i small compared to r so that for every $e > \frac{1}{2}$ there are infinitely many cases with $|k| < x^e$. On the other hand the numerical evidence found here seems to indicate that $e = \frac{1}{2}$ is truly the lower bound for e. This is the basis for the following conjecture:

CONJECTURE: *In $x^3 - y^2 = k \neq 0$, for every $e > \frac{1}{2}$ there are infinitely many cases with $|k| < x^e$, but there is an absolute constant C such that $|k| > Cx^{\frac{1}{2}}$.*

$C = \frac{1}{5}$ appears to work in all known cases and in Section 5 all values with $|k| < 2x^{\frac{1}{2}}$ are listed for $x < 700{,}000{,}000$, and indeed for certain selected larger values for x. If we write $x = u^2 + v$, $u > 0$, $-u < v \leqslant u$ as we may, taking u^2 to be the square nearest to $x > 0$, then it is proved in Section 4 that if $0 < |v| \leqslant u^{4/5}$ then

$$|k| > \frac{x^{3/5}}{48} - \frac{23}{576}.$$

The most striking small values for k which have been found are

$$\begin{aligned} (5234)^3 - (378{,}661)^2 &= -17 \\ (28{,}187{,}351)^3 - (149{,}651{,}610{,}621)^2 &= -1090. \end{aligned} \tag{1.9}$$

The second of these is the stronger and here $|k|$ is only slightly greater than $\frac{1}{5}x^{\frac{1}{2}}$.

I must thank William Bergquist, Harriet Hahn, and Jonathan Hall for assistance in programming the computer search. In Section 5 a sequence depending on a parameter M is defined and the computer search took M to the limit 20,000. As $x > (4M/3)^2$ this covers all values of $x \leqslant 711{,}111{,}111$ and certain other cases.

2. Definition of the Minimal Sequence

We shall study solutions in rational integers of the Diophantine equation

$$x^3 - y^2 = k. \tag{2.1}$$

If $k = 0$ then $x^3 = y^2$ and there are infinitely many solutions given by

$$x = u^2, \qquad y = u^3, \tag{2.2}$$

where u is an integral parameter.

If $k \neq 0$ it is well known that (2.1) has only a finite number of solutions in integers (x, y). Without loss of generality we may take $y \geqslant 0$. Then for a given value of k we test the finite set of y's satisfying $0 \leqslant y \leqslant |k|$ and see which of these yield integral values for x. For the purposes of this study we consider such solutions as elementary, and confine our attention to the non-elementary solutions in which $|k| < y$.

THEOREM 2.1 *If* $0 < |k| < y$ *in* $x^3 - y^2 = k$, *then* $x > 0$ *and* $|x^{3/2} - y| < \frac{1}{2}$. *Conversely if* $x > 0$ *and* $|x^{3/2} - y| < \frac{1}{2}$ *then* $|k| \leqslant y$.

Proof. If $x \leqslant 0$ then $|k| \geqslant y^2 \geqslant y$ and we do not have $|k| < y$. Hence $x > 0$. Then write $x^{3/2} - y = \varepsilon$. We have

$$(x^{3/2} - y)(x^{3/2} + y) = k \quad \text{or} \quad k = \varepsilon(x^{3/2} + y),$$

and as $|k| < y$ certainly $|\varepsilon| < 1$. Then also

$$k = \varepsilon(2y + \varepsilon) \quad \text{and} \quad |k| \leqslant y - 1$$

so that

$$|\varepsilon| \left| 2 + \frac{\varepsilon}{y} \right| = \left| \frac{k}{y} \right| \leqslant 1 - \frac{1}{y},$$

and so

$$2|\varepsilon| \leqslant 1 - \frac{1}{y} + \frac{\varepsilon^2}{y} \leqslant 1.$$

Thus $|\varepsilon| \leqslant \frac{1}{2}$ but with $|\varepsilon| \leqslant \frac{1}{2}$ then

$$2|\varepsilon| \leqslant 1 - \frac{1}{y} + \frac{1}{4y} < 1$$

whence $|\varepsilon| < \frac{1}{2}$. Conversely suppose that $x > 0$ and $|x^{3/2} - y| < \frac{1}{2}$. Then writing $x^{3/2} - y = \varepsilon$ we have

$$k = (x^{3/2} - y)(x^{3/2} + y) \quad \text{and } |k| = |\varepsilon|\,|2y + \varepsilon|$$

whence

$$|k| \leqslant y + \varepsilon^2 \leqslant y + \tfrac{1}{4},$$

and as k is an integer $|k| \leqslant y$. For cases with $|k| = y$ our equation (2.1) becomes $x^3 - y^2 = \pm y$, $x^3 = y(y \pm 1)$ so that y, $y \pm 1$ are consecutive integral cubes. The only possibilites here are $x = 0$, $y = 0$, $k = 0$; $x = 0$, $y = 1$, $k = -1$ and in the second of these we do not have $|x^{3/2} - y| < \frac{1}{2}$. Hence except for $x = y = k = 0$ we have $|k| < y$ when $|x^{3/2} - y| < \frac{1}{2}$. This proves all parts of the theorem.

Given $x > 0$ there is a unique integer y such that $|x^{3/2} - y| < \frac{1}{2}$, y being the integer nearest to $x^{3/2}$. (Note that $x^{3/2} - y = \pm \frac{1}{2}$ is impossible.) For a given x choose y as the integer nearest to $x^{3/2}$ and determine k from $x^3 - y^2 = k$. Taking $x = 1, 2, \ldots$, we then have what we call the *minimal sequence* of values (x, y, k) which by our theorem yields all non-elementary solutions of $x^3 - y^2 = k$. Listings of the minimal sequence have been made by the author, by R. M. Robinson, and possibly by others. We list here the minimal sequence for $1 \leqslant x \leqslant 100$.

Minimal sequence $1 \leqslant x \leqslant 100$

x	y	k	x	y	k	x	y	k	x	y	k
1	1	0	26	133	−133	51	364	155	76	663	−593
2	3	−1	27	140	83	52	375	−17	77	676	−443
3	5	2	28	148	48	53	386	−119	78	689	−169
4	8	0	29	156	53	54	397	−145	79	702	235
5	11	4	30	164	104	55	408	−89	80	716	−656
6	15	−9	31	173	−138	56	419	55	81	729	0
7	19	−18	32	181	7	57	430	293	82	743	−681
8	23	−17	33	190	−163	58	442	−252	83	756	251
9	27	0	34	198	100	59	453	170	84	770	−196
10	32	−24	35	207	26	60	465	−225	85	784	−531
11	36	35	36	216	0	61	476	405	86	798	−748
12	42	−36	37	225	28	62	488	184	87	811	782
13	47	−12	38	234	116	63	500	47	88	826	−804
14	52	40	39	244	−217	64	512	0	89	840	−631
15	58	11	40	253	−9	65	524	49	90	854	−316
16	64	0	41	263	−248	66	536	200	91	868	147
17	70	13	42	272	104	67	548	459	92	882	764
18	76	56	43	282	−17	68	561	−289	93	897	−252
19	83	−30	44	292	−80	69	573	180	94	911	663
20	89	79	45	302	−79	70	586	−396	95	926	−101
21	96	45	46	312	−8	71	598	307	96	941	−745
22	103	39	47	322	139	72	611	−73	97	955	648
23	110	67	48	333	−297	73	624	−359	98	970	292
24	118	−100	49	343	0	74	637	−545	99	985	74
25	125	0	50	354	−316	75	650	−625	100	1000	0

(2.3)

For $101 \leqslant x \leqslant 1000$ we shall not list the complete minimal sequence but only the subsequence for which $|k| \leqslant x$

x	y	k	x	y	k	x	y	k
101	1015	76	325	5859	244	625	15625	0
109	1138	−15	331	6022	207	654	16725	639
121	1331	0	336	6159	−225	675	17537	506
136	1586	60	351	6576	−225	676	17576	0
143	1710	107	356	6717	−73	677	17615	508
144	1728	0	361	6859	0	717	19199	212
145	1746	109	366	7002	−108	729	19683	0
152	1874	−68	377	7320	233	741	20171	−220
158	1986	116	399	7970	299	783	21910	587
169	2197	0	400	8000	0	784	21952	0
175	2315	150	401	8030	301	785	21994	589
190	2619	−161	422	8669	−113	799	22585	174
195	2723	146	441	9261	0	810	23053	191
196	2744	0	483	10615	362	841	24389	0
197	2765	148	484	10648	0	891	26596	755
225	3375	0	485	10681	364	899	26955	674
243	3788	−37	529	12167	0	900	27000	0
255	4072	191	560	13252	496	901	27045	676
256	4096	0	568	13537	63	909	27406	593
257	4120	193	575	13788	431	937	28682	−171
289	4913	0	576	13824	0	944	29004	368
312	5511	207	577	13860	433	961	29791	0
327	5644	277	584	14113	−65	978	30585	−873
323	5805	242	592	14404	−528			
324	5832	0						

3. Parametrization of the Minimal Sequence

Let x and y be a pair of integers in the minimal sequence, so that $x > 0$, $y > 0$, $|x^{3/2} - y| < \frac{1}{2}$. Let us choose $u > 0$ so that u^2 is the square nearest to x, more precisely taking u, v so that

$$x = u^2 + v, \qquad u > 0, \qquad -u < v \leqslant u. \tag{3.1}$$

It is easily seen that every positive integer x has a unique representation of the form (3.1). We use this representation to parametrize x, y and k in the minimal sequence $x^3 - y^2 = k$.

THEOREM 3.1 *If x, y with $x^3 - y^2 = k$ are in the minimal sequence, then there exist integers $u > 0$, v, M such that*

$$x = u^2 + v, \qquad u > 0, \qquad -u < v \leqslant u$$
$$y = u^3 + \tfrac{3}{2}uv + \tfrac{1}{2}M \tag{3.2}$$

where M is determined by the conditions

$$M \equiv uv \pmod 2$$
$$\left|\frac{M}{1} - \frac{3v^2}{8u} + \frac{v^3}{16u^3} - \frac{3v^4}{128u^5} + \frac{3v^5}{256u^7}\right| < \tfrac{1}{2} \tag{3.3}$$

and k is given by

$$k = \tfrac{3}{4}u^2v^2 - Mu^3 + v^3 - \tfrac{3}{2}Muv - \tfrac{1}{4}M^2 \tag{3.4}$$

Proof. Define Y by

$$Y = u^3 + \tfrac{3}{2}uv + \frac{3v^2}{8u} - \frac{v^3}{16u^3} + \frac{3v^4}{128u^5} - \frac{3v^5}{256u^7}. \tag{3.5}$$

Here Y consists of the initial terms in the binomial expansion of $(u^2 + v)^{3/2}$. If we put $e = Y^2 - x^3$, direct calculation yields

$$e = \frac{-7v^6}{512u^6} - \frac{3v^7}{256u^8} + \frac{33v^8}{16384u^{10}} - \frac{9v^9}{16384u^{12}} + \frac{9v^{10}}{65536u^{14}}. \tag{3.6}$$

Since $|v| \leqslant u$, it is immediate that $|e| \leqslant \frac{1}{32}$. Hence

$$Y^2 - y^2 = Y^2 - x^3 + x^3 - y^2 = e + k. \tag{3.7}$$

Put $\varepsilon = Y - y$. Then from (3.7)

$$\varepsilon = Y - y = (e + k)/(Y + y) \tag{3.8}$$

and since $Y > 0$

$$|\varepsilon| < (y - 1 + \tfrac{1}{32})/y < 1. \tag{3.9}$$

Hence as $Y + y = 2y + \varepsilon$

$$|\varepsilon| < (y - 1 + \tfrac{1}{32})/(2y + \varepsilon) < \tfrac{1}{2}. \tag{3.10}$$

If we now write

$$y = u^3 + \tfrac{3}{2}uv + \frac{M}{2} \tag{3.11}$$

this will determine M as an integer satisfying

$$M \equiv uv \pmod 2 \tag{3.12}$$

and if we substitute (3.5) and (3.11) into

$$|Y - y| = |\varepsilon| < \tfrac{1}{2} \tag{3.13}$$

we obtain

$$\left| \frac{M}{2} - \frac{3v^2}{8u} + \frac{v^3}{16u^3} - \frac{3v^4}{128u^5} + \frac{3v^5}{256u^7} \right| < \tfrac{1}{2}. \tag{3.14}$$

Thus we have completed the proof of (3.2) and (3.3) in the theorem. The formula (3.4) is the result of substituting the parametric forms of x and y in $x^3 - y^2 = k$. Our proof is now complete. The inequality (3.4) confines M to an open interval of length 2 and the condition $M \equiv uv \pmod 2$ determines the parity of M, and so M is uniquely determined by u, v and these conditions.

4. Further Parameters. Some Special Subsequences

It is convenient to define further integral parameters in terms of the original parameters u, v, M. We define N, F, T by the rules

$$\begin{aligned} N &= 3v^2 - 4Mu, \\ F &= 2Mv - 3Nu, \\ T &= 8M^2 - 9Nv. \end{aligned} \tag{4.1}$$

We may express k in terms of these parameters in the following ways

$$\begin{aligned} k &= (2Nu^2 - v^3 + 3Nv - 2M^2)/8, \\ k &= (-Fu + 4Nv - 3M^2)/12, \\ k &= [-4u(2Tu - MN) - 9Tv + 5N^2]/288v. \end{aligned} \tag{4.2}$$

The above relations lead to the following:

$$\begin{aligned} N &= (8k + v^3 - 3Nv + 2M^2)/2u^2, \\ F &= (-12k + 4Nv - 3M^2)/u = (-96k + 5Nv - 3T)/8u, \\ T &= (-288kv + 4MNu - 9Tv + 5N^2)/8u^2. \end{aligned} \tag{4.3}$$

There are also some identities on the parameters following directly from their definition

$$\begin{aligned} T^2 - 27N^3 &= 4M(2MT - 9NF), \\ 3F^2 - 4M^2N &= u(2MT - 9NF), \\ Tu - 3Fv &= -2MN. \end{aligned} \tag{4.4}$$

If we put $M = (3v^2 - N)/4u$ in the inequality (3.14) we obtain

$$\left| -\frac{N}{8u} + \frac{v^3}{16u^3} - \frac{3v^4}{128u^5} + \frac{3v^5}{256u^7} \right| < \tfrac{1}{2}. \tag{4.5}$$

This gives immediately

$$|N| \leqslant 4u + \frac{u}{2} + \frac{3}{16} + \frac{3}{32u}. \tag{4.6}$$

Since N and u are integers and $u \geqslant 1$ it follows that

$$|N| \leqslant \tfrac{1}{2}9u. \tag{4.7}$$

THEOREM 4.1 *If $u \geqslant 2$ and $1 \leqslant |v| \leqslant u^{\frac{1}{2}}$, then $|k| > u^2 + u + u \geqslant x$ except for the subsequences*:

$$\begin{aligned} &u = 2t, \quad v = -1, \quad t \geqslant 1, \\ &x = 4t^2 - 1, \quad y = 8t^3 - 3t, \quad k = 3t^2 - 1; \\ &u = 2t, \quad v = 1, \quad t \geqslant 1, \\ &x = 4t^2 + 1, \quad y = 8t^3 + 3t, \quad k = 3t^2 + 1. \end{aligned} \tag{4.8}$$

For these $|k| \geqslant \frac{3}{4}u^2 - 1$. (4.9)

Proof. The correctness of the theorem may be checked directly in the minimal sequence for $1 \leqslant x \leqslant 100$. Hence in our proof we may assume $u \geqslant 10$ as well as $1 \leqslant |v| \leqslant u^{\frac{1}{2}}$ and $|k| \leqslant u^2 + u$. Now

$$M = (3v^2 - N)/4u \tag{4.10}$$

whence using $|v| \leqslant u^{\frac{1}{2}}$ and (4.7) we find

$$|M| \leqslant \tfrac{3}{4} + \tfrac{9}{8} = \tfrac{15}{8} \tag{4.11}$$

Hence $M = -1, 0$, or 1. From (4.3),

$$
\begin{aligned}
|N| &= |8k + v^3 - 3Nv + 2M^2|/2u^2 \\
&\leqslant \frac{4u^2 + 4u}{u^2} + \frac{u^{3/2}}{2u^2} + \frac{27u^{3/2}}{4u^2} + \frac{1}{u^2} = 4 + \frac{4}{u} + \frac{29}{4u^{\frac{1}{2}}} + \frac{1}{u^2} \\
&\qquad + \frac{29}{4u^{\frac{1}{2}}} + \frac{1}{u^2}.
\end{aligned}
\tag{4.12}
$$

Since we have $u \geqslant 10$, this gives $|N| \leqslant 6$. If $M = \pm 1$, then as $uv \equiv M \pmod 2$ both u and v are odd, and so $3v^2 = 4Mu + N$ becomes modulo 8, $3 \equiv 4 + N \pmod 8$ whence $N = -1 \pmod 8$ and as $|N| \leqslant 6$, $N = -1$. But then $3v^2 + 1 = \pm 4u$, while $|3v^2 + 1| \leqslant 3u + 1 < 4u$ conflicting with $3v^2 + 1 = \pm 4u$. Hence $M = \pm 1$ leads to a conflict and so we must have $M = 0$. With $M = 0$ we have $3v^2 = N$ and as $v \neq 0$, $|N| \leqslant 6$, the only possibilities are $v = \pm 1$, $N = 3$, $M = 0$. As $uv \equiv M \pmod 2$ u must be even say $u = 2t$. Here $u = 2t$, $v = -1$, $M = 0$ gives the subsequence (4.8) while $u = 2t$, $v = +1$, $M = 0$ gives the subsequence (4.9). Thus our theorem is proved.

The following subsequence of the minimal sequence lists all cases with $1 \leqslant x \leqslant 1000$ for which $|k| \leqslant u^{3/2}$ but $k \neq 0$.

$0 < |k| \leqslant u^{3/2}$

x	y	k	u	v	M	N	F	T
2	3	−1	1	1	1	−1	5	17
32	181	7	6	−4	2	0	−8	32
40	253	−9	6	4	2	0	8	32
43	282	−17	7	−6	4	−4	36	−88
46	312	−8	7	−3	1	−1	15	−19
52	375	−17	7	3	1	−1	27	35
109	1138	−15	10	9	6	3	18	45
243	3788	−37	16	−13	8	−5	32	−73
356	6717	−73	19	−5	1	−1	47	−37
568	13537	63	24	−8	2	0	−16	32
584	14113	−65	24	8	2	0	16	32
937	28682	−171	31	−24	14	−8	72	−160

(4.13)

THEOREM 4.2 *For* $1 \leqslant |v| \leqslant u^{2/3}$, $u \geqslant 32$ *we have* $|k| \geqslant u^{3/2}$ *except for the sequence*:

$$u = 6t^2, \quad v = 4t, \quad M = 2, \quad N = 0, \quad F = 8t, \quad T = 32,$$
$$x = 36t^4 + 4t, \quad y = 216t^6 + 36t^3 + 1, \quad k = -8t^3 - 1, \tag{4.14}$$

where t is any integer with $|t| \geqslant 2$. Here we have $|k| \geqslant \sqrt{\tfrac{8}{27}}u^{3/2} - 1$.

Proof. All cases with $|k| \leqslant u^{3/2}$ and $u \leqslant 31$ are listed in (4.13). Hence the theorem assumes $u \geqslant 32$. From (4.1)

$$M = (3v^2 - N)/4u \tag{4.15}$$

and from (4.7), $|N| \leqslant 9u/2$. Hence as $|v| \leqslant u$, $u \geqslant 32$

$$|M| \leqslant \tfrac{4}{3}u + \tfrac{9}{8} < u. \tag{4.16}$$

From (4.3)

$$N = (8k + v^3 - 3Nv + 2M^2)/2u^2. \tag{4.17}$$

From this, using $|v| \leqslant u^{2/3}$, $|k| \leqslant u^{3/2}$, $|M| \leqslant u$ and $|N| \leqslant 9u/2$,

$$|N| \leqslant \frac{8}{u^{\frac{1}{2}}} + \frac{1}{2} + \frac{27}{2u^{1/3}} + 1. \tag{4.18}$$

As $u \geqslant 32$ this gives $|N| < 7$. Using this improved estimate for $|N|$ in (4.15) gives

$$|M| \leqslant \frac{3u}{4} + \frac{7}{4u} \tag{4.19}$$

and as M, and $u \geqslant 32$ are integers, $|M| \leqslant \tfrac{4}{3}u$. Taking $|M| \leqslant \tfrac{4}{3}u$ and $|N| \leqslant 7$ on the right hand side of (4.17) gives

$$|N| \leqslant \frac{8}{u^{1/2}} + \frac{1}{2} + \frac{27}{u^{4/3}} + \frac{9}{16}. \tag{4.20}$$

As $u \geqslant 32$, we have $|N| < 3$. But as $N = 3v^2 - 4Mu \equiv 3v^2 \pmod 4$ the only possibilities are $N = 0, -1$.

First let us consider cases with $N = -1$. Here

$$0 < M = (3v^2 + 1)/4u \leqslant (3u^{4/3} + 1)/4u < u^{1/3}. \tag{4.21}$$

Also from (4.2)

$$k = (-2u^2 - v^3 - 3v - 2M^2)/8. \tag{4.22}$$

Consequently

$$|k| \geqslant \tfrac{1}{4}u^2 - |v^3 + 3v + 2M^2|/8$$
$$\geqslant \tfrac{1}{4}u^2 - (u^2 + 3u^{2/3} + 2u^{2/3})/8$$
$$= \frac{u^2}{8} - \frac{5u^{2/3}}{8}.$$

Here

$$\frac{u^2}{8} - \frac{5u^{2/3}}{8} - u^{3/2} = \frac{u^{3/2}}{8}\left(u^{\frac{1}{2}} - 8 - \frac{5}{u^{5/6}}\right)$$

and as $u^{\frac{1}{2}} - 8 - (5u/^{5/6})$ is an increasing function of u it follows that $|k| \geqslant u^{3/2}$ for $u \geqslant 81$. Hence when $N = -1$, it remains to consider cases with $32 \leqslant u \leqslant 81$. As $4Mu = 3v^2 + 1 \equiv 4 \pmod 8$ it follows that Mu is odd. Also $0 < M < (81)^{1/3}$ and so $M = 1$ or 3. But 3 is not a divisor of $3v^2 + 1$ and so the only possibility is $M = 1$. We now have $3v^2 = 4u - 1 \leqslant 323$ and so v is odd, $|v| \leqslant 9$. Also as $u \geqslant 32$, $3v^2 = 4u - 1 \geqslant 127$ and so the only values to consider are $v = \pm 7$, $v = \pm 9$. With $v = \pm 7$, we have $u = 37$, $k = -388$ or -297 and with $v = \pm 9$ we have $u = 61$, $k = -1025$ or -386. In these cases we do not have $|k| \leqslant u^{3/2}$ and this disposes of the cases with $N = -1$.

When $N = 0$, (4.2) gives us

$$k = (-v^3 - 2M^2)/8. \tag{4.24}$$

Also $3v^2 = 4Mu$ and so $v = \pm\sqrt{(4Mu/3)}$ giving

$$k = -\left(\pm\left(\frac{4M}{3}\right)^{3/2} u^{3/2} + 2M^2\right)\Big/8 = \mp\left(\frac{M}{3}\right)^{3/2} u^{3/2} - \frac{M^2}{4}. \tag{4.25}$$

Here as v is even and $v \neq 0$, M is even. Also as $M = 3v^2/4u < \tfrac{3}{4}u^{1/3}$, it follows from (4.25) that $|k| > u^{3/2}$ unless $M = 2$. When $M = 2$, $3v^2 = 8u$, and so v must be a multiple of 4. Writing $v = 4t$ for an integer $t \neq 0$, we have $u = 6t^2$. As $M = 2$, $N = 0$ the rest of the formula (4.14) follows and all parts of our theorem are proved.

It has been shown by H. Davenport (1965) that if we put $x = f(t)$, $y = g(t)$ where $f(t)$ and $g(t)$ are polynomials, with real or complex coefficients, of degrees $2r$, $3r$ respectively, then the degree of $k = x^3 - y^2 = f^3(t) - g(t)$ is at least $r + 1$ except when $f^3(t) - g^2(t)$ is identically zero. If $f(t)$, $g(t)$ have integral coefficients and $f^2(t) - g^3(t) = h(t) \neq 0$ then $x^3 - y^2 = k$ is satisfied with $x = f(t)$, $y = g(t)$ and $k = h(t)$. If $h(t)$ is of degree $r + i$ where $i \geqslant 1$, then for sufficiently large t

$$|k| \leqslant Cx^{(r+i)/(2r)} \tag{4.26}$$

where C is a constant depending on the leading coefficients of $f(t)$ and $h(t)$. If there are infinitely many degrees $2r$ for integral $f(t)$, $g(t)$ such that $h(t)$ has degree $r+1$ (or $r+i$ with i small compared to r) then it will follow that for every $e > \frac{1}{2}$ there are infinitely many $|k| < x^e$. This is highly plausible, but not yet proved. In the opposite direction, numerical evidence seems to indicate that there is a constant C such that $|k| > Cx^{\frac{1}{2}}$ for all x, and indeed $C = \frac{1}{5}$ works in all cases known to the writer.

The formulae of Theorems 4.1 and 4.2 give instances with $x = f(t)$, $y = g(t)$ $k = x^3 - y^2 = h(t)$ of degrees $2r$, $3r$, $r+1$ for $r = 1$ and 2. Such formulae are also known for $r = 3$, 4, and 5 and are given here:

$$r = 3$$

$$\begin{gathered} x = 147{,}456t^6 + 6144t^4 + 160t^2 + 1 \\ y = 56{,}623{,}104t^9 + 3{,}538{,}944t^7 + 129{,}024t^5 + 2240t^3 + 21t \\ k = 1152t^4 + 39t^2 + 1, \quad u = 384t^3 + 8t, \quad v = 96t^2 + 1, \\ M = 18t, \quad N = 3, \quad F = -36t, \quad T = -27. \end{gathered} \tag{4.27}$$

$$r = 3$$

$$\begin{gathered} x = 147{,}456t^6 - 6144t^4 + 160t^2 - 1 \\ y = 56{,}623{,}104t^9 - 3{,}538{,}944t^7 + 129{,}024t^5 - 2240t^3 + 21t \\ k = -1152t^4 + 39t^2 - 1, \quad u = 384t^3 - 8t, \quad v = 96t^2 - 1 \\ M = 18t, \quad N = 3, \quad F = 36t, \quad T = 27. \end{gathered} \tag{4.28}$$

$$r = 4$$

$$\begin{gathered} x = 4t^8 + 24t^7 + 84t^6 + 200t^5 + 344t^4 + 456t^3 + 436t^2 + 296t + 112 \\ y = 8t^{12} + 72t^{11} + 360t^{10} + 1248t^9 + 3264t^8 + 6768t^7 + 11{,}328t^6 \\ + 15{,}456t^5 + 18{,}388t^4 + 14{,}984t^3 + 10{,}068t^2 + 4668t + 1196 \\ k = -1728t^5 - 6480t^4 - 16{,}416t^3 - 26{,}352t^2 - 26{,}784t - 25{,}488 \\ u = 2t^4 + 6t^3 + 12t^2 + 14t + 8, \quad v = 24t^3 + 48t^2 + 72t + 48, \\ M = 216t^2 + 216t + 216, \quad N = 1728t \\ F = 10{,}368t + 20{,}736, \quad T = 373{,}248. \end{gathered} \tag{4.29}$$

$$r = 5$$

t odd

$$\begin{gathered} x = 6561t^{10} + 1458t^7 + 135t^4 + 4t \\ y = 531{,}441t^{15} + 177{,}147t^{12} + 26{,}244t^9 + 1944t^6 + \tfrac{135}{2}t^3 + \tfrac{1}{2} \\ k = -(81t^6 + 14t^3 + 1)/4, \quad u = 81t^5 + 9t^2 \\ v = 54t^4 + 4t, \quad M = 27t^3 + 1, \quad N = 12t^2 \quad F = 8t, \quad T = 8. \end{gathered} \tag{4.30}$$

The following lemma is helpful in establishing certain lower bounds on $|k|$.

LEMMA: *If* $v \neq 0$, *then* $T \neq 0$.

Proof. Suppose $T = 0$. Then $8M^2 - 9Nv = 0$ and $N = 8M^2/9v$. Then $3v^2 - 4Mu = 8M^2/9v$ and so

$$27v^3 - 36Muv = 8M^2. \tag{4.31}$$

From this

$$(4M + 9uv)^2 = 9v^2(9u^2 + 6v). \tag{4.32}$$

Hence $9u^2 + 6v$ is an integral square, say z^2 with $z \geqslant 0$.

$$9u^2 + 6v = z^2. \tag{4.33}$$

But as $-u < v \leqslant u$ this gives

$$(3u - 1)^2 = 9u^2 - 6u + 1 < 9u^2 + 6v - z^2 < 9u^2 + 6u + 1 = (3u + 1)^2. \tag{4.34}$$

This gives $z^2 = (3u)^2 = 9u^2$ as the only possibility, but then $v = 0$, contrary to assumption.

As (4.30) gives an infinite sequence with $|k| < u^{6/5}$, in seeking a lower bound for $|k|$, we may assume $|k| < u^{6/5}$.

THEOREM 4.3 *For* $0 < |k| < u^{6/5}$, $u \geqslant 32$, *the following inequalities hold*:

$$|v| > u^{2/3}$$

$$\left|M - \frac{3v^2}{4u}\right| = \left|\frac{-N}{4u}\right| \leqslant \frac{1}{8},$$

$$\left|N - \frac{v^3}{2u^2}\right| < \frac{15}{32},$$

$$\left|T - \frac{MN}{2u}\right| \leqslant \frac{36|v|}{u^{4/5}} + \frac{121}{256} + \frac{81}{2u^{4/5}} + \frac{9}{2u},$$

$$3 \leqslant M \leqslant \frac{3|v|}{4},$$

$$|N| \leqslant \frac{2M}{3} \leqslant \frac{|v|}{2}. \tag{4.35}$$

Proof. From Theorem 4.1 and 4.2 we must have $|v| > u^{2/3}$. From our original definitions

$$\begin{aligned} N &= 3v^2 - 4Mu, \\ F &= 2Mv - 3Nu, \\ T &= 8M^2 - 9Nv. \end{aligned} \tag{4.36}$$

From (4.3)

$$\begin{aligned} N &= (8k + v^3 - 3Nv + 2M^2)/2u^2 = (32k + 4v^3 - 3Nv + T)/8u^2, \\ F &= (-12k + 4Nv - 3M^2)/u = (-96k + 5Nv - 3T)/8u, \\ T &= (-288kv + 4MNu - 9Tv + 5N^2)/8u^2. \end{aligned} \tag{4.37}$$

From (4.7) $|N| \leqslant 9u/2$, and so

$$\left| M - \frac{3v^2}{4u} \right| = \left| \frac{-N}{4u} \right| \leqslant \frac{9}{8}.$$

This gives

$$|M| \leqslant \frac{3u}{4} + \frac{9}{8} < u.$$

Also as $T = 8M^2 - 9Nv$ we have trivially

$$|T| \leqslant 8u^2 + 81u^2/2 = 97u^2/2.$$

We can use these trivial estimates on the right hand side of (4.37) to obtain better estimates:

$$\begin{aligned} |T| &\leqslant \frac{36|v|}{u^{4/5}} + \frac{9u^2}{4u} + \frac{873u^3}{8u^2} + \frac{405u^2}{8u^2}, \\ &\leqslant \frac{36|v|}{u^{4/5}} + \frac{891u}{8} + \frac{405}{8} < 116u. \end{aligned} \tag{4.38}$$

$$\left| N - \frac{v^3}{2u^2} \right| \leqslant \frac{4}{u^{4/5}} + \frac{27u|v|}{4u^2} + 1 < \frac{4}{16} + \frac{27}{4} + 1 = 8. \tag{4.39}$$

This last gives

$$|N| \leqslant \frac{|v|}{2} + 8 \leqslant \frac{u}{2} + \frac{u}{4} = \frac{3u}{4}$$

and so

$$\left|M - \frac{3v^2}{4u}\right| = \left|\frac{-N}{4u}\right|$$

gives

$$|M| \leqslant \frac{3|v|}{4} + \frac{3}{16}$$

and as M and v are integers

$$|M| \leqslant \frac{3|v|}{4}.$$

We use (4.37) again to improve the estimates

$$|T| \leqslant \frac{36|v|}{u^{4/5}} + \frac{9|v|}{32} + \frac{1044|v|}{8u} + \frac{45}{128}$$

$$< \frac{36|v|}{u^{4/5}} + \frac{9|v|}{32} + 131 \leqslant \frac{81|v|}{32} + 131. \tag{4.40}$$

Iterating this procedure

$$|T| \leqslant \frac{36|v|}{u^{4/5}} + \frac{9|v|}{32} + \frac{729|v|}{256u} + \frac{45}{128}$$

$$< \frac{36|v|}{u^{4/5}} + \frac{9|v|}{32} + \frac{819}{256} + \frac{36|v|}{u^{4/5}} + \frac{9|v|}{32} + 4. \tag{4.41}$$

Substituting in (4.37)

$$\left|N - \frac{v^3}{2u^2}\right| = \left|\frac{32k - 3Nv + T}{8u^2}\right| \leqslant \frac{4}{u^{4/5}} + \frac{9}{32} + \frac{9}{2u^{9/5}} + \frac{9}{256u} + \frac{1}{2u^2} < \frac{9}{16}. \tag{4.42}$$

Thus $|N| \leqslant \frac{1}{2}v + \frac{9}{16}$ and as N and v are integers $|N| \leqslant \frac{1}{2}v + \frac{1}{2}$. With this value we find

$$\left|N - \frac{v^3}{2u^2}\right| \leqslant \frac{4}{u^{4/5}} + \frac{3}{16} + \frac{3}{16u} + \frac{9}{2u^{9/5}} + \frac{9}{256u} + \frac{1}{2u^2} < \frac{15}{32} \tag{4.43}$$

and as N and v are integers and $|N| \leqslant \dfrac{|v|}{2} + \dfrac{15}{32}$

$$|N| \leqslant \frac{|v|}{2}. \tag{4.44}$$

Using (4.41) and (4.44) in (4.37) we have

$$\left|T - \frac{MN}{2u}\right| \leqslant \frac{36|v|}{u^{4/5}} + \frac{81}{2u^{4/5}} + \frac{81}{256} + \frac{9}{2u} + \frac{5}{32}. \tag{4.45}$$

This is the inequality of (4.35) for T. As $|M| \leqslant 3|v|/4$ and $|N| \leqslant |v|/2$, $u \geqslant 32$, a weaker form of (4.45) is

$$|T| \leqslant \frac{3}{16}|v| + \frac{36|v|}{u^{4/5}} + 4. \tag{4.46}$$

From (4.37)

$$\left|\frac{F}{3u}\right| = \left|\frac{2vM}{3u} - N\right| \leqslant \left|\frac{+4k}{u^2}\right| + \left|\frac{5Nv}{24u^2}\right| + \left|\frac{T}{8u^2}\right|$$

$$\leqslant \frac{4}{u^{4/5}} + \frac{5}{48} + \frac{1}{8u^2}\left(\frac{3|v|}{16} + \frac{36|v|}{u^{4/5}} + 4\right)$$

$$\leqslant \frac{4}{u^{4/5}} + \frac{5}{48} + \frac{3}{128u} + \frac{9}{2u^{9/5}} + \frac{1}{2u^2}$$

$$< \frac{4}{u^{4/5}} + \frac{11}{96} \leqslant \frac{4}{16} + \frac{11}{96} \leqslant \frac{35}{96}.$$

Here $\frac{4}{u^{4/5}} + \frac{11}{96} < \frac{1}{3}$ when $u \geqslant 40$ and so as M and N are integers $|N| < \frac{2}{3}M + \frac{1}{3}$, $u \geqslant 40$ gives

$$|N| \leqslant \frac{2M}{3}, \qquad u \geqslant 40. \tag{4.48}$$

But if $|v| = u$, then $v = u$, $3u^2 - 4Mu = N$ and as $|N| \leqslant \frac{1}{2}v$, $N = 0 \pmod{u}$ forces $N = 0$. Otherwise $|v| \leqslant u - 1$ and we have

$$|N| \leqslant \frac{2|u-1|}{3u}M + \frac{35}{96} = \frac{2M}{3} - \frac{2M}{3u} + \frac{35}{96}. \tag{4.49}$$

Since for $u \geqslant 32$, $|v| \geqslant u^{2/3}$,

$$M > \frac{3v^2}{4u} - \frac{1}{8} > \frac{3u^{1/3}}{4} - \frac{1}{8}$$

giving $M \geqslant 3$. Thus again (4.49) gives

$$|N| \leqslant \frac{2M}{3} - \frac{2}{u} + \frac{35}{96} < \frac{2M}{3} + \frac{1}{3}$$

for $32 \leqslant u \leqslant 39$ and so

$$|N| \leqslant \frac{2M}{3}, \qquad u \geqslant 32. \tag{4.50}$$

Thus all parts of Theorem 4.3 have been proved. We are now in a position to find a lower bound for $|k|$ whenever $|v| \leqslant u^{4/5}$.

THEOREM 4.4 *If* $0 < |k| < u^{6/5}$, *and if* $|v| \leqslant u^{4/5}$, $u \geqslant 32$, *then* $1 \leqslant |T| \leqslant 37$ *and*

$$|k| \geqslant \frac{2(|T| - \frac{1}{4})u^2}{72|v|} - \frac{|T|}{32} - \frac{5}{576}.$$

Hence

$$|k| \geqslant \frac{u^2}{48|v|} - \frac{23}{576}$$

and

$$|k| \geqslant \frac{u^{6/5}}{48} - \frac{23}{576}.$$

Proof. From Theorem 4.3 we estimate MN/u when $|v| \leqslant u^{4/5}$, $u \geqslant 32$.

$$\begin{aligned} |MN/u| &\leqslant \left(\frac{3v^2}{4u} + \frac{1}{8}\right) \cdot \left(\frac{|v^3|}{2u^2} + \frac{15}{32}\right) / u \\ &\leqslant \frac{3|v|^5}{8u^4} + \frac{|v^3|}{16u^3} + \frac{45v^2}{128u^2} + \frac{15}{256u} \\ &\leqslant \frac{3}{8} + \frac{1}{16u^{3/5}} + \frac{45}{128u^{2/5}} + \frac{15}{256u} \leqslant \frac{1}{2}. \end{aligned} \tag{4.51}$$

Thus from (4.35)

$$|T| < \frac{1}{4} + 36 + \frac{121}{256} + \frac{81}{2u^{4/5}} + \frac{9}{2u} < 40. \tag{4.52}$$

As $T = 8M^2 - 9Nv$ and $3v^2 - 4Mu = N$, $M \equiv uv \pmod 2$ it follows that if $T \equiv 0 \pmod 3$ then $T \equiv 0 \pmod 9$ whence $T = \pm 39$ is impossible. Also if $T \equiv 0 \pmod 2$ then $Nv \equiv 0 \pmod 2$ and so $N \equiv v \equiv 0 \pmod 2$ and also $M \equiv uv \equiv 0 \pmod 2$. Then $N \equiv 0 \pmod 4$ $v \equiv 0 \pmod 2$ whence $T \equiv 0 \pmod 8$. Thus $T = \pm 38$ is impossible. Hence $|T| \leqslant 37$ and by the lemma $T \neq 0$. From (4.2) we may write k in the form

$$k = \frac{-u^2}{72v}\left(2T - \frac{MN}{u}\right) - \frac{T}{32} + \frac{5\,N^2}{288\,v}. \tag{4.53}$$

Here

$$|N| < \frac{|v|^3}{2u^2} + \frac{15}{32}$$

gives immediately with $|v| < u^{4/5}$, $|v| \geqslant 11$

$$\frac{N^2}{|v|} \leqslant \frac{|v|^5}{4u^4} + \frac{15v^2}{32u^2} + \frac{225}{1024|v|} < \tfrac{1}{2} \tag{4.54}$$

and so (4.53) gives

$$|k| \geqslant \frac{u^2}{72|v|}\left(2|T| - \frac{1}{2}\right) - \frac{|T|}{32} - \frac{5}{576}. \tag{4.55}$$

For fixed u and v this gives the lowest bound when $|T| = 1$ and for fixed u when $|v| = u^{4/5}$, giving the inequalities of the theorem.

5. Cases with $k < 2\sqrt{x}$. The MN Sequence

We have $x = u^2 + v \leqslant u^2 + u$ and so $\sqrt{x} < u + \frac{1}{2}$. Consequently if $|k| < 2\sqrt{x} < 2u + 1$, then $|k| \leqslant 2u$. Thus for $u \geqslant 32$, $|k| \leqslant u^{6/5}$ and the estimates of Theorem 4.3 apply. But these can of course be improved.

THEOREM 5.1 *If $0 < |k| \leqslant 2u$, $u \geqslant 32$, then the following inequalities hold*:

$$|v| > u^{2/3}$$

$$\left|M - \frac{3v^2}{4u}\right| = \left|\frac{-N}{4u}\right| \leqslant \frac{1}{8}$$

$$\left|N - \frac{v^3}{2u^2}\right| = \left|\frac{4k}{u^2} - \frac{3\,Nv}{8u^2} + \frac{T}{8u^2}\right| \leqslant \frac{3}{16}\left|\frac{v}{u}\right|^2 + \frac{9}{u}$$

$$\left|F - \frac{5\,Nv}{8u}\right| < 25$$

$$\left|T - \frac{MN}{2u}\right| \leqslant \left|\frac{v}{u}\right|\left(72 + \frac{47}{128} + \frac{675|v|}{8u^2}\right) < \left|\frac{v}{u}\right|\left(\frac{145}{2} + \frac{675|v|}{8u^2}\right)$$

$$0 < M \leqslant 3|v|/4$$

$$|N| \leqslant 2M/3 \leqslant |v|/2. \tag{5.1}$$

Proof. Some of these come directly from Theorem 4.3. In the identity

$$T - \frac{MN}{2u} = \frac{-36kv}{u^2} - \frac{9Tv}{8u^2} + \frac{5N^2}{8u^2} \tag{5.2}$$

let us take $|k| \leqslant 2u$, $|N| \leqslant |v|/2$ and use the estimate (4.46) for T on the right. This gives

$$\left|T - \frac{MN}{2u}\right| \leqslant \left|\frac{v}{u}\right|\left(72 + \frac{27|v|}{128u} + \frac{81|v|}{2u^{9/5}} + \frac{9}{2u} + \frac{5}{32}\left|\frac{v}{u}\right|\right). \tag{5.3}$$

As $|v| \leqslant u$ and $u \geqslant 32$, this gives

$$\left|T - \frac{MN}{2u}\right| \leqslant \left|\frac{v}{u}\right|\left(72 + \frac{65}{128} + \frac{81}{2u^{4/5}}\right) \leqslant 75\left|\frac{v}{u}\right|. \tag{5.4}$$

This gives the improved estimate

$$|T| \leqslant \frac{3}{8}|N| + 75\left|\frac{v}{u}\right| \leqslant \frac{3}{16}|v| + 75\left|\frac{v}{u}\right|. \tag{5.5}$$

Using this on the right hand side of (5.2) gives

$$\begin{aligned}\left|T - \frac{MN}{2u}\right| &\leqslant \left|\frac{v}{u}\right|\left(72 + \frac{27}{128}\left|\frac{v}{u}\right| + \frac{675|v|}{8u^2} + \frac{5}{32}\left|\frac{v}{u}\right|\right)\\ &\leqslant \left|\frac{v}{u}\right|\left(72 + \frac{47}{128} + \frac{675|v|}{8u^2}\right) < \left|\frac{v}{u}\right|\left(\frac{145}{2} + \frac{675|v|}{8u^2}\right).\end{aligned} \tag{5.6}$$

This is the estimate of the theorem. A further identity is

$$F - \frac{5Nv}{8u} = \frac{-12k}{u} - \frac{3T}{8u}. \tag{5.7}$$

Using (5.5) we have

$$\left|F - \frac{5Nv}{8u}\right| \leqslant 24 + \frac{9}{128}\left|\frac{v}{u}\right| + \frac{225|v|}{8u^2} < 25 \tag{5.8}$$

the inequality of the theorem

From the identity

$$N - \frac{v^3}{2u^2} = \frac{4k}{u^2} - \frac{3Nv}{8u^2} + \frac{T}{8u^2} \tag{5.9}$$

we find, using 5.6

$$\left|N - \frac{v^3}{2u^2}\right| \leqslant \frac{8}{u} + \frac{3}{16}\left|\frac{v}{u}\right|^2 + \frac{|MN|}{16u^3} + \frac{145}{16u^2} + \frac{675|v|}{64u^4} \tag{5.10}$$

$$\leqslant \frac{3}{16}\left|\frac{v}{u}\right|^2 + \frac{8}{u} + \frac{3}{128u} + \frac{145}{16u^2} + \frac{675}{64u^2}$$

$$< \frac{3}{16}\left|\frac{v}{u}\right|^2 + \frac{9}{u}$$

which is the inequality of Theorem 5.1 This completes all parts of Theorem 5.1 not covered by Theorem 4.3.

The minimal sequence may be considered as an enumeration of cases by choosing $u > 0$ and for a given u all v's, $-u < v \leqslant u$. We shall construct a sequence (called the MN sequence) based on choosing $M > 0$ and for a given M, values of N with $-2M/3 \leqslant N \leqslant 2M/3$. From (5.5) the value of T is limited by the value of N. Hence in

$$8M^2 - 9Nv = T \tag{5.11}$$

given M and N ($N \neq 0$) we can divide $8M^2$ by $9N$ to obtain a quotient v and remainder T in a range specified by N.

From M, N, T, v we can calculate u from

$$u = (3v^2 - N)/4M. \tag{5.12}$$

If this gives an integral value for u, and if $M \equiv uv \pmod 2$, the further values of F, k, x, and y are easily calculated. We note that when $N = 0$, M is even and (5.5) gives $T = 8M^2 < 75$. The only possible value for M is $M = 2$ and these values yield the parametric solutions of (4.14) and only for $t = -1$ and $t = +1$ is $|k| \leqslant 2u$ corresponding to $x = 32$, $y = 181$, $k = 7$ and $x = 40$, $y = 253$, $k = -9$.

Given N and T, the formula

$$T^2 - 27N^3 = 4M(2MT - 9NF) \tag{5.13}$$

determines M as a divisor of $T^2 - 27N^3$ such that $8M^2 \equiv T \pmod{9N}$. This fails if $T^2 - 27N^3 = 0$, but if this holds, condition (5.5) that

$|\sqrt{27N^3}| < \frac{3}{8}|N| + 75$, can easily be shown to hold only when $N = 3$, $T = \pm 27$. These values are easily shown to lead to the parametric solutions of (4.27) and (4.28), but for no values of the parameter is $|k| \leqslant 2u$. But if N is large the factorization of $T^2 - 27N^3$ is not feasible, whereas a computer testing of the MN sequence is entirely practical.

We seek a bound on T entirely in terms of M and N. We have from the definition of F

$$\frac{v}{u} = \frac{3N + e}{2M}, \qquad e = \frac{F}{u}. \tag{5.14}$$

From (5.1)

$$|e| = \left|\frac{F}{u}\right| \leqslant \left|\frac{5Nv}{8u^2}\right| + \frac{25}{u} \leqslant \frac{5}{16} + \frac{25}{u} \leqslant \frac{35}{32}. \tag{5.15}$$

Also as $u \geqslant |v| \geqslant 4M/3$

$$|e| \leqslant \min\left(\frac{35}{32}, \frac{5}{16} + \frac{75}{4M}\right). \tag{5.16}$$

Note in particular that $|e| < 1$ if $M \geqslant 30$. Also identically

$$\frac{M}{u} = \frac{3v^2}{4u^2} - \frac{N}{4u^2}. \tag{5.17}$$

Hence

$$0 < \frac{M}{u} \leqslant \frac{3}{4}\left|\frac{v}{u}\right|^2 + \left|\frac{v}{8u^2}\right| \tag{5.18}$$

and so from (5.14)

$$\frac{M}{u} \leqslant \frac{3}{4}\left(\frac{3N + e}{2M}\right)^2 + \left|\frac{v}{8u^2}\right| \tag{5.19}$$

The bound

$$|T| \leqslant \frac{3|N|}{8} + 75\left|\frac{v}{u}\right|$$

yields

$$|T| \leqslant \frac{3|N|}{8} + 75 \tag{5.20}$$

and

$$|T| \leqslant \frac{3|N|}{8} + 75\left(\frac{3|N| + |e|}{2M}\right). \tag{5.21}$$

For $M \geqslant 30$ $|e| < 1$ and so (5.21) gives

$$|T| \leqslant \frac{3}{8}|N| + \frac{225|N| + 75}{2M} \leqslant \frac{33}{8}|N| + \frac{5}{4} \tag{5.22}$$

and so for $M \geqslant 30$ $|N| \geqslant 4$ we have $|T| < 9/2|N|$. Thus when $M \geqslant 30$, $|N| \geqslant 4$ T is the least remainder in absolute value in dividing $8M^2$ by $9N$.

For $M \geqslant 30$ we have $|e| < 1$ and as $M \leqslant \frac{3}{4}|v| \leqslant \frac{3}{4}u$, we must have $u \geqslant 40$ and the bound from Theorem 5.1 takes the form, using (5.14) and (5.19)

$$\begin{aligned} |T| &\leqslant \left|\frac{MN}{2u}\right| + \left|\frac{v}{u}\right|\left(72 + \frac{47}{128} + \frac{675|v|}{8u^2}\right) \\ &\leqslant \left|\frac{N}{2}\right|\left(\frac{3}{4}\left(\frac{3|N| + 1}{2M}\right)^2 + \left|\frac{v}{8u^2}\right|\right) + \left|\frac{v}{u}\right|\left(72 + \frac{47}{128} + \frac{675|v|}{8u^2}\right) \\ &= \frac{3|N|(3|N| + 1)^2}{32M^2} + \left|\frac{v}{u}\right|\left(72 + \frac{47}{128} + \frac{675|v|}{8u^2} + \frac{|N|}{16u}\right) \\ &\leqslant \frac{3|N|(3|N| + 1)^2}{32M^2} + \frac{(3|N| + 1)}{2M}\left(72 + \frac{51}{128} + \frac{1925}{32M}\right). \end{aligned} \tag{5.23}$$

The most useful results we state as a theorem

THEOREM 5.2 *Suppose* $0 < |k| \leqslant 2u$, $u \geqslant 32$. *Then*

$$0 < |N| \leqslant 2M/3, \qquad |T| \leqslant \frac{3|N|}{8} + 75. \tag{5.24}$$

Further if $M \geqslant 30$, *then*

$$0 < |T| \leqslant \frac{3|N|}{8} + \frac{225|N| + 75}{2M}. \tag{5.25}$$

and

$$0 < |T| \leqslant \frac{3|N|(3|N| + 1)^2}{32M^2} + \frac{3(|N| + 1)}{2M}\left(72 + \frac{51}{128} + \frac{1925}{32M}\right). \tag{5.26}$$

Also the following congruences hold:
If M is even then $N \equiv 0, 3, 4 \pmod 8$. If M is odd then $N \equiv -1 \pmod 8$. If $M \equiv 0 \pmod 3$ then $N \equiv 0 \pmod 3$. If T is even $T \equiv 0 \pmod 8$. Also $T \equiv 2 \pmod 3$ or $T \equiv 0 \pmod 9$.

Proof. The inequalities were established in (5.20), (5.22), and (5.23). From $N = 3v^2 - 4Mu$ and $M \equiv uv \pmod 2$ we see that if M is even, then $N \equiv 3v^2 \pmod 8$ and so $N \equiv 0, 3, 4 \pmod 8$. If M is odd then also u and v are odd and $N \equiv 3 - 4 \equiv -1 \pmod 8$. Also if $M \equiv 0 \pmod 3$ then $N = 3v^2 - 4Mu \equiv 0 \pmod 3$. As $T = 8M^2 - 9Nv$ we have $T \equiv 8M^2 \pmod 9$ and so $T = 2 \pmod 3$ or $T = 0 \pmod 9$. Also if $T = 0 \pmod 2$ then Nv is even and from $N = 3v^2 - 4Mu$, both v and N are even and indeed $N \equiv 0 \pmod 4$ whence $Nv \equiv 0 \pmod 8$ and $T \equiv 0 \pmod 8$.

In calculating the MN sequence it is not difficult using (5.21) along with the theorem to calculate by hand all values with $M \leqslant 30$. For $M \geqslant 30$ the computer was used, this being the IBM 360–75. As the ratios u/v, $3v/4M$, $2M/3N$ are nearly equal, it follows that if M is much larger than N, the corresponding values of v and u are proportionately larger, and enough so to be troublesome. Thus it is desirable to eliminate the small values of N as soon as possible. As $|T| \geqslant 1$, (5.26) shows that for a fixed N, there is an upper bound on the values for M. But for moderate values of M, T will be restricted and we may use $T^2 - 27N^3 = 4M(2MT - 9NF)$ to determine M from N and T. For example with $N = -1$ and $M \geqslant 30$ the inequality (5.26) gives $|T| < 6$. The only possible values are $T = -1$ and $T = 5$. But if $T = -1 = 8M^2 + 9v$, then $1 - 27(-1)^3 = 4M(-2M + 9F)$. Here $28 = 4M(-2M + 9F)$ gives $7 = M(-2M + 9F)$ and $M = 7$ (as $M = 1$ is too small). Then $8(49) + 9v = -1$ but $-393 = 9v$ does not give an integral value for v. If $T = 5$, then $25 + 27 = 4M(-2M + 9F)$ or $13 = M(-2M + 9F)$. Here take $M = 13$, and then $8(169) + 9v = 5$ or $9v = -1347$, which does not give an integral value for v. Similarly we can eliminate $N = 3, -4, 4$ for cases with $M \geqslant 30$. Hence we take $5 \leqslant |N| \leqslant \frac{2}{3}M$ for $30 \leqslant M \leqslant 2000$. In the same way if $M \geqslant 2000$ and $|N| \leqslant 100$ the only possible values for T are -1 and 5 and these values can be eliminated with only a moderate amount of effort. Only $N = -13$, $T = 5$, $M = 7418$ gives integral values for v, u, k, but here $|k|$ is much larger than $2u$.

If we take an MN sequence with $|N| \leqslant (2M/3)$ and require only $|T| < 9|N|/2$ then from (4.2) we see that $|k|$ is no greater than a number approximately

$$\left|\frac{Tu^2}{36v}\right| < \left|\frac{Nu^2}{8v}\right|$$

and as we can also show that $|T| < 9|N|/2$ forces N to be approximately

$v^3/2u^2$, it follows that the MN sequence gives only values of $|k|$ not greatly exceeding $v^2/16$. As the minimal sequence gives all values with $|k| < y$ where y is approximately u^3, we may consider the MN sequence with $|T| < 9|N|/2$.

In taking M to the limit 20,000 the following cases with $|k| < 2u$ have been found.

$|k| \leqslant 2u$

x	y	k	u	v	M	N	F	T
2	3	−1	1	1	1	−1	5	17
3	5	2	2	−1	0	3	−18	27
5	11	4	2	1	0	3	−18	−27
32	181	7	6	−4	2	0	−16	32
40	253	−9	6	4	2	0	16	32
46	312	−8	7	−3	1	−1	15	−19
109	1138	−15	10	9	6	3	18	45
2660	137,190	−100	52	−44	28	−16	32	−64
5234	378,661	−17	72	50	26	12	8	8
8158	736,844	−24	90	58	28	12	8	8
47,044	10,203,669	−337	217	−45	7	−1	21	−13
93,844	28,748,141	−297	306	208	106	48	32	32
421,351	273,505,487	−618	649	150	26	4	12	8
657,547	533,200,074	847	811	−174	28	−4	−12	8
117,188	40,116,655	−353	342	224	110	48	32	32
27,564,105	144,715,764,559	−6856	5250	1605	368	75	30	17
367,806	223,063,347	207	606	570	402	252	144	72
918,493	880,265,693	−1092	958	729	416	211	114	77
720,114	611,085,363	−225	849	−687	417	−225	117	63
28,187,351	149,651,610,621	−1090	5309	1870	494	116	28	8

For cases with $M \geqslant 500$ we do not list x or y, since these numbers become very large.

k	u	v	M	N	F	T
307	969	826	528	300	156	72
44678	29847	−4746	566	−60	−12	8
−14668	19602	4362	728	108	24	8
−14857	19764	4386	730	108	24	8
−192057	96177	10354	836	60	28	8
−20513	10296	3440	862	192	64	32
−3753	2241	2157	1557	999	621	405
−8569	10525	5761	2365	863	305	113
−30788	20710	−8388	2548	−688	192	−64
−315969	252900	33730	3374	300	40	8
−316844	253350	33770	3376	300	40	8
142463	100815	−23433	4085	−633	75	−1
22189	11845	−11348	8154	−5208	3096	−1728
−11492	12423	−11660	8208	−5136	3024	−1728
−2381192	1360926	129626	9260	588	58	8
−2383593	1361808	129682	9262	588	56	8

For the range $10{,}000 \leqslant M \leqslant 20{,}000$ there are no cases with $|k| \leqslant 2u$. There are, however, three cases in which $|k| \leqslant 3u$. These are:

k	u	v	M	N	F	T
1,299,225	481,552	101,445	16,028	2251	264	17
−10,759,129	4,782,240	354,258	19,682	972	72	8
−10,764,232	4,783,698	354,330	19,684	972	72	8

References

Baker, A. (1967–68). The Diophantine Equation $y^2 = x^3 + k$. *Phil. Trans. Roy. Soc. London Ser.* **A263**, 193–208.

Birch, B. J., Chowla, S., Hall, Marshall, Jr., Schinzel, A. (1965). "On the difference $x^3 - y^2$". *Norske Vid. Selsk. Forh.* **38,** 65–69.

Davenport, H. (1965). "On $f^3(t) - g^2(t)$". *Norske Vid. Selsk. Forh.* **38,** 86–87.

Mordell, L. J. (1913). The Diophantine equation $y^2 - k = x^3$. *Proc. London. Math. Soc.* **13,** 60–80.

The Diophantine Equation $x^3 - y^2 = k$

F. B. COGHLAN

Department of Mathematics, University of Manchester, England

AND

N. M. STEPHENS

Pembroke College, Oxford, England

1. Introduction

Initially, we set out to solve the four diophantine equations

$$x^3 - y^2 = 18, 72, 288, 648$$

in order to complete an outstanding step in one of the authors' Ph.D. thesis (Coghlan, 1967). This was done by using Skolem's method as applied by Ljunggren (1963) to the equations

$$x^3 - y^2 = 7, 15.$$

With the machinery set up we attempted to solve

$$x^3 - y^2 = k \tag{1}$$

for the outstanding values of k, $0 \leqslant k \leqslant 100$. (Hemer (1952) completed the solutions of (1) for $-100 \leqslant k \leqslant 0$.) In this paper we show that there are no other solutions of (1), $0 \leqslant k \leqslant 100$, apart from those previously known (see Hemer, 1952).

2. Determination of quartics

Equation (1) may be factorised over $Q(\sqrt{-k})$, giving rise to equations of the form

$$f(a, b) = 1 \tag{2}$$

where f is a homogeneous cubic. One of these cubics is reducible and is readily solved; henceforth we neglect solutions of (1) corresponding to this cubic. The other cubics are irreducible. If $k < 0$, Skolem's method can be applied; but if $k > 0$, this method cannot be applied directly, so that another approach is needed.

Equation (1) may also be factorised over $Q(\alpha)$ where $\alpha^3 = k$, giving rise to equations of the form

$$x - \alpha = \lambda\beta^2 \tag{3}$$

where β is an integer of $Q(\alpha)$ and λ is one of a finite list of elements of $Q(\alpha)$. This list, which contains 1, depends on k, the fundamental unit and class number of $Q(\alpha)$. Equation (3) is equivalent to an equation of the form

$$g(u, v) = 1 \tag{4}$$

where g is a homogeneous quartic. When λ in (3) is 1, (4) is reducible and readily solved. Otherwise (4) is irreducible. Some of these quartics may be excluded by congruence considerations. For of the remaining quartics corresponding to the values of k under consideration, a solution of (4) and hence of (1), x_0, y_0, say, was known. This meant that (3) could be written as

$$(x - \alpha)/(x_0 - \alpha) = \text{square of } Q(\alpha),$$

and (4) as

$$g(u, v) = u^4 - 6x_0u^2v^2 - 8y_0uv^3 - 3x_0{}^2v^4 = 1. \tag{5}$$

Clearly distinct solutions of (1) give rise to distinct equations of this type.

So, finally, our problem is that of showing that each such quartic equation arising has only the trivial solutions $(u, v) = (\pm 1, 0)$. This will mean that (1) has no other solutions apart from those mentioned in this section.

Following Ljunggren (1963), let $u = \theta$ be a root of

$$g(u, 1) = 0 \tag{6}$$

so that (5) may be written in the form

$$N(u - v\theta) = 1. \tag{7}$$

Since (6) has just two real roots, $Q(\theta)$ has two fundamental units, ε_1 and ε_2 say. In every case, $Q(\theta)$ has as its only roots of unity ± 1, so (7) is equivalent to

$$u - v\theta = \pm\, \varepsilon_1{}^m \varepsilon_2{}^n. \tag{8}$$

3. Integer basis of $Q(\theta)$

In order to find a pair of fundamental units of $Q(\theta)$, it is necessary to determine a basis for its integers. The polynomial $g(u, 1)$ has discriminant $-2^{12}\, 3^3\, k^2$, so that it is only necessary to determine a basis at those primes dividing $6k$. This is a finite but tedious task and can be greatly simplified by considering several cases at once.

For example, if $\phi = (\theta^2 - x_0)/2$, then

$$\phi^4 - 4x_0\phi^3 - 4k\phi - kx_0 = 0.$$

So that, when $2\|kx_0$, an integer basis at the prime 2 is $1, \phi, \phi^2, \phi^3$.
We omit the details.

4. Units of $Q(\theta)$

To find a pair of fundamental units of $Q(\theta)$ we use the method of Berwick (1932). Let θ', θ'' be the two real roots, and θ''' a complex root of $g(u, 1) = 0$. Let $1, w_1, w_2, w_3$ be an integer basis. By w_1' we mean the value of w_1 with θ' for θ, etc. Any unit of $Q(\theta)$ will thus have the form

$$\varepsilon = x + yw_1 + zw_2 + tw_3,$$

with $x, y, z, t \in Z$, and $\varepsilon' \varepsilon'' \varepsilon''' \bar{\varepsilon}''' = \pm 1$. $\varepsilon_1, \varepsilon_2$ are a pair of fundamental units if they are units such that:

$$|\varepsilon''_1| < 1, \quad |\varepsilon_1'''| < 1, \quad |\varepsilon_1'| \text{ is minimal};$$

$$|\varepsilon_2'| < 1, \quad |\varepsilon_2'''| < 1, \quad |\varepsilon_1''| \text{ is minimal}.$$

Such units we shall call 'minimal'.

The conditions $|\varepsilon_1''| < 1$, $|\text{Re}(\varepsilon_1''')| < 1$, $|\text{Im}(\varepsilon_1''')| < 1$, provide an upper and lower bound for the coefficients y, z, t of ε_1 in terms of x of the type

$$|y - g_1 x| < h_1. \tag{9}$$

Using these inequalities, a search on the computer was made testing, for each $x = 0, 1, \ldots, A$ and the corresponding y, z and t's, whether the norm of $\alpha = x + yw_1 + zw_2 + tw_3$ was ± 1 and whether $|\alpha''| < 1$, $|\alpha'''| < 1$. When a unit, $\varepsilon = x_0 + y_0w_1 + z_0w_2 + t_0w_3$, did occur, it was invariably minimal, but this could be checked by a further small search. For, suppose there is another unit, $\delta = x + yw_1 + zw_2 + tw_3$ such that $|\delta''| < 1$, $|\delta'''| < 1$, $|\delta'| < |\varepsilon'|$ and $x > x_0$. Then, using the inequalities (9) we have a bound for x, given by

$$|x(1 - g_1w_1' - g_2w_2' - g_3w_3')| < |\varepsilon'| + h_1|w_1'| + h_2|w_2'| + h_3|w_3'|.$$

A similar procedure was adopted for ε_2.

At the initial stage, the actual value taken for A was such that the computing time spent searching for each unit was small (about 5-10 seconds); essentially, this depended on the product $h_1h_2h_3$. In practice, this product varied wildly but for a large number of cases its value was about 2 (with a corresponding value for A of about 4000.) By the end of this stage, about half the fundamental units had been found.

Rather than increase the value of A (perhaps indefinitely) for the remaining cases, it seemed pertinent to obtain some bound, and therefore to find two independent units of the field. This was done, in each field, by forming the norms of a few integers with small coefficients and, by drawing conclusions about their ideal factorisations, suitably multiplying and dividing them. For this last step, a teletype console with direct access to a computer proved essential, for the coefficients of any resulting unit were usually very large.

From two independent units, it is easy to find two other independent units δ_1, δ_2, generating the same multiplicative subgroup of units, say U, such that

$$|\delta_1''| < 1, \quad |\delta_1'''| < 1, \quad |\delta_1'| \text{ is minimal in } U,$$

$$|\delta_2'| < 1, \quad |\delta_2'''| < 1, \quad |\delta_2''| \text{ is minimal in } U.$$

$|\delta_1'|$, $|\delta_2''|$ now provide suitable bounds. When A could be increased without affecting the accuracy of the computer program and without using an excessive amount of computing time (usually the first criterion was the deciding factor), a search was made and the fundamental unit found.

The quartic fields where only one minimal unit was known, and a bound for the second was known but beyond the range of computation, were dealt with on the lines of the following argument. Suppose we have two units δ_1 and ε_2 with

$$|\delta_1''| < 1, \quad |\delta_1'''| < 1,$$

$$|\varepsilon_2'| < 1, \quad |\varepsilon_2'''| < 1, \quad |\varepsilon_2''| \text{ minimal},$$

and suppose there exists a minimal unit ε_1 such that

$$|\varepsilon_1''| < 1, \quad |\varepsilon_1'''| < 1, \quad B < |\varepsilon_1'| < |\delta_1'|,$$

then for some $a, b \in Z$ we have

$$\delta_1 = \varepsilon_1^a \varepsilon_2^b.$$

It is easy to prove that $a > 0$ and $0 \leqslant b < a$, so that

$$a < \log |\delta_1'|/(B' + \log |\varepsilon_2'|).$$

This gives, providing B' is large enough, a finite number of possible pairs (a, b) some of which can be further eliminated by a refinement of the above argument. For each of the remaining pairs with $a > 1$, it was shown that $\delta_1 \varepsilon_2^{-b}$ was not an ath power by congruence considerations modulo a suitable integer. Thus $a = 1$, $b = 0$ and δ_1 is a minimal unit.

Finally, there were quartic fields where two independent units were known, but where we were unable to determine 2 minimal units. However, in these cases, these independent units gave sufficient information about any pair of fundamental units for us to be able to apply the method of Skolem (1934) (see section 5).

5. p–adic methods

We solve (8) by the p-adic method of Skolem (1934). This involves choosing a suitable prime p such that by expanding $\varepsilon_1{}^m\varepsilon_2{}^n$ p-adically, it can be infered that, since the coefficients of w_2 and w_3 are zero, $m = n = 0$.

In practice, one proceeds as follows. By considering $\varepsilon_1{}^m\varepsilon_2{}^n$ modulo p and perhaps other (generally small) primes, one tries to show that $a|m$, $b|n$ where a, b are the orders of $\varepsilon_1, \varepsilon_2$ modulo p respectively. One then continues, by induction, to show that $m \equiv n \equiv 0 \pmod{p^\lambda}$ for each positive integer λ, and hence to show that $m = n = 0$. For, if $m \equiv n \equiv 0 \pmod{p^\lambda}$ than $\varepsilon_1{}^m\varepsilon_2{}^n$ may be expanded p-adically using the facts that ap^λ/m, $\varepsilon_1{}^{ap^\lambda} = 1 + p^{\lambda+1}\alpha$ for some α; similarly for ε_2. If p has been well chosen, and if $\varepsilon_1{}^m\varepsilon_2{}^n$ has been expanded enough, it follows that, because the coefficients of w_2 and w_3 are zero, $m \equiv n \equiv 0 \pmod{p^{\lambda+1}}$.

In most cases, Skolem's p-adic method was successful with $p = 2$ or 3; in all cases with p dividing $6k$ (see Table).

In the few cases where fundamental units were not known, knowledge of a particular pair of independent units was sufficient to make the p-adic method work.

For example, when $k = 87$ there are two independent units δ_1, δ_2 where $\delta_1 \equiv 1 + \theta - \theta^3$, $\delta_2 \equiv 1 - \theta - \theta^2 \pmod 3$. G_3 (the multiplicative group of residues modulo 3) is a product of four cyclic groups $C_3 \times C_3 \times C_2 \times C_2$ generated by the residues of $\delta_1, \delta_2, 1 - \theta, -1$. The norm of $1 - \theta$ is congruent to $-1 \pmod 3$, and it can be shown, by considering the generators of G_4, that there are no integers of norm $\equiv -1 \pmod 4$. Hence, there are no units of norm -1. So if, $\varepsilon_1, \varepsilon_2$ is a pair of fundamental units then the subgroup of G_3 generated by the residues of $\varepsilon_1, \varepsilon_2, -1$ is contained in that generated by the residues of $\delta_1, \delta_2, -1$ and is thus the same. Therefore, there is a pair of fundamental units $\varepsilon_1, \varepsilon_2$ with $\varepsilon_1 \equiv \delta_1, \varepsilon_2 \equiv \delta_2 \pmod 3$. This gives enough information about $\varepsilon_1, \varepsilon_2$ to apply Skolem's method 3-adically to (8) in this case.

6. Table

We list in Table I all those solutions of (1) for the outstanding values of k, for $0 \leqslant k \leqslant 100$ and $k = 288, 648$. For all those solutions giving rise to an irreducible quartic we also indicate whether the units found were fundamental (F) or just independent but sufficient for the p-adic method to be applied (*I*). The last column lists the prime p for which the p-adic method of Skolem was used.

TABLE I. *Solutions of* $x^3 - y^2 = k$

k	x	$\pm y$	F or I	p
18	3	3	*F*	2
23	3	2	*F*	3
25	5	10	*F*	2
26	3	1	*F*	2
	35	207	*F*	2
28	4	6	*F*	7
	8	22		
	37	225		
39	4	5	*F*	3
	10	31	*I*	3
	22	103	*I*	3
45	21	96	*F*	2
47	6	13	*I*	3
	12	41	*F*	3
	63	500		
53	9	26	*F*	2
	29	156	*F*	2
55	4	3	*F*	11
	56	419		
60	4	2	*F*	2
	136	1586		
61	5	8	*I*	3
63	4	1	*F*	3
	568	13537		
71	8	21	*F*	2
72	6	12	*F*	2
79	20	89	*F*	3
87	7	16	*I*	3
89	5	6	*F*	2
95	6	11	*I*	3
100	5	5		
	10	30	*F*	2
	34	198		
288	9	21	*F*	2
648	9	9	*F*	2
	18	72	*F*	2
	22	100		
	54	396	*F*	3
	97	955		
	1809	76941	*F*	2

The machine computation was initially done on the ICL 1905E at the University of East Anglia, and latterly on ATLAS at the Atlas Computer

Laboratory. We would like to express our thanks to the staff concerned for advice and assistance in the running of the programs.

Like all computations of this nature, there is the question of the reliability of the calculations. Every care was taken to preserve good accuracy in the machine computations and to double check hand calculations. As a further check we have heard that independent results for some k have been obtained by students of Ljunggren ($k = 18, 23, 25, 26, 28$) and by Finkelstein and London ($k = 18, 25, 100$).

References

Berwick, W. E. H. (1932). Algebraic Number-Fields with two independent units. *J. London Math. Soc.* **34,** 360–378.

Coghlan, F. B. (1967). Elliptic curves with conductor $N = 2^m 3^n$, Ph.D. thesis, Manchester University.

Hemer, O. (1952). On the diophantine equation $y^2 - k = x^3$. Dissertation, Uppsala University.

Ljunggren, W. (1963). On the diophantine equation $y^2 - k = x^3$. *Acta Arithmetica.* **VIII,** 451–463.

Skolem, T. (1934). Ein Verfahren zur Behandlung gewisser exponentialer Gleichungen und diophantischer Gleichungen, 8 de. Skand. Mat. Kongress, Stockholm, 163–188.

A Non-Trivial Solution of the Diophantine Equation $9(x^2 + 7y^2)^2 - 7(u^2 + 7v^2)^2 = 2$

OSKAR HERRMANN

University of Heidelberg, Heidelberg, Germany

My problem is the construction of a solution of a very special diophantine equation, yet this equation is closely related to the Hilbert Tenth Problem, first posed by Hilbert (1900) in his famous lecture on mathematical problems. Many of these problems are settled, some by solving, some by proving the unsolvability. In Hilbert's own words the Tenth Problem is: "Eine diophantische Gleichung mit irgendwelchen Unbekannten und ganzen rationalen Koeffizienten sei vorgelegt: Man soll ein Verfahren angeben, nach welchem sich mittels einer endlichen Anzahl von Operationen entscheiden lässt, ob die Gleichung in ganzen rationalen Zahlen lösbar ist." In a recent paper Martin Davis stated the following theorem: if the diophantine equation

$$9(x^2 + 7y^2)^2 - 7(u^2 + 7v^2)^2 = 2 \tag{1}$$

has no non-trivial solution (i.e. a solution different from $x = \pm 1, y = 0, u = \pm 1, v = 0$), then the Hilbert Tenth Problem is unsolvable in the sense of recursive number theory. The proof of this theorem is based on a paper of Julia Robinson (1952). In that paper diophantine predicates and predicates of exponential order of growth are investigated. According to Martin Davis (1963) the hypothesis "There is a diophantine predicate of exponential order of growth", which was conjectured by Julia Robinson, implies the unsolvability of the Tenth Problem. The unsolvability of the diophantine equation (1) implies according to Martin Davis (1968) the truth of the hypothesis of Julia Robinson. Trying to prove the unsolvability of the diophantine equation (1) I investigated properties of a solution and finally discovered a solution. Thus in the event no light is shed on Hilbert's Tenth Problem. The very extensive computations were performed on the Siemens 2002 of the Astronomisches Recheninstitut Heidelberg.

The diophantine equation (1) is obviously equivalent to the following system of diophantine equations:

$$(3A)^2 - 7B^2 = 2, \tag{2}$$

$$A = x^2 + 7y^2, \tag{3'}$$

$$B = u^2 + 7v^2. \tag{3''}$$

Let $\alpha = 3A + B\sqrt{7}$ be an integer of the quadratic field with discriminant 28. The equation (2) is equivalent to $N\alpha = 2$. A simple computation mod 3 shows that every integer of this field with norm 2 has a rational part divisible by 3. The class number of this field is 1. There is only one ideal in this field with norm 2. The fundamental unit of this field is $\varepsilon = 8 + 3\sqrt{7}$. Hence, every integer α with $N\alpha = 2$ is of the form

$$\alpha = \pm (3 + \sqrt{7})(8 + 3\sqrt{7})^n.$$

The equation (3′) and the equation (3″) imply that A and B are non-negative integers. Thus finally all non-negative solutions of (2) are given by

$$3A_n + B_n\sqrt{7} = (3 + \sqrt{7})(8 + 3\sqrt{7})^n, \quad n \geqslant 0.$$

By representation of the field $Q(\sqrt{28})$ by matrices of order 2, we find that

$$\begin{pmatrix} A_n \\ B_n \end{pmatrix} = \begin{pmatrix} 8 & 7 \\ 9 & 8 \end{pmatrix}^n \begin{pmatrix} 1 \\ 1 \end{pmatrix},$$

and the recursion

$$\begin{aligned} A_{n+1} &= 8A_n + 7B_n, \\ B_{n+1} &= 9A_n + 8B_n, \\ A_0 &= B_0 = 1. \end{aligned} \tag{4}$$

Next, since $Q(\sqrt{-7})$ has class number 1, a number N is representable as $r^2 + rs + 2s^2$ if and only if every prime divisor of N with odd multiplicity is not a quadratic non-residue modulo 7. Using this and some computation modulo 2 we obtain:

an odd number N is representable as $x^2 + 7y^2$ if and only if every prime divisor of N with odd multiplicity is not a quadratic non-residue modulo 7. (5)

The corresponding result with 7 replaced by 3 is true, but not for 1 1, 19 43, 67, or 163. We require also some detailed results on A_n and B_n.

LEMMA. *Let p be prime, and all congruences below taken modulo p. If $A_n \equiv 0$,* *then*

$$A_{2n} \equiv -1,\ B_{2n} \equiv +1,\ A_{2n+1} \equiv B_{2n+1} \equiv -1,\ A_{4n+2} \equiv B_{4n+2} \equiv +1. \quad (6.1)$$

If $B_n \equiv 0$, *then*

$$A_{2n} \equiv +1, \qquad B_{2n} \equiv -1, \qquad A_{2n+1} \equiv B_{2n+1} \equiv +1. \quad (6.2)$$

If $A_n \equiv B_n \equiv 1$, *then*

$$A_{nk} \equiv B_{nk} \equiv +1 \textit{ for all } k. \quad (6.3)$$

If $A_n \equiv 1$, $B_n \not\equiv 1$, *then*

$$B_n \equiv -1, \qquad A_{n+1} \equiv B_{n+1} \equiv 1. \quad (6.4)$$

If $A_n \not\equiv 1$, $B_n \equiv 1$, *then*

$$A_n \equiv -1, \qquad A_{n+1} \equiv B_{n+1} \equiv -1, \qquad A_{2n+2} \equiv B_{2n+2} \equiv 1. \quad (6.5)$$

The sequence of pairs (A_n, B_n) *considered modulo* p *is periodic. If* $p \neq 3$, *then the primitive period* $\pi(p)$ *is a divisor of* $p - (p/7)$. (6.6)

$$A_n \equiv B_n \equiv 1 \pmod 2. \quad (6.7)$$

$$A_n \equiv (-1)^n(1 + 2n), \qquad B_n \equiv (-1)^n \pmod 3. \quad (6.8)$$

$$A_n \equiv 1, \qquad B_n \equiv 1 + 2n \pmod 7. \quad (6.9)$$

The proofs of (6.1) to (6.5) are easy, and (6.6) is a special case of a theorem of Legendre ($p = 3$ is exceptional since it occurs in the coefficients of A^2 in (2)); (6,7) to (6.9) are trivial.

The construction of a solution of (1) consists of two steps. In the first step, by a method of sieving, numbers n, such that (3′) and (3″) were probably solvable, were determined. Beginning with the set of numbers $\{n|0 \leqslant n < 10000\}$ a number n was eliminated if $n \equiv 1$, 2 or 6 (7). In these cases B_n is not a quadratic residue mod 7 by (6.9) and so (3″) is not solvable by (5). Next, for every prime p with $(p/7) = -1$ less than 40000, the sequence of pairs A_n, B_n mod p up to an index n_0 was computed, where n_0 was the least number satisfying one of the following conditions:

(i) $A_n \equiv B_n \equiv \pm 1\ (p)$,

(ii) $n = 10000$,

(iii) $A_nB_n \equiv 0\ (p)$.

In the first case, there is no number n such that A_n or B_n is divisible by p. In the second case, there is no number $n < 10000$, such that A_n or B_n is divisible by p. In the third case, all the zeros of A_n resp. $B_n \bmod p$ are given by

$$n = n_0 + (2n_0 + 1)k; \quad k = 0, 1, 2, \ldots.$$

If any of these numbers was not already eliminated, then the numbers A_n, $B_n \bmod p^2$ and $\bmod p$ were computed. A number n was then eliminated if A_n or B_n was divisible by p but not by p^2. If A_n or B_n was divisible by p^2, the computation was in principle to be repeated $\bmod p^3$ and $\bmod p^4$, but in order to save time, this computation was dropped, and the number p was punched out. This procedure was initially very efficient, but lost its efficiency very soon, so it that was necessary to investigate the remaining numbers n separately.

The first remaining number was 26. Some of the numbers A_n, B_n are

$$\begin{aligned}
A_0 &= 1\\
B_0 &= 1\\
A_1 &= 15\\
B_1 &= 17\\
A_2 &= 239\\
B_2 &= 271\\
A_3 &= 3809\\
B_3 &= 4319\\
A_6 &= 15\ 418831\\
B_6 &= 17\ 483311\\
A_{26} &= 17\ 231429\ 089624\ 614166\ 470862\ 182959\\
B_{26} &= 19\ 538604\ 045167\ 506118\ 097869\ 511631
\end{aligned}$$

By (6.1) the primitive period $\pi(p)$ of a divisor of A_{26} is a divisor of 106 and the primitive period of a divisor of B_{26} is a divisor of 53. So, the primitive period of a prime divisor of $A_{26} B_{26}$ is a divisor of 106. Let $\pi(p) = 1$ or $\pi(p) = 2$. Then $A_2 \equiv B_2 \equiv 1 \pmod p$. So p is a prime divisor of $B_2 - A_2 = 32$. Hence, $p = 2$ and $\pi(p) = 1$. A_{26} and B_{26} are odd numbers, so the primitive period of any divisor of A_{26} or B_{26} is divisible by 53.

By (6.6) we find $53 | \pi(p) | p - (p/7)$ and

$$p \equiv (p/7) \pmod{53}.$$

Neither A_{26} nor B_{26} has a prime divisor satisfying this congruence up to 10^8.

In order to factorize A_{26} and B_{26} or to prove A_{26} and B_{26} to be primes by means of a theorem of Lehmer (1939) it is necessary to factorize $A_{26} - 1$ and $B_{26} - 1$. Let p a common divisor of $A_{26} - 1$ and $B_{26} - 1$ then $\pi(p)$ is a divisor of 26. Hence, $p = 2$ or, by (6.6), $p \equiv (p/7) \pmod{13}$. Then (6.4) and (6.5) imply that primitive period $\pi(p)$ of a prime p, which divides either $A_{26} - 1$ or $B_{26} - 1$, is a divisor of 54. Hence, $\pi(p)$ is a divisor of 6 or a multiple of 9. The primes p, with $\pi(p)|6$ are easy to determined: these are the common prime divisors of $A_6 - 1 = 15\,418830$ and $B_6 - 1 = 17\,483310$. So, possible prime divisors of $A_{26} - 1$ and $B_{26} - 1$ are:

(a) $p = 2, 3, 5, 17$,

(b) primes p with $p \equiv (p/7) \pmod{53}$,

(c) primes p with $p \equiv (p/7) \pmod{9}$.

Using this information it was possible to factorize $A_{26} - 1$ and $B_{26} - 1$ by a very short computation. The factorizations are:

$$A_{26} - 1 = 2 . 7 . 17 . 131 . 4049 . 117701 . 159839 . 414991 . 17483311,$$
$$B_{26} - 1 = 2 . 3^5 . 5 . 19 . 71 . 131 . 117701 . 17483311 . 22110582149.$$

These are if fact prime decompositions. This may be checked by using Lehmer's table for the factors less than ten millions. There are two remaining factors. The smaller one is a common divisor of $A_{26} - 1$ and $B_{26} - 1$. There is no prime $p \equiv (p/7) \pmod{53}$, $p < 4200$ which is a divisor of 17483311, so this is a prime. The primality of the last divisor may be checked by using (c) instead of (b) in the same way.

Now, it is a simple computation to show, that the residue class of 7 mod A_{26} has the order $A_{26} - 1$ in the residue class group mod A_{26} and the residue class of 7 mod B_{26} has the order $B_{26} - 1$ in the residue class group mod B_{26}. The existence of primitive roots of this order shows that A_{26} and B_{26} are primes. (Martin Davis has checked this independently).

We conclude: A_{26} and B_{26} are primes and quadratic residues mod 7. So, there is in both cases only one prime facter with odd multiplicity and this prime factor is a quadratic residue and the equations (3′) and (3″) are solvable. This finishes the proof of the existence of a solution of (1).

According an oral commumication of Martin Davis there are similar equations to state in respect of every imaginary quadratic field $Q(\sqrt{-d})$ with class number 1 and d not a square, with the same application to Hilbert's problem. The last condition excludes the Gaussian field. So, according the Theorem of Heegner and Stark there are 8 values of d:

$$d = 2, 3, 7, 11, 19, 43, 67 \text{ and } 163.$$

In the case $d = 3$ the corresponding equation is

$$3(x^2 + 3y^2)^2 - (u^2 + 3v^2)^2 = 2. \tag{7}$$

The recursion-formula of the corresponding Pell-type equation is

$$A_{n+1} = 2A_n + B_n,$$
$$B_{n+1} = 3A_n + 2B_n,$$
$$A_0 = B_0 = 1.$$

The corresponding sieve shows the pair $(A_6, B_6) = (2131, 3691)$ to be the first pair of candidates for a solution. In fact it is, and

$$A_6 = 16^2 + 3 . 25^2 \text{ and } B_6 = 4^2 + 3 . 35^2.$$

So the equation (7) has a solution, and indeed with explicit x, y, u, and v. The solution of the corresponding equations in the other cases with odd d is much more laborious, for numbers of more than eighty digits have to be factorized, and (5) replaced by more a complicated lemma. These results may be published later after completing the computation.

References

Davis, M. (1963). Extensions and corollaries of recent work on Hilbert's Tenth Problem, *Illinois J. Math.* **7,** 264–250.

Davis, M. (1968). One Equation to rule them all. The Rand Corporation. Memorandum RM–5494–PR.

Hilbert, D. (1900). Mathematische Probleme. Vortrag, gehalten auf dem internationalem Mathematiker-Kongress zu Paris, 1900.

Lehmer, D. H. (1939). A Factorisation Theorem applied to a Test for Primality. *Bull. Amer. Math. Soc.* **45,** 132–137.

Matiyasevic, Yu. V. (1970). The Diophantine of enumerable sets. *Dokl. Akad. nauk. SSSR.* **191,** 279–282.

Robinson, J. (1952). Existential definability in arithmetic. *Trans. Amer. Math. Soc.* **72,** 437–449.

Note added in proof:

In an important recent paper Mitiyasevic (1970) gave a proof of the unsolvability of the Tenth problem, but the method developed in order to solve this diophantine equation is of general interest only.

On the Fermat Quotient

J. BRILLHART, J. TONASCIA, AND P. WEINBERGER

University of Arizona, Tucson, Arizona,
Johns Hopkins University, Baltimore, Maryland and
University of California, Berkeley, California, U.S.A.

1. Introduction

(a) At various times over the last 7 years the authors, in collaboration with D. H. and Emma Lehmer, have conducted a search for *odd* solutions N of the congruence

$$a^{N-1} \equiv 1 \pmod{N^2} \tag{1}$$

for the bases $a = 2(1)99$. This computing, which was done almost completely on idle time, was carried out on the IBM 7090-94 and the CDC 6400 at the University of California, Berkeley and the IBM 7090 at Stanford University*. The purpose of this search was both to extend the search limits of other investigators (Fröberg, 1958; Kravitz, 1960; Pearson, 1963; Riesel, 1964; Hausner and Sachs, 1963; and Kloss, 1965) and to discover any composite solutions of (1) for the value of a considered.

The first part of this paper is devoted to a discussion of the search for prime solutions of (1), while the second part is given over to an investigation of the existence and nature of composite solutions.

The numerical results of these two investigations are given in Tables 1 and 2 below.

(b) It is a matter of some interest to note that quite different problems lead to (1) as a condition for their solution. Best known among these, of course, is the criterion of Wieferich (1909) in the first case of Fermat's Last Theorem:

If p is an odd prime, $p \nmid xyz$, for which $x^p + y^p = z^p$ has a non-trivial solution, then $2^{p-1} \equiv 1 \pmod{p^2}$. (see Lehmer (1941) for further references).

Another instance is given by Inkeri (1964), who has connected (1) with Catalan's conjecture:

* Use of the IBM 7090 at Stanford was made possible by the Department of Computer Sciences under grant No. NSF–GP948.

If p and q are primes, then

$$x^p - y^q = 1 \tag{2}$$

has no non-trivial solutions, except in the case $p = 2, q = 3$, *where* $x = \pm 3$, $y = 2$ *are solutions.*

In particular, he has shown the following:

Let p and q be primes >3 *with* $p \equiv 3 \pmod 4$ *and* $q \nmid h(p)$, *the class number of the quadratic field* $Q(\sqrt{-p})$. *If* (2) *has non-trivial solutions, then* $p^{q-1} \equiv 1 \pmod{q^2}$, *and* $x \equiv 0 \pmod{q^2}$, $y \equiv -1 \pmod{p^{2p-1}}$.

In a different direction we have the theorem:

If g is a primitive root of a prime p, but not a primitive root of p^2, *then* $g^{p-1} \equiv 1 \pmod{p^2}$. (See Nagell, 1964; Shanks, 1962. Also, see Shanks (1963) for a relation of this congruence to the smallest positive primitive root). Closely related to this question is the fact that the expansions of $1/p$ and $1/p^2$ to base b have the same period if and only if $b^{p-1} \equiv 1 \pmod{p^2}$. (See Dickson, 1952).

Recently Warren and Bray (1967) have observed that if q is a prime for which $q^2 \mid (2^p - 1)$, p an odd prime, then $2^{q-1} \equiv 1 \pmod{q^2}$. (See Brillhart and Johnson (1960), p. 366).

For further matters relating to (1) and its history see Dickson (1952), Beeger (1913), Hasse (1957), and the other references at the end of this paper.

2. The Search for Prime Solutions

When the present investigation was begun in 1962, the highest published search limit for $a = 2$ was 200183 due to Pearson (1963). In our initial report to the Number Theory Conference at Boulder Colorado in 1963 we gave the search limit of 2^{22} for $2 \leqslant a \leqslant 10$, and the new solutions $p = 1006003$ for $a = 3$, and $p = 534851$ and 3152573 for $a = 6$.

Shortly thereafter in November, 1963 Hausner and Sachs (1963) published (independently) the result that no new solutions exist below 10^6 for $a = 2$. In the following year Riesel (1964) extended the values of a to 150, finding all solutions $p < 500{,}000$ for $2 \leqslant a \leqslant 10$ and $p < 10{,}000$ for $11 \leqslant a \leqslant 150$.

In 1965 an unpublished table of solutions of (1) for $a \leqslant 100$ with the search limit 10^6 was sent to us by Dr. Sachs. This table serves as the basis of Table 1 below. (These results are presented here with the kind permission of Dr. Sachs.) In the same year Kloss (1965) published the results of his search for solutions for the *prime* bases $\leqslant 43$ to various search limits between 5×10^6 and 11×10^6, with the exception of the limit 31059000 for $a = 2$. From his investigation came several solutions of (1), which verified the solutions in the unpublished table of Sachs, with three solutions larger than 10^6 being $p = 1747591$ for $a = 13$, $p = 2481757$ for $a = 23$, and $p = 1025273$ for $a = 41$.

The present search has now established the following limits:

$a = 2,$	$p < 3 \times 10^9$	$a = 14, 15, 17,$ 18, 19, 20	$p < 2^{27}$
$a = 3,$	$p < 2^{30}$		
$a = 5,$	$p < 2^{29}$		
$a = 6, 7, 10, 11,$ 12, 13	$p < 2^{28}$		

For those bases for which no solutions larger than the base are known, we have the search limits:

$a = 21, 34$	$p < 2^{29}$
$a = 29, 47, 50$	$p < 2^{28}$
$a = 22, 23, 24, 26, 28$	$p < 2^{26}$
$a = 42, 51, 60, 61, 66, 72, 73, 74, 82, 88, 89, 90, 97, 99$	$p < 2^{25}$

For all other bases we have the limit of 10^6 due to Sachs.

The new prime solutions of (1) we have discovered are $p = 53471161$ for $a = 5$, $p = 56598313$ for $a = 10$, $p = 1284043$ for $a = 18$, $p = 63061489$ for $a = 19$, $p = 9377747$ and 122959073 for $a = 20$, $p = 1595813$ for $a = 22$, and $p = 13703077$ for $a = 23$. These results are given in Table 1 along with all other known odd prime solutions.

It is worth mentioning that the probability of p being a solution of $a^{p-1} \equiv 1 \pmod{p^2}$, a given, is roughly $1/p$ since the congruence $x^{p-1} \equiv 1 \pmod{p^2}$ has $p - 1$ solutions. (See Dickson, 1952). The rarity of solutions for a given $x = a$ implied by this probability is clearly borne out in Table 1.

Among the solutions p in Table 1 are those which also satisfy

$$a^{p-1} \equiv 1 \pmod{p^3}, \qquad 2 \leqslant a \leqslant 100. \tag{3}$$

These are marked with an asterisk. Of these, only $p = 3$ satisfies a similar congruence $(\bmod\, p^4)$ for $a = 80$ and 82, $2 \leqslant a \leqslant 100$.

We observe the error in Dickson (1952), (for H. Hertzer) and Beeger (1913), where $p = 11$ is claimed as a solution of (3) for $a = 3$. We also note the curious partition $66161 = 2^{16} + 5^4$ in the case $a = 6$.

It is proper to point out here that Table 1 does not include those a which are powers. That we may omit these without leaving out any essential information is clear from the following.

THEOREM 1: *If p is a prime, $p \nmid n$, $n \geqslant 2$, and $\alpha \geqslant 2$, then $(a^n)^{p-1} \equiv 1 \pmod{p^\alpha}$ if and only if $a^{p-1} \equiv 1 \pmod{p^\alpha}$.*

Proof: The "if" direction is immediate. In the other direction, let a belong to

TABLE I. Solutions of $a^{p-1} \equiv 1 \pmod{p^2}$, p an odd prime
(a not a power, $a < 100$)

a	p	a	p
2	1093 3511	40	11 17 307 66431
3	11 1006003	41	29 1025273
5	20771 40487 53471161	42	23*
6	66161 534851 3152573	43	5 103
7	5 491531	44	3 229 5851
10	3 487 56598313	45	1283 131759
11	71	46	3 829
12	2693 123653	47	
13	863 1747591	48	7 257
14	29 353	50	7
15	29131	51	5 41
17	3 46021 48947	52	461
18	5 7* 37 331 33923 1284043	53	3* 47 59 97
19	3 7* 13 43 137 63061489	54	19 1949
20	281 46457 9377747 122959073	55	3* 30109
21		56	647
22	13 673 1595813	57	5* 47699 86197
23	13 2481757 1370377	58	131
24	5 25633	59	2777
26	3* 5 71	60	29
28	3* 19 23	61	
29		62	3 19 127 1291
30	7 160541	63	23 29 36713 401771
31	7 79 6451	65	17 163
33	233 47441	66	
34		67	7 47 268573
35	3 1613 3571	68	5* 7 19 113* 2741
37	3 77867	69	19 223 631
38	17 127	70	13 142963
39	8039	71	3 47 331
72		87	1999 48121
73	3	88	
74	5	89	3 13
75	17 43 347 31247	90	
76	5 37 1109 9241 661049	91	3 293
77	32687	92	727 383951
78	43 151 181 1163 56149	93	5 509 9221 81551
79	7 263 3037	94	11 241 32143 463033
80	3* 7 13 6343	95	2137 15061
82	3* 5	96	109 5437 8329
83	4871 13691	97	7
84	163 653 20101	98	3 28627
85	11779	99	5 7 13 19 83
86	68239		

$e \pmod{p^\alpha}$. Then

$$e \mid n(p-1) \text{ and } e \mid p^{\alpha-1}(p-1).$$

But since $p \nmid n$, then $p \nmid e$. Hence $e \mid (p-1)$, so $a^{p-1} \equiv 1 \pmod{p^\alpha}$.

This Theorem shows, except possibly for the primes dividing n (which for Table 1 are <6), that the prime solutions for base a^n are the same as those for a itself.

3. Programming Developments

The program to search for prime solutions of (1) consisted of two parts:

(i) the production of a sequence of trial p's

(ii) the testing of each p in (1).

The sequence in (i) can be generated efficiently in a computer in the way one usually generates a sequence of trial divisors for factoring a given number, namely, through the use of an increment table in the memory, which is used over and over again to produce a sequence of numbers not divisible by certain small primes. (This table is actually the difference table for the numbers that remain when the integers are sieved by the small primes which are used).

In the IBM 7094, for example, only the primes $\leqslant 13$ can be used in this way, for $\phi(2.3.5.7.11.13) = 5760$ memory locations are required for the increment table. To have included 17 as well would have required a table 16 times as large ,which would exceed the memory available in the 7094.

Although the sequence produced in this way includes composite as well as prime numbers, the method is efficient timewise, since to test a trial p further is to test it in

$$a^{p-1} \equiv 1 \pmod{p} \tag{4}$$

anyway, (We note the time required to search, say, to 2^{22} for a given a was approximately 30 minutes).

When the sequence in (i) is used in factoring a number M, the monotonicity of the trial divisors assures us any divisor d of M will be prime, if each such divisor, when discovered, is removed as often as it will divide M. This clearly is not true in the present case, however, so composite solutions of (1) may well be discovered. Thus, each of the solutions discovered for a given a must be tested for primality. That this is very simple will follow when we show the interesting property (Sec 4) that composite solutions of (1) can only have prime divisors which are already solutions of (1).

The testing in (ii) is itself broken into two parts: (a) determining if p is a pseudoprime, (b) testing pseudoprime p in (1). (We remark here that the term "pseudoprime" refers to a number p satisfying (4) for a given base $a \geqslant 2$.

A prime is thus a pseudoprime under this definition.) In (a) the calculation of $a^{p-1} \pmod{p}$ is carried out by *single-precision* residue methods, which are possible since p is less than 2^{35}, the word-size in the computer. In this calculation the usual procedure of using the bits of $p - 1$ to prescribe the sequence of squarings and multiplications by a is followed. To shorten this build-up, the largest initial power of $a < 2^{35}$ is chosen from a table in memory on the basis of the leading bits of $p - 1$

If the result of this calculation is not 1, then p cannot satisfy (1), and the next value of p is considered. If the result is 1, then p is a pseudoprime, and we proceed to (b).

In testing p in (1), we unfortunately cannot use any of the arithmetic of (a). We must therefore re-compute a^{p-1}, this time using the *double-precision* modulus p^2 (as soon as $p > 2^{17.5}$). The double-precision, however, can be accomplished by single-precision in the following way:
If R is any residue in the calculation, $0 \leqslant R < p^2$, then we wish to find $R^2 \pmod{p^2}$. We first compute $R^2 = Q_1 p + R_1$, $0 \leqslant R_1 < p$, and then $Q_1 = Q_2 p + R_2$, $0 \leqslant R_2 < p$. Combining we have $R^2 \equiv R_2 p + R_1 \pmod{p^2}$, where the expression on the right is the desired remainder, since $0 \leqslant R_2 p + R_1 \leqslant (p-1)p + p - 1 = p^2 - 1 < p^2$. (It is clear that this process is merely finding the last two digits of R^2 in a number system with base p).

4. Composite Solutions

(a) The following theorem provides a strong characterization of the primes which divide a composite solution of (1).

THEOREM 2: *If* $a^{N-1} \equiv 1 \pmod{N^2}$, $a \geqslant 2$, $N \geqslant 2$, *and if* $p^\alpha | N$, *then* $p^{2\alpha} | (a^{p-1} - 1)$.

Proof: Setting $N = p^\alpha m$, we conclude directly that $a^{p^\alpha m - 1} \equiv 1 \pmod{p^{2\alpha}}$. Also, if a belongs to $e \pmod{p^{2\alpha}}$, then $e | p^{2\alpha-1}(p-1)$ and $e | (p^\alpha m - 1)$. The latter implies $p \nmid e$. Thus $e | (p-1)$, so that $a^{p-1} \equiv 1 \pmod{p^{2\alpha}}$.†

The information contained in this theorem is of two kinds. For one thing, the theorem shows a prime divisor p of a composite solution N of (1) must itself be a solution of (1), In addition, it shows the maximum power of p that can divide N is limited by the highest power of p that divides $a^{p-1} - 1$. In particular, if $p^\beta \| (a^{p-1} - 1)$, then $1 \leqslant \alpha \leqslant [\beta/2]$, the greatest integer in $\beta/2$. It follows then that $\beta = 2$ or 3 and $\alpha = 1$ for all primes in Table 1 (except for $p = 3$ with $a = 80$ or 82, where $\beta = 4$ and $\alpha = 1$ or 2), so the composite solutions of (1), divisible only by primes in Table 1, must necessarily be square-free if $a \neq 80$ or 82.

† This theorem was initially discovered by D. H. Lehmer for the case $\alpha = 1$

(b) It is clear from (a) that the most effective way to find composite solutions of (1) is to construct all possible square-free products N from the primes in Table 1 for a given a and then test each such N in (1). This testing can be greatly shortened, however, through the use of the following theorem, which will exclude with very little calculation all the trial N's not satisfying (1).

THEOREM 3: *Let* $N = \Pi_i p_i^{\alpha_i}$, *where* $a^{p_i - 1} \equiv 1 \pmod{p_i^2}$ *and* $1 \leqslant \alpha_i \leqslant [\beta_i/2]$, $p_i^{\beta_i} \| (a^{p_i - 1} - 1)$. *Also, for some* $p^\alpha \| N$ *let* $d = (p - 1, N/p^\alpha - 1)$. *If* $a^{N-1} \equiv 1 \pmod{N^2}$, *then* $a^d \equiv 1 \pmod{p^{2\alpha}}$.

Proof. From Theorem 2 we conclude for the chosen p that $a^{p-1} \equiv 1 \pmod{p^{2\alpha}}$ which implies $a^{p^\alpha} \equiv a \pmod{p^{2\alpha}}$. Hence, $1 \equiv a^{N-1} \equiv (a^{p^\alpha})^{N/p^\alpha} . \; a^{-1} \equiv a^{(N/p^\alpha) - 1} \pmod{p^{2\alpha}}$. Then if a belongs to $e \pmod{p^{2\alpha}}$, $e \mid (p-1)$ and $e \mid (N/p^\alpha - 1)$. Thus, $e \mid d$, from which the theorem follows.

In applying Theorem 3 we use the weakened form of the conclusion

$$a^d \equiv 1 \pmod{p}. \tag{5}$$

This easy-to-use necessary condition will usually eliminate most of the trial N's which do not satisfy (1). Of course, it may happen for certain p that (5) is actually satisfied, so N is not eliminated. In this case we must select another p (if possible), compute its d, and through the use of (5) see if N is eliminated.

Example. Let $a = 78$ and take $N = 43 . 151 . 1163 . 56149$. (See Table 1). Choose $p = 43$. Then

$$d = (42, 151 . 1163 . 56149 - 1) = (42, 9860494336) = 14.$$

But $78^{14} \equiv 1 \pmod{43}$, so N is *not* eliminated. If, however, we take $p = 151$, then $d = (150, 2807955340) = 10$ and $78^{10} \equiv 64 \not\equiv 1 \pmod{151}$, which shows N is not a solution of $78^{N-1} \equiv 1 \pmod{N^2}$.

In this example, we note since the conclusion of Theorem 3 allows for a modulus of $p^{2\alpha}$, it might seem that one should use a modulus of 43^2 in a further attempt to eliminate N without changing to another p and recomputing its d. That this will never work follows from

THEOREM 4: *If* a *belongs to* $e \pmod{p}$ *and* $a^{p-1} \equiv 1 \pmod{p^2}$, *then* a *belongs to* $e \pmod{p^2}$.

Proof. From $a^e \equiv 1 \pmod{p}$ we obtain

$$a^e = 1 + kp. \tag{6}$$

Then

$$1 \equiv a^{p-1} = (a^e)^{(p-1)/e} = (1 + kp)^{(p-1)/e} \equiv 1 + kp(p-1)/e \pmod{p^2}.$$

Hence, $p|k$ and from (6), $a^e \equiv 1 \pmod{p^2}$.

But no smaller power of a can yield $1 \pmod{p^2}$, since such a power would then contradict the fact that a belongs to $e \pmod{p}$.

By means of the above tests we readily determine all the composite solutions of (1) whose prime divisors lie in Table 1. These solutions are listed in Table 2. (We note from Riesel's table (1964) there are two composite solutions $N = 7.13$ for $a = 146$ and $N = 11.41$ for $a = 148$).

TABLE II. Solutions of $a^{N-1} \equiv 1 (\mathrm{mod}\ N^2)$, N odd and composite

a	N
26	$3 \cdot 5$, $3 \cdot 5 \cdot 71$
68	$7 \cdot 19$
80	3^2
82	3^2, $3^2 \cdot 5$
99	$5 \cdot 7$, $5 \cdot 13$, $7 \cdot 13 \cdot 19$

(c) As in the search for prime solutions, we can essentially disregard those a which are powers.

THEOREM 5: *If* $(a^n)^{N-1} \equiv 1 \pmod{N^2}$, $n \geqslant 2$, *and* $p^\alpha | N$, *but* $p \nmid n$, *then* $a^{p-1} \equiv 1 \pmod{p^{2\alpha}}$.

Proof. Let $N = p^\alpha m$. Thus, $(a^n)^{p^\alpha m - 1} \equiv 1 \pmod{p^{2\alpha}}$. Let a belong to e (mod $p^{2\alpha}$). Then $e | n(p^\alpha m - 1)$ and $e | p^{2\alpha - 1}(p-1)$. Now $p \nmid e$, since both $p \nmid n$ and $p \nmid (p^\alpha m - 1)$. Therefore, $e | (p-1)$, so that $a^{p-1} \equiv 1 \pmod{p^{2\alpha}}$.

We might remark in case $p | n$, that if $p^\beta \| (a^{p-1} - 1)$ and $p^\sigma \| n$, then $1 \leqslant \alpha \leqslant [(\beta + \sigma)/2]$. The testing of N in this case is carried out as before.

5. Acknowledgements

We would like to thank the directors of the computer centers at Stanford University and the University of California, Berkeley for their support of this work. We would also like to express our appreciation to Daniel Shanks, whose suggestions have helped to further this work. Finally, we wish to thank D. H. and Emma Lehmer for their generous contributions of time and effort to this investigation.

References

Abel, N. H. (1828). *J. für. Math.* V. **3**, 212.
Abel, N. H. (1881). *Oeuvres*, **1**, 619.

Beeger, N. G. W. H. (1913). Quelques remarkes sur les congruences $r^{p-1} \equiv 1(\bmod p^2)$ et $(p-1)! \equiv -1(\bmod p^2)$, *Mess. Math.* **43,** 72–85.

Beeger, N. G. W. H. (1922). On a new case of the congruence $2^{p-1} \equiv 1(\bmod p^2)$, *Mess. Math.* **51,** 149–150.

Beeger, N. G. W. H. (1939). On the congruence $2^{p-1} \equiv 1(\bmod p^2)$ and Fermat's Last Theorem. *Nieuw Archief voor Wiskunde.* **20,** 51–54.

Beeger, N. G. W. H. (1925). On the congruence $2^{p-1} \equiv 1(\bmod p^2)$ and Fermat's Last Theorem. *Mess. Math.* **55,** 17–26.

Brillhart, J. and Johnson, G. D. (1960). On the factors of certain Mersenne Numbers. *Math. Comp.* **14,** 365–369.

Bryant, S. (1967). Groups, graphs, and Fermat's Last Theorem, *Amer. Math. Monthly.* **74,** 152–156.

Dickson, L. E. (1952). "History of the Theory of Numbers", Ch. IV, Ch. XXVI. Chelsea, New York.

Fröberg, C. E. (1958). Some computations of Wilson and Fermat remainders, *MTAC.* **12,** 281.

Hardy, G. H. and Wright, E. M. (1965). "An Introduction to the Theory of Numbers". 104–105. Oxford University Press.

Hasse, H. (1957). Über die Charakterführer zu einem Arithmetischen Funktionen Körper vom Fermatschen Typus. *Wissenschaftliche Veröffentlichungen der Nationalen Technischen Universität, Athens,* **12,** 3–50.

Hausner, M. and Sachs, D. (1963). On the congruence $2^p \equiv 2 \ (\bmod p^2)$, *Amer. Math. Monthly,* **70,** 996.

Hertzer, H. (1908). Über die Zahlen der Form $a^{p-1}-1$, wenn p eine Primzahl, *Archiv der Math. u. Physik.* **13,** 107.

Jacobi, K. G. J. (1828). Beantwortung der Aufgabe *s.* 212 dieses Bandes. *J. für Math.* **3,** 301–302.

Inkeri, K. and Hyyrö, S. (1961). On the congruence $3^{p-1} \equiv 1(\bmod p^2)$ and the Diophantine equation $x^2 - 1 = y^p$, *Ann. Univ. Turku, Ser. A,* **1,** 1–4.

Inkeri, K. (1964). On Catalan's problem. *Acta Arith.* IX, 285–290.

Kloss, K. E. (1965). Some number-theoretic calculations. *J. of Research, Nat. Bur. Standards,* 69B, **4,** 335–336.

Kravitz, S. (1960). The congruence $2^{p-1} \equiv 1(\bmod p^2)$ for $p < 100000$. *MTAC* **14,** 378.

Kruyswijk, D. (1966). On the congruence $u^{p-1} \equiv 1(\bmod p^2)$. *Math. Centrum Amsterdam Afd., Zuivere Wisk.,* ZW-003, 7.

Lehmer, D. H. and Lehmer, E. (1941). On the first case of Fermat's Theorem. *Bull. Amer. Math. Soc.* **47,** 139–142.

Lehmer, D. H. (1941). "Guide to Tables in the Theory of Numbers". Nat. Research Council, Washington D.C. (b_4), 10.

Meissner, W. (1913). Über die Teilbarkeit von $2^p - 2$ durch das Quadrat der Primzahl $p = 1093$. *Akad. d. Wiss., Berlin, Sitzungsb.* **35,** 663–667.

Muskat, J. B. (1964). Criteria for Solvability of Certain Congruences. *Canadian J. Math.* **16,** 350.

Nagell, T. (1964). "Introduction to Number Theory". Chelsea, New York.

Oblath, R. (1939). Über die Zahl $x^2 - 1$, *Mathematica, B VIII,* 161–172.

Pearson, E. H. (1963). On the congruences $(p-1)! \equiv -1$ and $2^{p-1} \equiv 1(\bmod p^2)$. *Math. Comp.* **17,** 194–195.

Riesel, H. (1964). Note on the congruence $a^{p-1} \equiv 1(\bmod p^2)$. *Math. Comp.* **18,** 149–150.

Rotkiewicz, A. (1965). Sur les nombres de Mersenne dépourvus de diviseurs carrés et sur les nombres naturels n, tels que $n^2|2^n - 2$. *Mat. Vesnik*. **2**(17), 78–80.
Shanks, D. (1962). "Solved and Unsolved Problems in Number Theory", Spartan Books, Washington D.C.
Shanks, D. (1963). Review 72F, Math. Comp., V.17, 463–464.
Warren, L. and Bray, H. (1967). On the square-freeness of Fermat and Mersenne Numbers, *Pacific J. Math.*, **V.22**, 563–564.
Wieferich, A. (1909). Zum letzen Fermatschen Theorem. *J. für Math.* **136,** 293–302.

The Inhomogeneous Minima of Some Totally Real Cubic Fields

J. R. SMITH

R.N.E.C., Plymouth, Devon, England

Let $\{1, L(\theta), Q(\theta)\}$ be an integral basis of a totally real cubic field $R(\theta)$ and $\{1, L(\phi), Q(\phi)\}$, $\{1, L(\psi), Q(\psi)\}$ the corresponding bases of the conjugate fields. L is a linear and Q a quadratic polynomial. The inhomogeneous minimum, M, of $R(\theta)$ is defined by

$$M = \sup \inf \left| \left(x + yL(\theta) + zQ(\theta)\right) \left(x + yL(\phi) + zQ(\phi)\right) \left(x + yL(\psi) + zQ(\psi)\right) \right|,$$

where the sup is taken over all real x_1, y_1, z_1, and the inf is over x, y, z congruent modulo 1 to x_1, y_1, z_1.

$R(\theta)$ is Euclidean if and only if $M < 1$ for all rational x, y, z and Godwin (these proceedings) is interested chiefly in whether or not $M < 1$ for the fields considered. In this note we investigate all non-cyclic fields (type $K(3, 0)$) of discriminant $\Delta \leqslant 1957$ and some cyclic fields (type $K(3C, 0)$) and give the actual minima where known. The method used is that of Barnes and Swinnerton-Dyer (1952) as extended by Samet (1954). A computer (Elliott 503) was used to find the "uncovered regions" discussed by Godwin (these proceedings). The results obtained concerning non-cyclic fields are given in Table I. 34 other fields were considered (those of discriminants 761, 785, 788, 837, 892, 940, 993, 1016, 1076, 1101, 1129, 1229, 1257, 1300, 1304, 1373, 1384, 1396, 1425, 1436, 1489, 1492, 1509, 1524, 1556, 1573, 1593, 1620, 1708, 1765, 1772, 1901, 1940 and 1994) and were found to be Euclidean. The results obtained concerning cyclic fields are given in Table II. Two other fields were considered (those of discriminants 3721 and 4489) and were found to be Euclidean.

Table I

Δ	M
148	$\frac{1}{2}$*
229	$\frac{1}{2}$
257	$\frac{1}{3}$
316	$\frac{1}{2}$
321	$\frac{1}{3}$
404	$\frac{1}{2}$
469	$\frac{1}{2}$
473	$\frac{1}{3}$
564	$\frac{1}{2}$
568	$\frac{1}{2}$
621	$\frac{1}{2}$
697	$\frac{13}{31}$
733	$\frac{1}{2}$
756	$\frac{1}{2}$
985	1
1345	$\frac{7}{5}$
1825	$\frac{7}{5}$
1929	1
1937	1
1957	2

*Due to Clarke (1951)

Table II

Δ	M
49	$\frac{1}{7}$*
81	$\frac{1}{3}$*
169	$\frac{5}{13}$
361	$\frac{8}{19}$
961	$\frac{16}{31}$
1369	$\frac{31}{37}$
1849	$\frac{22}{43}$
5329	$\frac{9}{8}$

*Due to Davenport (1947)

References

Barnes, E. S. and Swinnerton-Dyer, H. P. F., (1952). The inhomogeneous minima of binary quadratic forms. *Acta Math.* **87,** 259–322.

Clarke, L. E., (1951). On the product of three non-homogeneous linear forms. *Proc. Cambridge Phil. Soc.* **47,** 260–265.

Davenport, H., (1947). On the product of three non-homogeneous linear forms. *Proc. Cambridge Phil. Soc.* **43,** 137–152.

Samet, P. A., (1954). The product of non-homogeneous linear forms I. *Proc. Cambridge Phil. Soc.* **50,** 372–379.

Computations Relating to Cubic Fields

H. J. GODWIN

Royal Holloway College, Egham, Surrey, England

This paper is concerned with two problems—the existence of a Euclidean algorithm in certain algebraic fields (especially in self-conjugate cubic fields) and the tabulation of cubic fields with signature one.

We denote a field of degree n and signature s by $K(n,s)$: self-conjugate cubic fields are denoted by $K(3C,0)$. If $K(n,s)$ has integral basis $\omega_1{}^{(1)}, \ldots, \omega_n{}^{(1)}$ with conjugates $\omega_i{}^{(j)}$ (where $\omega_i{}^{(n-2s+2j)} = \bar{\omega}_i{}^{(n-2s+2j-1)}$ for $j = 1, \ldots, s$) then

$$F(\mathbf{x}) = F(x_1, \ldots, x_n) = N(x_1\,\omega_1{}^{(1)} + \ldots + x_n\,\omega_n{}^{(1)}) =$$

$$\prod_{j=1}^{n} (x_1\,\omega_1{}^{(j)} + \cdots + x_n\,\omega_n{}^{(j)})$$

is the norm-form of $K(n,s)$. We denote $\min|F(\mathbf{y})|$ taken over all vectors $\mathbf{y} \equiv \mathbf{x} \pmod 1$ by $M(\mathbf{x})$: $K(n,s)$ is Euclidean if and only if $M(\mathbf{x}) < 1$ for all rational vectors $\mathbf{x}$. If $M = \max M(\mathbf{x})$ over all vectors $\mathbf{x}$ then $K(n,s)$ is Euclidean if $M < 1$. No case is known where $\max M(\mathbf{x})$ is attained at a non-rational $\mathbf{x}$, but the impossibility of this happening has not been proved.

The number M can be given a geometrical significance as follows: for a given K, denote the region $|F(\mathbf{x}-\mathbf{m})| \leqslant K$ by $R(\mathbf{m})$ and let R be the union of the $R(\mathbf{m})$ with $\mathbf{m}$ taking all values with rational integral co-ordinates (i.e. the union of regions $|F(\mathbf{x})| \leqslant K$ with the origin translated to the points of the integral lattice). If $K < M$, R does not include the whole of space: if $K \geqslant M$, it does.

Instead of considering the whole of space it is sufficient to consider a region S, such that the union of S and its translates, when the origin is translated to the points of the integral lattice, is the whole of space. If S and its translates are disjoint then S is called a fundamental region—an example is $0 \leqslant x_i < 1$. We can then express the problem of the existence of Euclid's Algorithm as that of finding the least K for which a given fundamental region is covered by regions $R(\mathbf{m})$ centred at the points $\mathbf{m}$ of the integral lattice. For $K(2,1)$ the region $R(\mathbf{m})$ is an ellipse and the problem is easy: for all other types of field the region is unbounded and the possibility exists of a covering of the fundamental region being obtained by using parts of regions with centres remote from the fundamental region.

The present state of knowledge is as follows:

$K(2,1)$ There are 5 Euclidean fields (see, e.g. Hardy and Wright, 1938).

$K(2,0)$ There are 16 Euclidean fields (see Barnes and Swinnerton-Dyer (1952) for references to, and corrections of, earlier work: also Godwin, 1955).

$K(3,1)$ The number of Euclidean fields is finite (Cassels, 1952).

$K(3C,0)$ The number of Euclidean fields is finite (Heilbronn, 1950).

$K(3,0)$ Examples of Euclidean fields and references to earlier work are given in Godwin (1965a).

$K(4,2)$ The number of Euclidean fields is finite (Cassels, 1952).

$K(4,0)$ and $K(5,0)$ Examples of Euclidean fields are given in Godwin (1965b).

We now consider in particular the case of $K(3C,0)$. Since the existence of Euclid's Algorithm is a sufficient (but not a necessary) condition for unique factorization, we need consider only those fields with class-number one: these are known to have discriminant d^2, where d is 9 or a prime congruent to 1 modulo 6. Heilbronn (1950) shows that the number of Euclidean $K(3C,0)$ is finite by showing that for sufficiently large d we can find positive integers n and n' which are cubic residues modulo d, but which both factorize as the product of two coprime non-cubic residues, and are such that $d = n + n'$. The proof depends on the orders of remainder terms in character sums, and no explicit bound for d is given. The proof that the existence of n and n' implies that the field is non-Euclidean is as follows.

We suppose that the field is Euclidean and can then factorize d as $(\delta)^3$, where (δ) is a principal ideal. Since n is a cubic residue modulo d, there exists m such that $m^3 \equiv n (\operatorname{mod} d)$: consider x_1, x_2, x_3 such that $x_1\omega_1 + x_2\omega_2 + x_3\omega_3 = (m/\delta)$. We can find an integer θ such that $|N((m/\delta) - \theta)| < 1$, i.e. such that $|N(m - \delta\theta)| < |N(\delta)| = d$. Now $N(m - \delta\theta) \equiv m^3 (\operatorname{mod} \delta)$ and so $N(m - \delta\theta) \equiv m^3 (\operatorname{mod} d)$, i.e. $N(m - \delta\theta) \equiv n (\operatorname{mod} d)$: if $|N(m - \delta\theta)| < d$ then $N(m - \delta\theta) = n$ or $-n'$. But by their construction neither n nor $-n'$ is the norm of an integer of the field: the contradiction shows that the field is non-Euclidean. Thus, in our terminology, Heilbronn finds an $\mathbf{x}$ of particular type for which $M(\mathbf{x}) \geqslant 1$: if his criterion fails to show that a field is non-

Euclidean there may still be $\mathbf{x}$'s of other kinds giving $M(\mathbf{x}) \geqslant 1$, and this, in fact, happens in the case $d = 73$ (see Smith, 1969).

Smith has developed programs which amplify Heilbronn's result in two ways. First he has investigated all d less than 10^4, and application of Heilbronn's criterion shows that $K(3C, 0)$ is not Euclidean for $d < 10^4$ except for $d = 7, 9, 13, 19, 31, 37, 43, 61, 67$, and possibly for $d = 103, 109, 127, 157$. (The last four cases remain undecided because of the extensive computation involved.) It seems unlikely that any Euclidean field with $d > 10^4$ exists.

We now describe more fully the method used by Smith to settle the individual cases, since it is applicable in principle to fields of all types. It is essentially that of Barnes and Swinnerton-Dyer (1952) (see also Godwin, 1955), using a computer to search for covering regions.

For a given K we attempt to cover a fundamental region by portions of a number of regions $R(\mathbf{m})$. If we succeed we know that $M \leqslant K$: if we fail, leaving a set of uncovered portions of the fundamental region, then either $M > K$ or we have not considered enough covering regions. We now use the technique which Barnes and Swinnerton-Dyer used for $K(2, 0)$: we transform space by a transformation which sends lattice points into lattice points and the region $R(\mathbf{0})$ into itself. (For example, if ε is a unit of the field, we may replace $x_1 \omega_1 + \ldots + x_n \omega_n$ by $\varepsilon (x_1 \omega_1 + \ldots + x_n \omega_n)$.) The transform of a fundamental region A may be dissected into a number of pieces which can be translated to parts of A by translating the origin to various lattice points. Let S denote the uncovered part of A, and S' the uncovered part of A as reassembled after transformation. If θ does not belong to S' then, for some $\mathbf{m}_1$ and $\mathbf{m}_2$, $\theta + \mathbf{m}_1$ is covered by $TR(\mathbf{m}_2)$, the transform of $R(\mathbf{m}_2)$. But $TR(\mathbf{m}_2)$ is $R(\mathbf{m}_3)$ since the regions $R(\mathbf{m})$ are unchanged and the integral lattice is transformed into itself. Hence θ is covered by $R(\mathbf{m}_3 - \mathbf{m}_1)$ although $\mathbf{m}_3 - \mathbf{m}_1$ may not appear in the list of $\mathbf{m}$ used in the original attempt at covering. Hence the part of A which is really uncovered belongs to $S \cap S'$. We repeat this procedure, narrowing the uncovered part of A still further: since the transformation will be a contraction mapping on at least one co-ordinate, the uncovered part of A is reduced, in the limit, when different transformations and their inverses are used, to a set of isolated points. Finally, to show that $M > K$ at these isolated points, we use congruence considerations.

Smith's program for investigating the covering of a fundamental region consists of compiling a list of suitable lattice points $\mathbf{m}_1, \mathbf{m}_2 \ldots$ (suitable because at least one co-ordinate lies within the range of those in A), subdividing A and testing if a given subdivision lies in one of the $R(\mathbf{m}_i)$. If a subdivision is not covered it is further subdivided and the process repeated on each of these portions. Some economy of effort is obtained by trying first for any given subdivision the $R(\mathbf{m})$ which covered a neighbouring one. After a certain number of repetitions either the whole of A has been shown to be covered (so that M is known to be not greater than K), or a list of small uncovered regions has been obtained: these are then examined individually by using transformations of space as described above. If $M(\mathbf{x})$ is either M, at a few points, or is appreciably less than M (i.e. if the first minimum of the norm-form is well isolated) then the procedure is not too tedious: if there are values of $M(\mathbf{x})$ close to M at many points, then a large list of uncovered portions of A is obtained, and the work becomes prohibitive.

Smith has also applied this method to $K(3, 0)$ and found a number of Euclidean fields additional to those in Godwin (1965a), and the non-Euclidean field with discriminant 985; this has $M = 1$ and so is "just non-Euclidean" (as is the $K(2, 0)$ with discriminant 65).

The second problem discussed here is the tabulation of $K(3, 1)$, i.e. the determination of all such fields with discriminant greater than some given value. I. O. Angell, in work currently in progress for the degree of Ph.D. of London University, has tabulated the fields with discriminant greater than $-20{,}000$, and found that there are 3169 of them. This number includes 27 pairs, 58 triads, and 22 tetrads of fields with the same discriminant. (In the case of $K(3, 0)$, Godwin and Samet (1959) found that there are 830 fields with discriminant less than 20,000: these include just four pairs (all $K(3C, 0)$) with the same discriminants.)

The method used is based on Godwin (1957): every field with discriminant D can be generated by a number α, with conjugates $\beta + i\gamma$, such that $(\alpha - \beta)^2 + 3\gamma^2 < \sqrt{|D|}$. We may suppose that $0 < \alpha < 1$ and thus obtain bounds for β, γ, and for the coefficients of the cubic having $\alpha, \beta \pm i\gamma$ as zeros. Having enumerated these cubics we can test that they give fields of the required type, and with discriminants greater than $-20{,}000$.

In some cases, particularly with small values of $|D|$, the same field is given by a number of different polynomials: if and only if the polynomials whose respective zeros are $\alpha, \beta \pm i\gamma$ and $\alpha', \beta' \pm i\gamma'$ give the same field, then there exist rational integers r, s, t, n such that $\alpha' = (r\alpha^2 + s\alpha + t)/n$, with similar equations for the complex roots. n is known from the indices of α and α', and r, s, t can be calculated approximately from the approximate values of α, α', etc. If the values of r, s, t so found are really integers, then one can test that the transformation

$$\alpha' = \frac{r\alpha^2 + s\alpha + t}{n}$$

(using integral values for r, s, t) does really send one cubic into the other. If r, s, t are not approximately integers then the fields are apparently distinct, and this may be checked, e.g. by factorizing various primes in the fields and finding differences between the ways in which they factorize.

Angell is now finding the fundamental units of the fields which he has tabulated, by a program based on Voronoi's algorithm (Delone and Faddeev (1940) pp. 215–220: see also Delone (1923)). This constructs, in a systematic way, a chain of transformations of a cubic lattice which eventually recurs, thus giving rise to a transformation of the lattice into itself, which corresponds to a unit of the field. Angell has also adapted a method, given in Godwin (1960) for $K(3, 0)$, by proving that, if $D < -86$, then the unit $(\xi, \eta \pm i\zeta)$, for which $(\xi - \eta)^2 + \zeta^2$ is least, is the fundamental unit (where we choose between ξ and $(1/\xi)$ so that $|\xi| < 1$). In many cases units can be found readily by inspection, and this result enables one to check fairly easily whether or not they are fundamental.

References

Barnes, E. S. and Swinnerton-Dyer, H. P. F., (1952). The inhomogeneous minima of binary quadratic forms (I). *Acta Math.* **87,** 259–323.

Cassels, J. W. S., (1952). The inhomogeneous minima of binary quadratic, ternary cubic and quaternary quartic forms. *Proc. Cambridge Phil. Soc.* **48,** 72–86 and 519–520.

Delone, B. N., (1923). Interprétation géométrique de la généralisation de l'algorithme des fractions continues donnée par Voronoi. *Comptes Rendus.* **176,** 554–556.

Delone, B. N. and Faddeev, D. K., (1940). Theory of irrationalities of the third degree. *Acad. Sci. U.R.S.S. Trav. Inst. Math. Stekloff,* **11.**

Godwin, H. J., (1955). On the inhomogeneous minima of certain norm-forms. *J. London Math. Soc.* **30,** 114–119.

Godwin, H. J., (1957). On quartic fields of signature one with small discriminant. *Quart. J. of Math.* (*Oxford*) (2) **8,** 214–222.

Godwin, H. J., (1960). The determination of units in totally real cubic fields. *Proc. Cambridge Phil. Soc.* **56,** 318–321.

Godwin, H. J., (1965a). On the inhomogeneous minima of totally real cubic norm-forms. *J. London Math. Soc.* **40,** 623–627.

Godwin, H. J., (1965b). On Euclid's algorithm in some quartic and quintic fields. *J. London Math. Soc.* **40,** 699–704.

Godwin, H. J. and Samet, P. A., (1959). A table of real cubic fields. *J. London Math. Soc.* **34,** 108–110.

Hardy, G. H. and Wright, E. M., (1938). "An introduction to the theory of numbers", Oxford.

Heilbronn, H., (1850). On Euclid's algorithm in cubic self-conjugate fields. *Proc. Cambridge Phil. Soc.* **46,** 377–382.

Smith, J. R., (1969). On Euclid's algorithm in some cyclic cubic fields. *J. London Math. Soc.* **44,** 577–582.

The Products of Three and of Four Linear Forms

H. P. F. SWINNERTON-DYER

Cambridge University, England

The work described here was begun as joint work with the late Professor Davenport, and is not yet complete. I hope that a full account of it will appear, under both our names, in the forthcoming volume of *Acta Arithmetica* dedicated to his memory.

The problem of the homogeneous product of n linear forms (in the Geometry of Numbers) can be stated in two natural ways; they are clearly equivalent, but each of them has its advantages:—

(i) *What can be said about the lattices admissible for the region* $|x_1 x_2 \dots x_n| < 1$*—that is, lattices* Λ *with no lattice point other than the origin in the region?*

(ii) *What can be said about the sets of* n *linear forms*

$$L_i = a_{i1} x_1 + a_{i2} x_2 + \dots + a_{in} x_n$$

such that $|L_1 L_2 \dots L_n| \geqslant 1$ *for all sets of rational integers* $x_1, \dots, x_n$ *not all zero?*

Let $\Delta = |\det \Lambda| = |\det(a_{ij})|$; then the most obvious problem is to find the lattice constant of the region $|x_1 x_2 \dots x_n| < 1$, in other words to find $\inf \Delta$ as Λ runs through all admissible lattices. More generally one can consider sets of n linear forms L_i such that

$$\alpha = \inf' |L_1 L_2 \dots L_n| > 0; \tag{1}$$

one then asks what are the possible values of $\Delta\alpha^{-1}$, and what can be said about the corresponding lattices. There is some scope for normalization, for we can make an arbitrary integral unimodular transformation on the x_i without altering $\Delta\alpha^{-1}$; and we can also multiply the L_i by arbitrary non-zero constants. Thus in the case where the infimum in (1) is attained—and this turns out to be the interesting case—one can normalize so that $\alpha = 1$ and is attained at $(1, 0, 0, \dots, 0)$, and so each $a_{i1} = 1$.

One reason for investigating this problem is its use for determining class numbers of algebraic number fields:

THEOREM. *Let K be a totally real algebraic number field, of degree n over the rationals, and let d be the discriminant of K. Then every ideal class in K contains an integral ideal of norm at most $C^{-1} d^{\frac{1}{2}}$, where C is the lattice constant of* $|x_1 x_2 \ldots x_n| < 1$.

Except in special cases, one actually finds class numbers in this way, and this involves factorizing all rational primes up to $C^{-1} d^{\frac{1}{2}}$ in K. If one could replace C by a larger constant C_1 this would diminish the labour involved in the process; and this can be done provided one has detailed enough information about the lattices with $\Delta\alpha^{-1} < C_1$—or, which is the same thing, about the admissible lattices with determinant less than C_1.

Conversely, such fields give a means of obtaining admissible lattices. For let K be a totally real algebraic number field of degree n, and let

$$\alpha_1 = 1, \quad \alpha_2, \ldots, \alpha_n$$

be elements linearly independent of K over the rationals. Let

$$L_1 = x_1 + \alpha_2 x_2 + \ldots + \alpha_n x_n$$

and let $L_2, \ldots, L_n$ be the algebraic conjugates of L_1; then

$$L_1 L_2 \ldots L_n = \mathrm{Norm}_{K/Q}(L_1)$$

is a polynomial in the x_i with rational coefficients. Hence it cannot take arbitrarily small non-zero values for integral x_i, because it is rational with bounded denominator; and it cannot take the value zero because no L_i can do so, by the linear independence of the α_i. Hence $\alpha > 0$ in (1) and is attained. The corresponding lattice can be called a norm-form lattice.

For $n = 2$ the theory is essentially complete, and is due to Markoff; for a full account see for example Cassels, "Introduction to Diophantine Approximation". The possible values of $\Delta\alpha^{-1}$ are

$$\sqrt{5}, \sqrt{8}, \tfrac{1}{5}\sqrt{221} = 2{\cdot}97321 \ldots \text{ and so on;}$$

the smallest limit point of values of $\Delta\alpha^{-1}$ is 3. All the values below 3 arise from norm-form lattices, but this is not true for 3; an example to the contrary is

$$L_1 = x_1 + \tfrac{1}{2}(3 + \sqrt{5})\, x_2, \qquad L_2 = x_1 + \tfrac{1}{2}(-3 + \sqrt{5})\, x_2 \tag{2}$$

for which the minimum is $\alpha = 1$, attained only at $(1, 0)$. There are uncountably many essentially different lattices with $\Delta\alpha^{-1} = 3$.

For $n = 3$ Davenport proved that the two smallest values of $\Delta\alpha^{-1}$ are 7 and 9, arising from norm-forms in the cubic fields of discriminant 49 and 81 respectively. There is no machinery analogous to the continued fraction algorithm on which Markoff's work depends; and no admissible lattices other than norm-form ones are known. It is certainly impossible to construct them by analogy with (2), for Cassels and Swinnerton-Dyer have shown that if L_1 is defined over a totally real cubic field and $\alpha \neq 0$ then L_2 and L_3 must be multiples of the algebraic conjugates of L_1. This raises interesting problems, the more so because if all admissible lattices are of norm-form type then the values of $\Delta\alpha^{-1}$ can have no finite limit point; and then $C^{-1}\, d^{\frac{1}{2}}$ in the Theorem above could be replaced by some $o(d^{\frac{1}{2}})$. This is a matter on which algebraic number theory computations could be expected to shed some light.

As theoretical attacks on these problems seem to yield very little, it seemed worthwhile to find the next few values of $\Delta\alpha^{-1}$ computationally. In its crudest form, this represents a search through a region of nine-dimensional space (given by the values of the a_{ij}); and the method falls into two parts:—

(i) Reject almost all the region, by examining the values of $L_1\, L_2\, L_3$ for particular sets of integers x_1, x_2, x_3.

(ii) If Λ has been shown to be close to an identified admissible lattice Λ_0, show that if Λ is admissible it must be equivalent to Λ_0.

An isolation theorem of the sort described in (ii) is essential, and it has to be provided in a constructive form by the theorists; for the numerical calculations are necessarily only approximate and so one must have a way of refining approximate to exact answers. Fortunately Cassels and Swinnerton-Dyer have given a strong isolation theorem for the case when Λ_0 is a norm-form lattice, and this meets our needs. (The isolation theorem is much stronger than anything that is true when $n = 2$, which is one reason for believing that the cases $n = 2$ and $n = 3$ behave quite differently.)

It remains to consider (i), and the first step is to normalize the L_i. We do not know in advance that α is attained; but we can at least assume that $L_1\, L_2\, L_3$ is very close to α at $(1, 0, 0)$ and the most convenient form of this statement is

$$1-10^{-12} < \alpha \leqslant 1, \qquad a_{11} = a_{21} = a_{31} = 1.$$

Given this, we can require that a_{12} and a_{13} lie in the interval $(0, 1)$ by replacing x_1 by $x_1 + m_2\, x_2 + m_3\, x_3$ for some integers m_2, m_3; and we can then make an integral unimodular transformation on x_2, x_3 only to ensure that the quadratic form

$$\tfrac{1}{2}\{(L_1 - L_2)^2 + (L_2 - L_3)^2 + (L_3 - L_1)^2\} = Ax_2^{\,2} + 2Bx_2\, x_3 + Cx_3^{\,2}$$

of determinant $AC - B^2 = \frac{3}{4}\Delta^2$ is reduced. This means at any rate

$$A \leqslant \Delta, \qquad AC \leqslant \Delta^2. \tag{3}$$

But A cannot be too small, because if A is small then a_{12}, a_{22} and a_{32} are close together and we can make $|L_1 L_2 L_3| < \alpha$ by choosing $x_1 = -1$ or $0, x_2 = 1, x_3 = 0$. A more detailed argument along these lines gives $A > 7 - 10^{-10}$; in fact if $\alpha = 1$ then $A \geqslant 7$ and this is best possible. After (3), C is now bounded above by a bound depending on Δ; and so all the a_{ij} are bounded. This is the only use that is made of the reduction of the quadratic form.

Obviously none of the a_{ij} is a rational integer, because none of the L_i can vanish for integral x_j. So at this stage we divide the problem into cases, a "case" being identified by the specification of the six integer parts $[a_{ij}]$; if an upper bound for Δ is chosen, this restricts us to finitely many cases. (Of course there are trivial equivalences among cases, which reduces the work considerably.) At a later stage of the calculation it is necessary to split cases up into subcases, in which one or more of the a_{ij} is restricted to lie in an interval of length 1/12; this splitting is done automatically by the computer during the calculation described below, but it must not be done too often because one cannot afford to have too many cases to deal with.

While a case or subcase is being dealt with, the current information about the a_{ij} is described by three convex polygons P_i such that (a_{i2}, a_{i3}) lies in P_i. At the start of a case, each P_i is a unit square; the object of the calculation is to whittle this down to nothing or at least to a very small region. A typical step is as follows, where to simplify the notation we assume that we are trying to reduce P_1. Choose integers x_1, x_2, x_3 not all zero; this gives upper bounds for the corresponding values of $|L_2|$ and $|L_3|$, and these upper bounds are easy to calculate since they are attained at well-determined vertices of P_2 and P_3 respectively. Since

$$|L_1 L_2 L_3| \geqslant \alpha > 1 - 10^{-12},$$

this gives a lower bound for $|L_1|$ and therefore a certain strip S in which (a_{12}, a_{13}) cannot lie. This may enable one to reduce P_1; for example, if P_1 is the square in the figure below, then P_1 is diminished if S is S', but not if S is S'' or S'''. However, if S'' can be made a little larger because contractions of P_2 or P_3 have given smaller upper bounds for $|L_2|$ or $|L_3|$, or if a small piece at the bottom right hand corner of P_1 can be removed for other reasons, then S'' becomes useful and leads to a substantial reduction of P_1.

The computer is programmed to choose a finite cyclic list of triples (x_1, x_2, x_3) and carry out this step with each triple and each L_i in turn, going on until working once through the list gives no substantial improvement in any of the P_i. If the resulting regions are very small (less than 10^{-6} in each direction) they are printed out and the resulting lattice (which has always turned out to be a norm-form lattice) identified and isolated by hand. If the regions are not small, they are subdivided and a new list of triples (x_1, x_2, x_3) chosen; the process is then repeated. The triples are chosen so that the associated strips are not too narrow; this means that either x_1, x_2, x_3 are small or else two of the L_i are nearly equal—so that x_2/x_3 must be an approximation to one of three real numbers. The smaller the P_i, the larger the x_j can usefully be; this is why the list of triples is rechosen after each subdivision.

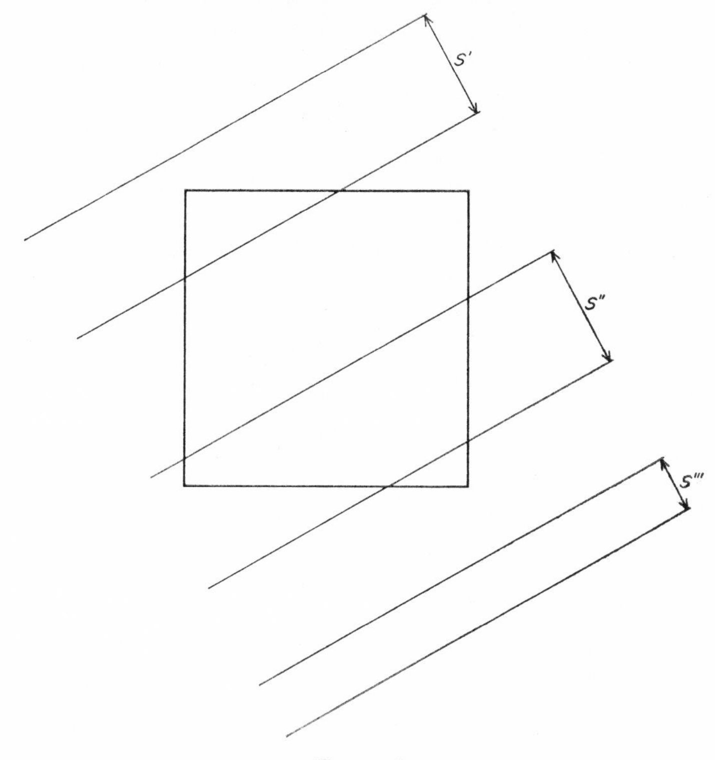

FIGURE 1

These calculations have been carried out for all possible $\Delta\alpha^{-1} \leqslant 17$, and subject to machine errors the permissible values of $\Delta\alpha^{-1}$ are

$$7, 9, \sqrt{148}, 63/5, 13, 14, 351/25, 189/13, 133/9, \sqrt{229}, 259/17,$$
$$559/35, \sqrt{257}, 273/17, \tfrac{1}{2}\sqrt{1078}, 117/7, \tfrac{1}{9}\sqrt{22736}, \tfrac{1}{3}\sqrt{2597}.$$

Each of these values comes essentially from only one lattice, except that 13 comes both from the lattice of integers of the real cubic field with $d = 13^2$ and from a non-principal ideal in one of the two real cubic fields with $d = 91^2$. The values of $\Delta\alpha^{-1}$ do not seem to be tending to a finite limit; and this impression is very much reinforced by looking at the associated lattices themselves. I doubt if there is much to be gained by further calculations.

The case $n = 4$ is of interest, and is just about within the bounds of what is feasible. Not even the minimum value of $\Delta\alpha^{-1}$ is known, though it is generally conjectured that it is $\sqrt{725}$ and corresponds to the lattice of integers of the totally real quartic field of least discriminant. Calculations are under way, and should be completed within a few months. The method is in principle the same as in the case $n = 3$, but there is one additional technical difficulty. In the case $n = 3$ it is easy to describe the polygons P_i and manipulate them; for the sides of P_i have a natural ordering and the vertices of P_i are just the intersections of two consecutive sides under this ordering. In the case $n = 4$, however, the convex regions involved are polyhedra in three dimensions; there is no natural ordering of the faces, and there are no simple and efficient algorithms for going from the set of faces to the set of vertices, for example. The techniques of linear programming are not very helpful. This sort of problem seems to turn up in a good many contexts (see for example Larmouth's paper, p.827, where some exceptionally unhandleable convex polyhedra occur), and some good algorithms would be very valuable.

The Enumeration of Perfect Forms

J. LARMOUTH

Computer Laboratory University of Cambridge, England

This paper describes an attempt which was made to automate the production of all perfect forms in a given number of variables, using the procedure described by Voronoi (1908), and following the work of Barnes (1957).

We are working with quadratic forms $\mathbf{x}' A \mathbf{x}$ in the n variables $x_1, x_2, \ldots, x_n$, and are concerned only with positive definite forms. Let the minimum of the form as $\mathbf{x}$ ranges over all integer values be M, say.

$$M = \min_{\mathbf{x} \neq \mathbf{0} \in \mathbf{Z}} \mathbf{x}' A \mathbf{x}$$

Then if the value M is taken at the points $\pm \mathbf{m}_k$, $k = 1, 2, \ldots, s$, we say that the form is *perfect* if it is uniquely determined by M and the s minimal vectors $\mathbf{m}_k$.

We wish to discover representatives of each equivalence class of perfect forms for $n \geqslant 7$. Barnes (1957) completely solved the problem for $n = 6$ working by hand, and using the approach described by Voronoi (1908). This method forms a correspondence between a perfect form and a region R in $\frac{1}{2}n(n+1)$ space. The region R is defined by s edge forms, and the procedure depends on finding all the faces of this region, and from them *neighbouring* perfect forms.

Throughout the work equivalent faces and equivalent forms must be rejected. Two forms A and B are equivalent if either one is a multiple of the other, or if there exists a transformation T (with integer elements and determinant $+1$) such that

$$T' A T = B$$

We see that the minima of A and B are the same, and that the minimal vectors are related by T. It follows that if we form the distance matrix with (i, j)th element

$$\mathbf{m}_i A \mathbf{m}_j$$

Table I

New forms for $n = 7$

s	$Det(A)$, $M=2$	Form
28	392/81	$3\phi - x_1 x_2 - 2(x_3 x_4 + x_3 x_5 + x_4 x_6 + x_5 x_6)$
28	71344/15625	$5\phi - 3x_1 x_2 - 2(x_3 x_4 + x_3 x_5 + x_3 x_6 + x_4 x_5 + x_4 x_6 + x_5 x_6)$
28	10240/2187	$3\phi - 2x_1 x_2 - (x_3 x_4 + x_3 x_5 + x_3 x_6 + x_3 x_7 + x_4 x_5 + x_4 x_7 + x_5 x_6 + x_5 x_7 + x_6 x_7)$
28	3584/729	$3\phi - x_1 x_2 - x_4 x_5 - 2(x_3 x_4 + x_3 x_5 + x_4 x_6 + x_5 x_7)$
32	1024/243	$3\phi - x_1 x_2 + x_2(x_2 + x_3 + x_4 + x_5 + x_6 + x_7) - (x_3 x_4 + x_3 x_5 + x_3 x_6 + x_4 x_5 + x_4 x_6 + x_5 x_6)$
29	10336/2187	$3\phi - x_1 x_2 - x_4 x_5 - x_3 x_7 - 2(x_3 x_4 + x_3 x_5 + x_4 x_6 + x_5 x_7)$

Here $\phi = \Sigma\Sigma x_i x_j$ and the summation is over all $i \geqslant j$.

Table II

New forms for $n = 8$

s	$Det(A)$, $M=2$	Form
36	2244/625	$5\phi - x_1 x_2 - 4(x_3 x_4 + x_3 x_5 + x_3 x_6)$
38	112/27	$3\phi - x_1 x_2 - 2(x_3 x_4 + x_3 x_5 + x_4 x_6 + x_5 x_6)$
37	345/64	$2\phi - x_1 x_2 - (x_3 x_4 + x_3 x_5 + x_3 x_6 + x_4 x_5 + x_4 x_7)$
36	3248/729	$3\phi - x_1 x_2 - 2(x_3 x_4 + x_3 x_5 + x_3 x_6 + x_4 x_7 + x_4 x_8)$
36	1449/256	$2\phi - x_1 x_2 - (x_3 x_4 + x_3 x_5 + x_3 x_6 + x_4 x_7 + x_5 x_8)$
36	370048/78125	$5\phi - 3x_1 x_2 - 2(x_3 x_4 + x_3 x_5 + x_4 x_5 + x_6 x_7 + x_6 x_8)$
38	65856/15625	$5\phi - 3x_1 x_2 - 2(x_3 x_4 + x_3 x_5 + x_3 x_6 + x_4 x_5 + x_4 x_6 + x_5 x_6)$
36	354368/78125	$5\phi - 3x_1 x_2 - 2(x_3 x_4 + x_3 x_5 + x_3 x_6 + x_4 x_5 + x_4 x_6 + x_7 x_8)$
36	1425/256	$2\phi - x_1 x_2 - (x_3 x_4 + x_3 x_5 + x_3 x_6 + x_4 x_5 + x_4 x_7 + x_5 x_8)$
39	1209/256	$2\phi - x_1 x_2 - (x_3 x_4 + x_3 x_5 + x_3 x_6 + x_4 x_5 + x_4 x_7 + x_6 x_8)$

Here $\phi = \Sigma\Sigma x_i x_j$ and the summation is over all $i \geqslant j$.

then we can test for equivalence by looking for a permutation of the distance matrix of one form into that of the other.

The task of finding the faces of the region R can be greatly assisted by using the techniques of linear programming.

A computer program (about 8000 orders, hand coded) was written to completely automate Voronoi's algorithm. Three different approaches were tried to the problem of generating all neighbours, but none proved sufficiently fast.

The first approach was to build up a face one edge-form at a time. The second was to build up a set of vectors lying *off* a face, and the third was a recursive technique which found all faces by choosing one face and finding all its subfaces, then generating all the neighbouring faces.

Although the work has not succeeded in producing a complete enumeration for $n = 7$, I have obtained a large number of inequivalent forms for $n = 7$ and $n = 8$. I list in table I 6 forms for $n = 7$, and in table II 10 forms for $n = 8$. I believe that these forms are not equivalent to any published forms.

A set of standard Fortran routines have been written to duplicate that part of the above work involved in verifying that two forms are inequivalent, and these are being used to maintain a tape of inequivalent forms.

References

Barnes, E. S., (1957). The complete enumeration of extreme senary forms. *Phil. Trans. Roy. Soc.* **A249,** 461–506.

Voronoi, G., (1908). Sur quelques propriétés des formes quadratiques positives parfaites. *J. reine angew. Math.* **133,** 97–178.

Two Problems in Diophantine Approximation

A. M. COHEN

University of Wales Institute of Science and Technology, Cardiff, Wales

1. The first problem concerns the estimation of lattice constants of three dimensional star bodies, which has been discussed in Cohen (1962). Briefly, it is as follows:—

Let S be a given star body having equation

$$f(\mathbf{x}) \leqslant 1,$$

and let **a**, **b** and **c** be three points on the boundary of S, i.e.,

$$f(\mathbf{a}) = f(\mathbf{b}) = f(\mathbf{c}) = 1.$$

Then

$$[l, m, n] = l\mathbf{a} + m\mathbf{b} + n\mathbf{c}, \qquad l, m, n \text{ integers},$$

defines a point on a lattice, Λ, with basis **a**, **b**, **c**. If $[l, m, n]$ lies outside or on S for all $(l, m, n) \neq (0, 0, 0)$ then we say that the lattice Λ is S–admissible and has lattice determinant

$$d(\Lambda) = \det(\mathbf{a}, \mathbf{b}, \mathbf{c}).$$

The problem of finding the lattice constant of S, $\Delta(S)$, is equivalent to finding

$$\min d(\Lambda) \qquad (\text{all admissible } \Lambda).$$

The problem is thus one of optimization and we can use the relevant techniques of numerical analysis, such as steepest descent, to obtain estimates for $\Delta(S)$.

One difficulty encountered is that of admissibility. For a finite star body, e.g. a convex body, we have only to test the admissibility of a limited number of lattice points but, for an infinite body such as $|xyz| \leqslant 1$, we require to search through all possible l, m, n. This is clearly impracticable. However, the difficulty can be overcome if we are able to assume that the star body is boundedly reducible for we know that in this case there exists a bounded body, K, such that

$$\Delta(S) = \Delta(S \cap K).$$

By suitable choice of K we can estimate $\Delta(S \cap K)$ and hence $\Delta(S)$. This method has been successfully employed, with K as the sphere $x^2 + y^2 + z^2 \leqslant r^2$, to generate critical lattices (by computer) for the bodies $|xyz| \leqslant 1$ and $|x|(y^2 + z^2) \leqslant 1$ with determinants 7 and $\sqrt{23}/2$ respectively (see Cassels (1959)).

We now give some results which arose as a result of numerical computation.†

(a) The convex body S: $|x|^3 + |y|^3 + |z|^3 \leqslant 1$.

For this body the lattice Λ with basis

$$\mathbf{a} = 2^{-1/3}(1, 1, 0)$$
$$\mathbf{b} = 2^{-1/3}(1, 0, -1)$$
$$\mathbf{c} = 2^{-1/3}(1 + \theta, -\theta, \theta)$$

is admissible, where θ satisfies the cubic equation

$$\theta^3 + \theta^2 + \theta - \tfrac{1}{3} = 0.$$

Further,

$$d(\Lambda) = \tfrac{1}{2}(1 + 3\theta) = 0.8797.$$

It follows that

$$\Delta(S) \leqslant 0.8797,$$

and the author hopes to establish that equality holds in this last result. It may be remarked that we know from Minkowski's convex body theorem that $\Delta(S) \geqslant 0.7075$.

(b) The body S: $|x| \max(y^2, z^2) \leqslant 1$.

The lattice constant of this body is of interest in connection with another problem. For, Davenport and Mahler (1946) showed that if C_2 is the lower bound of all numbers C with the property that every pair of real numbers θ_1, θ_2 admit infinitely many approximations satisfying

$$|p_1 - \theta_1 q| < (C/q)^{\frac{1}{2}}, \qquad |p_2 - \theta_2 q| < (C/q)^{\frac{1}{2}} \qquad (p_1, p_2, q \text{ integers})$$

then

$$C_2 = 1/\Delta(S).$$

† Bigg (1963) and Spohn (1969) give results relating to other bodies.

The best theoretical results obtained for $\Delta(S)$ are those of Cassels (1955) and Davenport (1952) who show respectively that

$$\Delta(S) \leqslant 3.5,$$

and

$$\Delta(S) \geqslant 46^{1/4} \doteqdot 2.6043.$$

Thus computational investigations are of interest in finding out more about $\Delta(S)$. The best numerical result obtained was that the lattice Λ with basis

$$\mathbf{a} = (0.8614,\ 1.0774,\ -0.0059),$$
$$\mathbf{b} = (1.5334,\ -0.1425,\ -0.8076),$$
$$\mathbf{c} = (1.1043,\ -0.9516,\ 0.7191),$$

and

$$d(\Lambda) = 2.8915,$$

was an admissible lattice for $S \cap K$, where K is the body $K: x^2 + y^2 + z^2 \leqslant 9.57$. Since all the nine points lying on $S \cap K$ lie on S it is tempting to conjecture that $\Delta(S) \leqslant 2.8915$. However, a closer examination of the data reveals that the point $[1, -2, 2]$ lies on $x^2 + y^2 + z^2 = 9.5794$ and well inside S. Thus the conjecture does not appear to have a solid foundation.

The same technique can also be used to estimate the lattice constants of four and higher dimensional bodies and the following problems are of interest.

(i) S_4: $|x_1 x_2 x_3 x_4| \leqslant 1$. What is $\Delta(S_4)$?
(ii) S: $x_1{}^2 + x_2{}^2 \leqslant 1$, $x_3{}^2 + x_4{}^2 \leqslant 1$. Is $\Delta(S) = \frac{3}{4}$? †
(iii) S: $|xyz(x + y + z)| \leqslant 1$. What is $\Delta(S)$ and are there more than three points of the critical lattice on the boundary of S ?

We note that, for (i), Noordzij (1967) has shown that

$$\sqrt{500} < \Delta(S_4) \leqslant \sqrt{725}.$$

An admissible lattice for this body has the basis

$$\mathbf{x}_r = (\theta_1{}^{r-1}, \dots, \theta_4{}^{r-1}), \qquad r = 1, 2, 3, 4,$$

† Dr. T. G. Murphy, Trinity College, Dublin, has found the answer to this question to be in the affirmative. The minimal determinant is attained by the product of two lattices giving the minimal determinants of the two circles. However, there do exist other lattices with the same determinant not expressible as products.

where $\theta_1, \theta_2, \theta_3, \theta_4$ are the roots of the polynomial equation

$$x^4 - x^3 - 3x^2 + x + 1 = 0.$$

This fact, and other results for lower dimensions, suggests the following approach for obtaining an upper bound for the lattice constant of $S_n: |x_1 x_2 \dots x_n| \leqslant 1$, namely, the determination for each n of an irreducible equation of the form

$$x^n + \alpha_1 x^{n-1} + \dots + \alpha_{n-1} x \pm 1 = 0 \qquad (\alpha_i \text{ integers})$$

having all the roots $\theta_1, \dots, \theta_n$ real and distinct and such that $|\Pi(\theta_1 - \theta_j)|, i < j$, is a minimum. A lattice Λ with basis points $\mathbf{x}_r = (\theta_1^{r-1}, \dots, \theta_n^{r-1})$ $(r = 1, \dots, n)$ is then S_n–admissible and

$$\Delta(S_n) \leqslant \min \left| \prod_{i<j} (\theta_i - \theta_j) \right|.$$

Thus for $n = 5$ we find

$$x^5 + 2x^4 - 5x^3 - 2x^2 + 4x - 1 = 0$$

is an irreducible equation having roots -3.2287, -1.0882, $.3728$, $.5462$, 1.3979 and consequently

$$\Delta(S_5) \leqslant \sqrt{14641} = 121.\dagger$$

In the other direction Godwin (1950) has shown that $\Delta(S_5) \geqslant 57.02$.

2. The second problem consists of estimating the constants γ_1 and γ_2 for which there exists a linear form $l\theta + m\phi + n$ such that

$$\max (l^2, m^2)|l\theta + m\phi + n| \geqslant \gamma_1,$$

and

$$(l^2 + m^2)|l\theta + m\phi + n| \geqslant \gamma_2,$$

for all $(l, m) \neq (0, 0)$. We know from a theorem in Cassels (1957) that these constants and linear forms exist and we give here a method to show that

$$\gamma_1 \doteqdot 0.2021, \qquad \gamma_2 \doteqdot 0.2807.$$

These values are probably best possible and are attained with $\theta \doteqdot 0.4566$, $\theta \doteqdot 0.2021$ and $\theta \doteqdot 0.4470$, $\phi \doteqdot 0.3067$ respectively.

† Hunter (1957) shows that the least discriminant of a totally real quintic field is 14641, so that the value 121 given here is the best possible upper bound by this method.

The method of evaluation is as follows. We first observe that we need only consider θ and ϕ satisfying $0 \leqslant \phi \leqslant \theta \leqslant \frac{1}{2}$. We shall call this the region A of the θ, ϕ plane. Next, we note that by suitable choice of the l, m, n we can find three strips

$$f(l_1, m_1)|l_1 \theta + m_1 \phi + n_1| \leqslant \delta, \quad (1)$$
$$f(l_2, m_2)|l_2 \theta + m_2 \phi + n_2| \leqslant \delta, \quad (2)$$
$$f(l_3, m_3)|l_3 \theta + m_3 \phi + n_3| \leqslant \delta, \quad (3)$$

which just cover A where $f(l, m)$ denotes $l^2 + m^2$ or $\max(l^2, m^2)$. For instance, with $\delta = \delta_1 = 4/13$, $f(l, m) = l^2 + m^2$ the strips

$$|\phi| \leqslant \delta_1, \qquad 2|\theta - \phi| \leqslant \delta_1, \qquad 4|-2\theta + 1| \leqslant \delta_1,$$

just cover A (figure 1). Clearly if δ is reduced below δ_1 then a gap will appear

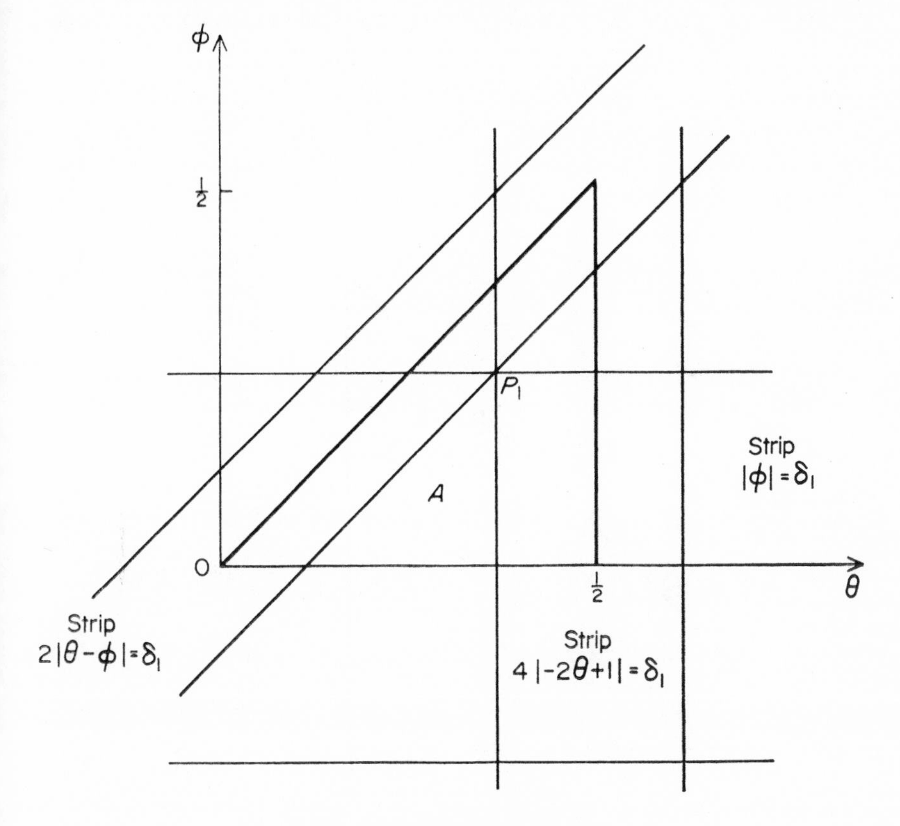

FIG. 1. Covering of the region A

in the region of the point P_1 (θ_1, ϕ_1) and A will no longer be covered. We next search for integers l^*, m^*, n^* which are as small as possible and which satisfy

$$f(l^*, m^*)|l^*\theta_1 + m^*\phi_1 + n^*| < \delta_1.$$

These integers can always be found since, by their construction, θ_1 and ϕ_1 are always rational. Now consider the intersection of the strips

$$f(l^*, m^*)|l^*\theta + m^*\phi + n^*| \leqslant \delta, \quad (4)$$

$$f(l_1, m_1)|l_1\theta + m_1\phi + n_1| \leqslant \delta, \quad (1)$$

$$f(l_2, m_2)|l_2\theta + m_2\phi + n_2| \leqslant \delta. \quad (2)$$

It can be shown—see Cohen (1963)—that these will cover the region A for some value of $\delta < \delta_1$, call it δ_2. Similarly, by combining (4) with (1) and (3) or with (2) and (3) we can determine two other numbers δ_3 and δ_4 which are less than δ (figure 2). We choose the largest δ held in the computer store

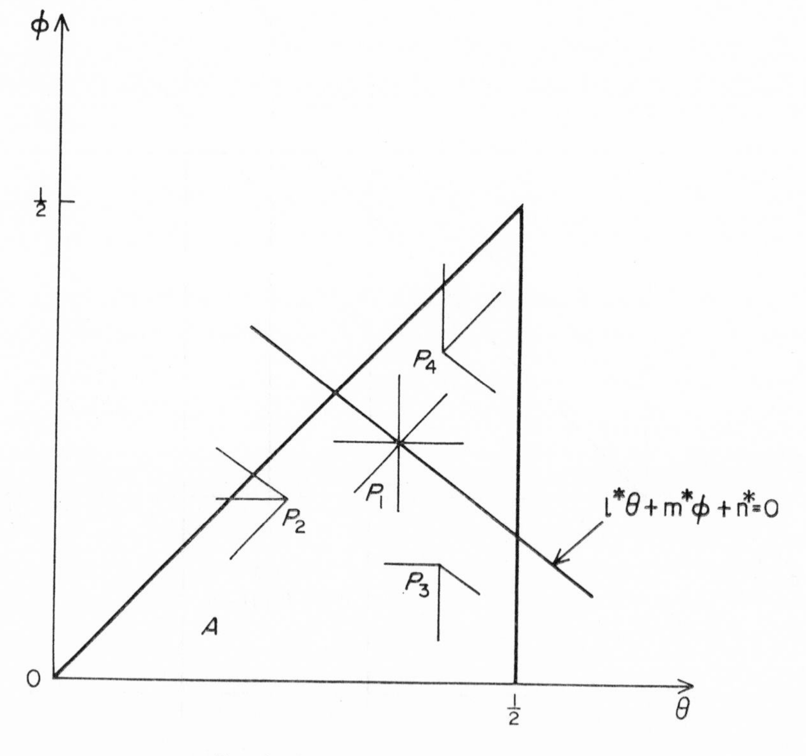

FIG. 2. Generations of three new strips

and repeat the process, terminating when the δ's appear to converge. This is the case when large l, m, n appear in the solution or, alternatively, when θ and ϕ become restricted to an extremely small region of A. Note that by taking the initial three strips parallel to to the edges of A we ensure all θ and ϕ belong to A. At the end of the process we have another problem on our hands, namely, to discover more about the computed numbers θ, ϕ and δ. The author does not wish to say anything about this, except that θ and ϕ cannot both belong to quadratic fields, and that this is another problem in which computers can assist.

Thanks are due to Professor K. Mahler, F.R.S., and the late Dr. C. B. Haselgrove for respectively suggesting the problems in sections 1 and 2. I am also grateful for constructive comments delivered during the discussion of this paper.

References

Bigg, M. D., (1963). *Proc. Cambridge Phil. Soc.* **59,** 523–530.

Cassels, J. W. S., (1955). Simultaneous Diophantine Approximation. *J. London Math. Soc.* **30,** 119–121.

Cassels, J. W. S., (1957). "An Introduction to Diophantine Approximation". Cambridge tract No. 45.

Cassels, J. W. S., (1959). "An Introduction to the Geometry of Numbers". Springer Verlag, Berlin.

Cohen, A. M., (1962). The Numerical Determination of Lattice Constants. *J. London Math. Soc.* **37,** 185–8.

Cohen, A. M., (1963). Ph.D. Thesis, University of Manchester.

Davenport, H., (1952). Simultaneous Diophantine Approximation. *Proc. London Math. Soc.* (3) **2,** 406–416.

Davenport, H. and Mahler, K., (1946). Simultaneous Diophantine Approximation. *Duke Math. J.* **13,** 105–111.

Godwin, H. J., (1950). On the Product of Five Homogeneous Linear Forms. *J. London Math. Soc.* **25,** 331–339.

Hunter, J., (1957). *Proc. Glasgow Math. Soc.* **3,** 57–67.

Noordzij, P., (1967). Uber das Produkt von vier reelen, homogenen, linearen Formen, *Monatsh. Math.* **71,** 436–445.

Spohn, W. G., (1969). *Math. Comp.* **23,** 141–149.

The Improbable Behaviour of Ulam's Summation Sequence

M. C. WUNDERLICH

Northern Illinois University, De Kalb Illinois, U.S.A.

In the proceedings of the 1963 Number Theory Conference at Boulder, S. Ulam posed the problem of determining the asymptotic density of the following sequence $U = \{u_k\}$: We define $u_1 = 1$ and $u_2 = 2$. If U is defined up to $n-1$, n is in the sequence if and only if n can be expressed uniquely as the sum of two distinct elements in U. The first 10 terms of U are 1, 2, 3, 4, 6, 8, 11, 13, 16, 18. In 1967, P. Muller computed the first 20,000 terms of the sequence. In addition to showing that the density of U appeared to resemble rather closely the density of the primes, he observed two other curiosities; namely that there are no consecutive elements of U greater than 47 and also that an unusually large number of sequence elements (over 60%) differ from another element by 2. In this paper we shall employ a heuristic argument to describe the behaviour of two functions; $p(k)$ which is the probability that k is in U and $s(k)$ which is the expected number of ways that k can be expressed as a sum of distinct elements of U. We will then produce some computational evidence which shows that U as well as two other similarly defined sequences deviates sharply from its expected behaviour. There appears to be a connection between the reasons for this deviation and the curiosities noted by Muller.

As n becomes large, we can expect the functions $p(k)$ and $s(k)$ to satisfy

$$s(n) = \sum_{k=1}^{[(n-1)/2]} p(k)\,p(n-k) \tag{1}$$

and

$$p(n) = s(n)/e^{s(n)}. \tag{2}$$

(1) follows from the fact that the probability that both k and $n-k$ are elements of U is the product $p(k)\,p(n-k)$ assuming of course that the two

* All computations for this paper were done on the IBM 360/50 computer at Northern Illinois University, DeKalb, Illinois.

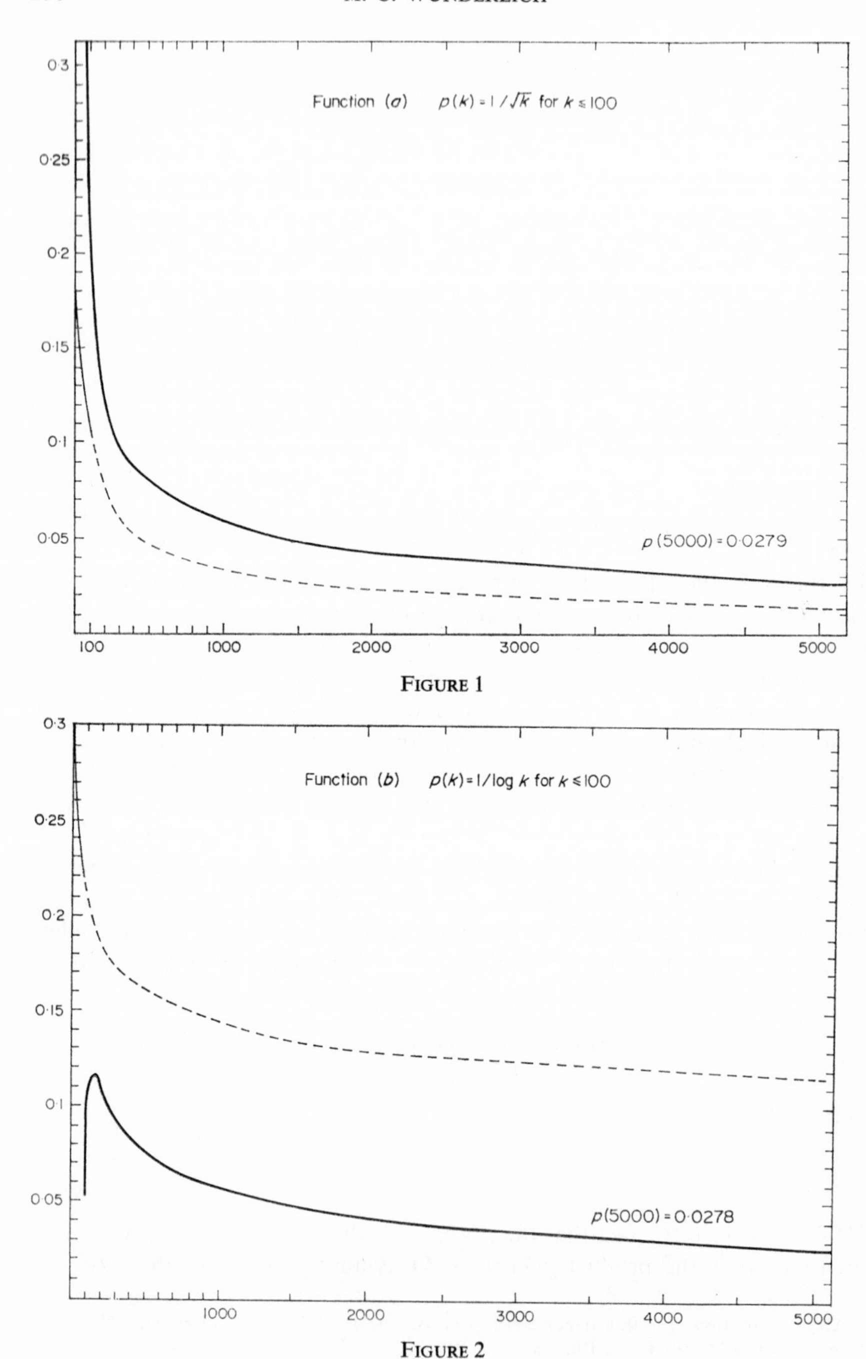

FIGURE 1

FIGURE 2

events are statistically independent. This means that in an interval about n of length r, we would expect a distribution of $r.s(n)$ sums of distinct sequence elements. We shall assume for the moment that each of the $r^{r.s(n)}$ distributions is equally probable. A standard result in the theory of probability tells us that the probability that a number in that interval is represented in exactly one way as a sum is

$$\frac{r.s(n)}{r}\left(1-\frac{1}{r}\right)^{r.s(n)-1} \tag{3}$$

$$= s(n)\left[\left(1-\frac{1}{r}\right)^{r}\right]^{s(n)-(1/r)}$$

Letting $r \to \infty$ and noting that

$$\lim_{r\to\infty}\left(1-\frac{1}{r}\right)^{r} = \frac{1}{e},$$

we obtain (2). We will return later to a discussion of the assumptions made to obtain (1) and (2).

The author was unable to find a function which satisfies the recursions (1) and (2). However, it is clear that if we assign values to $p(k)$ for k less than some integer M, the recursion relationships define $p(k)$ and $s(k)$ for all k. Computations were performed for a variety of starting functions for $M = 100$ and the results of those computations are displayed in Figs. 1–4.

(a) and (b). The starting functions of $1/\sqrt{n}$ and $1/\log n$ for $n < 100$ were chosen for no particular reason. They are too small and too large respectively to satisfy the recursion as is indicated by the nature of the jump discontinuities at $n = 100$.

(c) $p(n) = 0{\cdot}26$ for $n < 100$ was chosen because there are 26 elements in U which are less than 100.

(d) $p(n) = 0{\cdot}2187$ was found experimentally as the constant function for which no jump occurs at $n = 100$. For all three of these starting functions, the value of $p(5000)$ is remarkably similar.

(e) On the other hand, a starting function which is very small like $1/n$ produces an oscillating function which appears to be unstable.

(f) This starting function is linear between 0 and 100, has the property that

$$\sum_{k=1}^{100} p(k)$$

is 26 in order to emulate U, and has no discontinuity at $n = 100$. There is only one function possessing this property and it was obtained by computa-

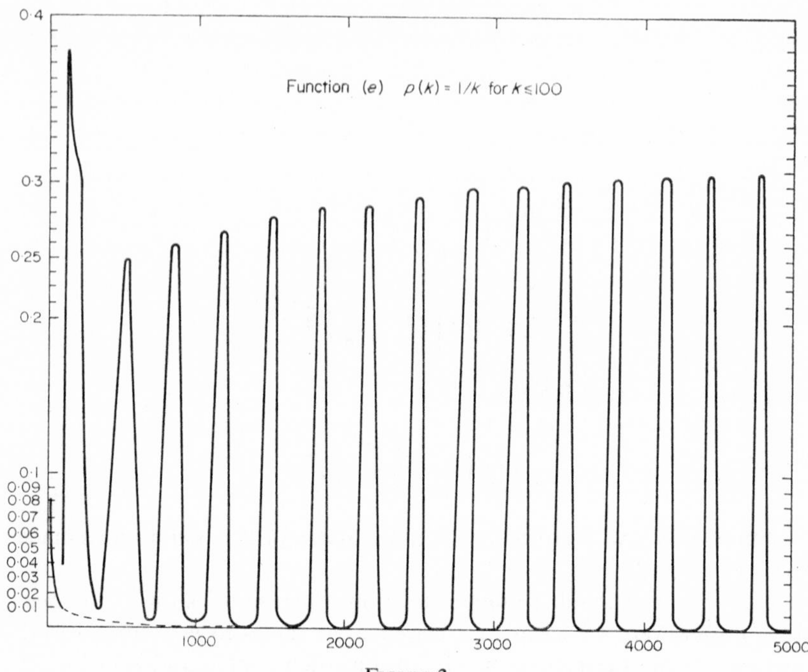

FIGURE 3

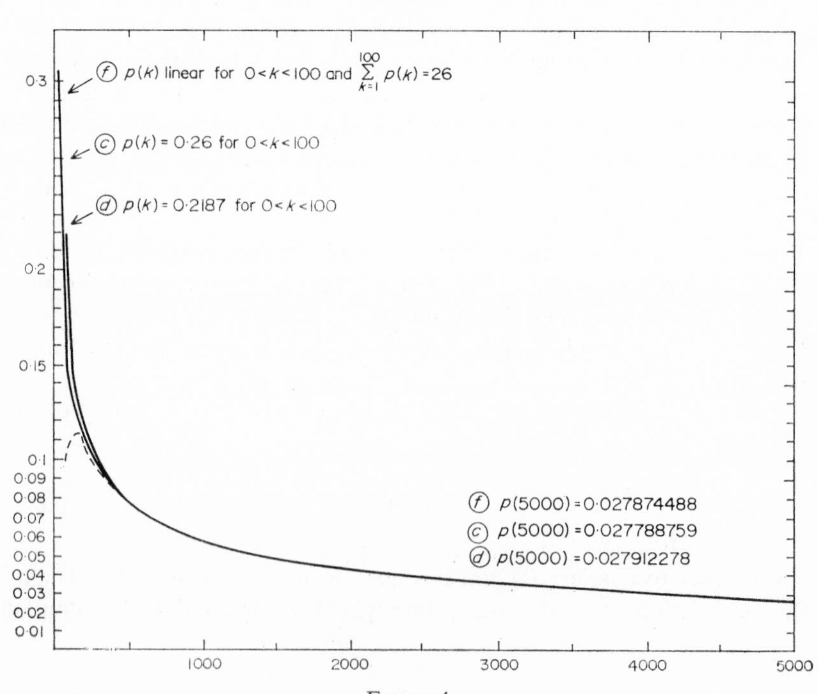

FIGURE 4

TABLE I

Function	$p(5000)$	$s(5000)$	$S = \sum_{k=1}^{5000} p(k)$
a	0·02792	5·2336	249·274
b	0·02783	5·2375	249·159
c	0·02789	5·2349	249·575
d	0·02791	5·2338	249·327
e	0·000000111	18·95423	417·737
f	0·02787	5·2355	249·706

tional methods. As one can see, it eventually differs very little from the other functions, with the exception of $1/n$.

Table I lists for each function the values of $p(5000)$, $s(5000)$ and

$$S = \sum_{k=1}^{5000} p(k).$$

One would expect the number of elements in U less than 5000 to be near S and this value is in the neighborhood of 250 for all the sequences except for (e). Therefore, on the basis of what is admitted to be a highly heuristic argu-

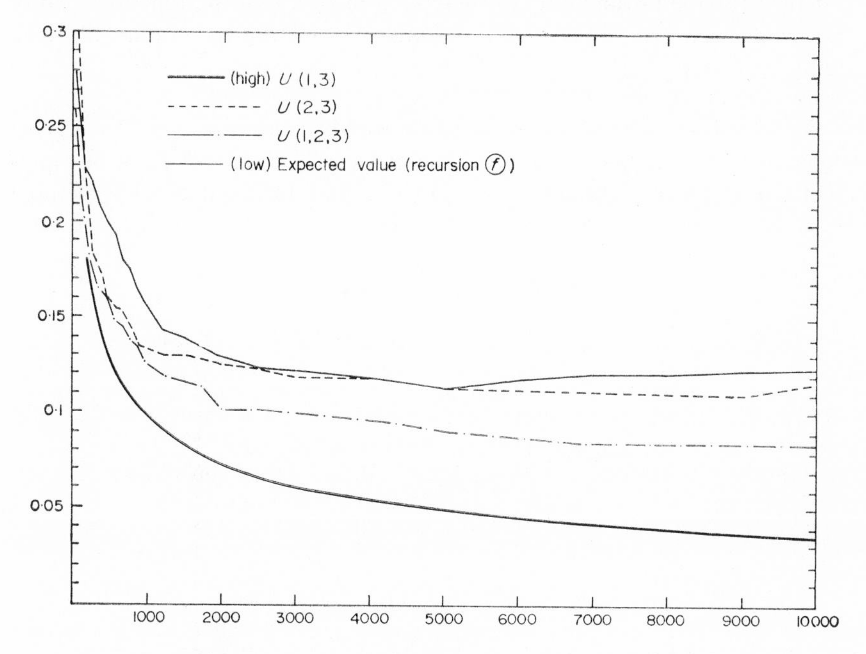

FIGURE 5

ment, we assert that the probable number of elements in Ulam's sequence which are less than or equal to 5000 is approximately 250.

Before discussing our computational evidence concerning U, we shall generalize our definition and introduce some additional terminology. If A is a finite set of distinct positive integers, and k is a positive integer, we will define the infinite sequence $U_k(A)$ as follows: If $n \leqslant \max\{i \mid i \in A\}$, $n \in U_k(A)$ if and only if $n \in A$. If $n > \max\{i \mid i \in A\}$ $n \in U_k(A)$ if and only if n can be uniquely expressed as the sum of k distinct elements of $U_k(A)$. In view of this definition, $U = U_2(\{1, 2\}) = U_2(\{1, 2, 3\})$. We have computed the sequences $U_2(\{1, 2, 3\})$, $U_2(\{1, 3\})$, and $U_2(\{2, 3\})$ up to 10,000 in order to test the heuristic results given above. (For notational conveniences, we will refer to them as $U = U_2(\{1, 2, 3\})$, $U(1, 3) = U_2(\{1, 3\})$ and $U(2, 3) = U_2(\{2, 3\})$.) It is difficult to extract a probability function analogous to $p(k)$ from the computed sequences, so we have instead displayed in Fig. 5 the three density functions $N_k(U_i)/k$ where $N_k(U_i)$ is the counting function of U_i and U_i ranges over the three sequences computed. These functions are compared with the "expected value" function

$$\sum_{j=1}^{k} p(j)$$

for the probability function $p(j)$ described in (f). (All the functions except (e) would have worked as well.) As one can see from the graphs, all three sequences have densities which deviate from the expected value as well as from each other.

Table II begins to show us why this erratic behaviour occurs. For a given summation sequence U, let $n(j)$ be the number of ways that j can be represented as a sum of distinct elements in U, in other words, U consists of those

TABLE II

E	$U(1, 2, 3)$	$U(2, 3)$	$U(1, 3)$
	32·336	57·000	71·048
$t(0)$	188	198	179
$t(1)$	72	105	122
$t(2)$	42	15	25
$t(3)$	26	11	7
$t(4)$	30	12	6
$t(5)$	12	7	7
$t(6)$	4	10	3
$t(7)$	7	4	4
$t(8)$	7	5	2
$t(9)$	8	4	3
$t(10)$	6	7	7
$\sum_{j>10} t(j)$	598	622	635

j for which $n(j) = 1$. For each sequence tabulated in Table II, E is the mean value of $n(i)$ for i in the interval (9000,10000). i.e.,

$$E = 0{\cdot}001 \sum_{i=9001}^{10000} n(i).$$

We let $t(k)$ be the number of i in that interval for which $n(i) = k$. We note that for each summation sequence tabulated, E is rather large. This represents the expected number of ways that a number in that interval can be represented as a sum. Assuming that the distribution of the sum is normal, we would expect $t([E])$ to be the largest value of all the t's and yet, $t(0)$ is by far the largest. This suggests that the sums do not distribute themselves randomly.

We next turn to a study of the sum distribution function $n(j)$ generated by U. For $j = 1000$, the function displays an easily recognizable pattern. The values of $n(j)$ for $j = 99981, 99982, \ldots, 10000$ are the following: 70, 63, 1, 132, 1, 73, 21, 5, 43, 0, 31, 3, 19, 22, 0, 65, 4, 57, 44, 0. The numbers appear to oscilate regularly between large values and small values with a period of either 2 or 3. The following computation demonstrated this pattern, The sequence $n(j)$, $j = 1001, \ldots, 10000$, was used to generate another sequence T of relative maximum points. $t \in T$ if and only if $n(t-1) < n(t)$ and $n(t) \geqslant n(t+1)$. So the numbers in T identify the "peaks" of the sums $n(i)$. Then let D be the sequence of successive differences of elements of T, i.e. $d_i = t_i - t_{i-1}$. D essentially describes the periods between peaks. As was expected, all the numbers in D were either 2's or 3's, and the average value was approximately $2 \cdot 44289$. We will from now on refer to this number as p the period of the distribution. To further demonstrate this phenomenon, consider the set of integers $A_\rho = \{\langle np + \rho\rangle\}$ as n ranges over the set of non-negative integers, ρ is a real number satisfying $0 < \rho \leqslant p$, and $\langle x\rangle$ means the "nearest integer to x". i.e., $\langle x\rangle = [x + \frac{1}{2}]$. If our hill–valley hypothesis is correct and our value of p is accurate, one would expect that for some values of p, A_ρ would hit all the hills and for others, A_ρ would hit all the valleys. Table III lists for fifty values of ρ the corresponding mean value of $\{n(i) \mid i \in A_\rho, i \leqslant 10000\}$ which is labeled M_ρ. The valleys appear to occur for $\rho = \frac{1}{2}p$ and the hills at $\rho = 0$.

One can now see, at least qualitatively, why the summation sequence is so dense. Most of the sums are distributed along the peaks of the distribution function so that a small number of sums are left to be distributed in the valleys, thus producing more one's in the distribution. It would be interesting to empirically obtain a distribution function, use it to modify the recursion relationships (1) and (2) and see if the modified recursion produces a probability function closer to the actual density of U. This would take more

TABLE III

ρ	M_ρ	ρ	M_ρ	ρ	M_ρ	ρ	M_ρ
·05	35·94	·7	11·64	1·35	3·27	2·0	28·82
·1	35·06	·75	10·28	1·4	4·71	2·05	31·79
·15	34·03	·8	9·30	1·45	6·50	2·1	34·13
·2	33·04	·85	8·44	1·5	8·52	2·15	35·86
·25	32·05	·9	7·51	1·55	10·51	2·2	36·89
·39	31·01	·95	6·34	1·6	12·37	2·25	37·40
·35	29·68	1·0	5·05	1·65	13·87	2·3	37·52
·4	27·81	1·05	3·75	1·7	15·20	2·35	37·36
·45	25·30	1·1	2·71	1·75	16·46	2·4	37·04
·5	22·31	1·15	1·99	1·8	18·00	2·45	36·55
·55	19·17	1·2	1·68	1·85	20·04	2·5	35·84
·6	16·11	1·25	1·77	1·9	22·66		
·65	13·60	1·3	2·28	1·95	25·70		

computing than has been possible at this time. The periodic nature of $n(j)$ also explaines the fact that no consecutive elements of U occur which are greater than 47. The period of $n(j)$ being 2·44289 makes it impossible for consecutive numbers to both lie in a valley, so to speak, and elements of U always come from valleys. This would also lead us to conjecture that sufficiently large elements of U can never differ by six, a conjecture not tested out as yet. The preponderance of twins separated by one integer remains a mystery. It simply means that in terms of our sum distribution function that an unusually large number of consecutive valleys contain a one. The other two sequences $U(1, 3)$ and $U(2, 3)$ also produced hill–valley sum distribution functions with different periods, but we did not analyse them.

We can also readily see why this hill–valley phenomenon propagates itself once it gets started. If most numbers in U are the form $\langle np + \frac{1}{2}p\rangle$, then most sums of two numbers in U will be of the form $\langle mp\rangle$ thus building the peaks even higher. One would expect this same phenomenon to occur for all summation sequences $U_k(A)$ where k is even, although if k gets too large, one would expect more diffusion to take place. For k odd, it is not clear what would occur. Muller has studied $U_3(\{1, 2, 3\})$ and has found that the successive differences of that sequence were eventually periodic. The exact statement of Muller's theorem is as follows:

THEOREM (Muller). *Let u_n be the nth term of $U_3(1, 2, 3)$. Then for $k > 5$,*

$$t_{3k} = 25k - 45$$

$$t_{3k+1} = 25k - 43$$

$$t_{3k+2} = 25k - 21.$$

The question which remains completely open is whether summation sequences in general tend to behave in this way. Is there an instability inherent in this type of progress which tends to produce sum distribution functions having the hill–valley property? The three we studied do and a more extensive computer study will yield an empirical answer to the question. It would probably require the application of more sophisticated methods from the theory of probability to show that the improbable behaviour of Ulam's summation sequence is indeed probable for a large class of summation sequences.

References

Muller, P. N. (1966). On Properties of Unique Summation Sequences. Masters Thesis, State University of New York at Buffalo.

Non-repetitive Sequences

P. A. B. PLEASANTS

University College, Cardiff, Wales

The sequences considered here (which may be finite or infinite in length) have terms belonging to a finite set of symbols which we may take to be the integers $1, 2, \ldots, n$. Such a sequence is said to have a repetition if two adjacent chunks of it (which may be of any finite length) are equal. For $n = 2$ any sequence with more than three terms has a repetition, but for $n = 3$ it is a surprising fact that there exist infinite sequences with no repetitions. This was first proved by de Bruijn in the 1930's and later by Peltesohn and Sutherland, but neither of these proofs was published. Proofs have since been published by Morse and Hedlund (1944), Hawkins and Mientka (1956), Leech (1957), and Pleasants (1970).

Erdös (1961) has raised the question what happens when the meaning of 'repetition' is extended to include pairs of adjacent chunks of a sequence which are merely permutations of each other (that is, which are such that each symbol occurs in one chunk as many times as it occurs in the other). We call a sequence *non-repetitive* if it has no repetitions in this extended sense. The following are non-repetitive sequences for $n = 1$, 2 and 3 that are as long as possible:

$$n = 1,\ 1;$$
$$n = 2,\ 121;$$
$$n = 3,\ 1213121.$$

At first sight this pattern suggests that the maximum length of a non-repetitive sequence on n symbols is $2^n - 1$, but for $n = 4$ there are non-repetitive sequences longer than this and, in fact, a non-repetitive sequence on four symbols with over 1,600 terms has been found by M. R. Bird using the Chilton Atlas computer. The computer constructed this sequence a term at a time, trying each symbol in turn and going back to alter preceding terms whenever there was no continuation. The fact that no more than the last 25 terms had to be altered at any stage in the construction shows that such sequences can really be found very easily and makes it almost certain that there is an infinite non-repetitive sequence for $n = 4$. So far, however, no such infinite sequence has been found. For $n = 5$ infinite non-repetitive sequences do exist, and I shall sketch the construction of such a sequence and indicate how the method might be used to give an infinite non-repetitive sequence for $n = 4$ with the help of a computer. (The details of the construction are given by Pleasants (1970).)

The basis of the construction is finding five blocks of terms, $B_1, \ldots, B_5$, with the property that when they are substituted for the five symbols in any non-repetitive sequence the resulting sequence is still non-repetitive. Suitable blocks are given by taking $B_1 = 213151314151412$ and obtaining the other blocks successively by adding 1 (mod 5) to each term of the previous block; so that $B_2 = 324212\ldots$, $B_3 = 435323\ldots$, etc. For $i = 1, \ldots, 5$, B_i contains seven i's and two of each other symbol. Let $S = ij\ldots$ be any sequence on five symbols, and let $S' = B_iB_j\ldots$ be the sequence got by replacing each symbol in S by its corresponding block. We have to show that if two adjacent chunks, C_1 and C_2, of S' are permutations of each other then S itself has a repetition. The following diagram illustrates the situation. (By omitting, if necessary, terms at the beginning and end of S, we may suppose that each block in S' intersects C_1 or C_2.).

The first step is to cancel blocks contained in C_1 with equal blocks contained in C_2 as far as possible. When this has been done the remaining parts of C_1 and C_2 will still be permutations of each other, and it can be shown that the preponderance of one symbol in each block is large enough to ensure that to each remaining block contained in C_1 there corresponds an equal block overlapping C_2, and vice versa. Since there are at most two blocks that overlap C_2 but are not contained in it at most two of the blocks contained in C_1 remain, and similarly at most two of the blocks contained in C_2 remain. The proof now reduces to a bounded amount of checking, as it is enough to show that to any repetition in a sequence made up of seven† or fewer complete blocks there corresponds a repetition in the parent sequence. If this can be done then, when the cancelled blocks are replaced, the repetition in S' will correspond to a repetition in S. The checking is not difficult to do by hand and has the desired result. (The fact that the blocks can be transformed into one

† This number is obtained by crudely estimating a maximum of two complete blocks in each of C_1 and C_2, one block overlapping both C_1 and C_2, one overlapping the beginning of C_1, and one overlapping the end of C_2. By a more detailed argument the number can be reduced to four.

another by permuting the symbols considerably reduces the amount of checking that is necessary, and so does the partial symmetry of the blocks.)

An infinite non-repetitive sequence on five symbols can now be built up by starting with a sequence of only one term (which is trivially non-repetitive) and repeatedly applying the process of replacing the symbols by blocks. In this way we get a chain of non-repetitive sequences in which each sequence is an extension (in both directions) of the previous one, and in the limit this gives a sequence that extends to infinity in both directions. This infinite sequence is non-repetitive, since any repetition it had would be contained in all sequences of the chain from some point on. Each sequence of the chain can be embedded in the next sequence in several different ways, and different choices of the set of embeddings give rise to different limiting sequences. Consequently the set of non-repetitive sequences on five symbols that can be constructed in this way has the cardinal of the continuum.

It may be possible to use the method just described to construct an infinite non-repetitive sequence on four symbols. A computer could be used to find a non-repetitive block with equal numbers of 2's, 3's and 4's but with an excess of 1's, and three other blocks could be obtained from this block, as before, by applying to the symbols the permutations in a subgroup of the permutation group S_4. A cancellation argument of the kind used above would then reduce the proof to a bounded amount of checking which could also be done by computer. To keep the amount of checking down to a reasonable level an excess of five or six 1's would be needed in the initial block and consequently the block would probably have to contain thirty or more terms (judging from the amount of discrepancy between the symbols in non-repetitive sequences that have been computed). Finding blocks, even of this length, by hand is very tedious.

The main difficulty in this programme will be finding a suitable block. Apart from the condition on the numbers of symbols it contains the block has to satisfy other essential conditions—it has to begin and end with the same symbol, its first three terms have to be different, and its last three terms have to be equal to its first three terms in some order. One could form such a block a term at a time, adding a 1 whenever possible and trying to keep level the numbers of occurrences of the other symbols. One would then hope that a block satisfying all the conditions occurred at some stage and apply the checking process to it when it did. If this failed different blocks might be found by specifying the first few terms in different ways. Each block satisfying the conditions that was found would give four chances of success, as the other blocks could be chosen in three different ways corresponding to the four subgroups of S_4 of order four. If the method succeeds the computer will have played an essential part in proving an interesting result that is not in itself finitely computable.

Added in proof: Professor G. A. Hedlund has pointed out to me that the result described in the first paragraph of this chapter had already been published by Thué, A. (1906). Über unendliche Zeichereihen. *Viderskass elskabets Skrifter, I. Mat.-nat. Kl., Christiania.*
Thué, A. (1912). Über die gegenseitige Lage gleicher Teile gewisser Zeichenreihen. *Viders kasselskabets Skrifter, I. Mat.-nat. Kl., Christiania.*

References

Erdős, P. (1961). Some unsolved problems. *Magyar Tud. Mat. Akad. Kutató Int. Közl.* **6,** 221–254 (in particular, p. 240).
Hawkins, D. and Mientka, W. E. (1956). On sequences which contain no repetitions. *Math. Student.* **24,** 185–187.
Leech, J. (1957). A problem on strings of beads. *Math. Gaz.* **41,** 277–278.
Morse, M. and Hedland, G. A. (1944). Unending chess, symbolic dynamics and a problem in semigroups. *Duke Math. J.* **11,** 1–7.
Pleasants, P. A. B. (1970). Non-repetitive sequences. *Proc. Camb. Phil. Soc.* **68,** 267–274.

On Sorting by Comparisons

R. L. GRAHAM

Bell Telephone Laboratories, Incorporated,
Murray Hill, New Jersey, U.S.A.

1. Introduction

A problem which frequently occurs in the study of sorting algorithms can be phrased as follows. Suppose we are given a partition of the set $\{1, 2, ..., m + n\}$ into two disjoint sets $A = \{a_1 < \cdots < a_m\}$ and $B = \{b_1 < ... < b_n\}$. We wish to determine the set A (and hence, B) by asking the minimum number of questions of the type: "Is $a_i > b_j$ or is $a_i < b_j$?" The answer to each question is known before the next question is asked. For each strategy S of asking questions, there is some choice of A and some set of answers which will require a maximum number of questions to determine A; we let $M(S)$ denote this maximum number of questions. Let $h(m, n)$ denote $\min_S M(S)$ where S ranges over all possible strategies which eventually determine A.

It is well-known (and easily shown) that $h(1, n) = 1 + [\log_2 n]$. In this paper we describe $h(2, n)$ as well as a related more general function. It will be seen, in particular, that the *least* value of n for which $h(2, n) = t \geqslant 2$ is exactly $\lceil \frac{12}{7} \cdot 2^{(t-2)/2} \rceil$ if t is even and $\lceil \frac{17}{7} \cdot 2^{(t-3)/2} \rceil$ if t is odd. The proofs of the assertions are rather lengthy and will not appear here.

2. A Generalization

For the case $m = 2$, write $A = \{\alpha < \beta\}$, $B = \{b_1 < ... < b_n\}$. Suppose that in addition to knowing the complete ordering within A and B we also have partial information concerning the ordering between A and B. Specifically, assume that for some i and j, $1 \leqslant i, j \leqslant n + 1$, we know $\alpha < b_i$ and $\beta > b_{n+1-j}$, where we use the convention that $\alpha < b_{n+1}$ indicates that we know nothing about the relative order of α and any element of B (with $\beta > b_0$ defined similarly). Let $f_n(i, j)$ denote $\min_S M(S)$ where S ranges over all strategies which eventually determine A. The main result of this paper is the determination of $f_n(i, j)$. Note that $f_n(n + 1, n + 1)$ is by definition equal to $h(2, n)$.

3. A Recursion for $f_n(i,j)$

We first note that $f_n(1,1) = 0$ for $n \geqslant 1$. Define $f_0(1,1) = 0$. Consider the situation which defines $f_n(i,j)$. That is, we know

$$\alpha < \beta, \quad b_1 < \ldots < b_n, \quad \alpha < b_i, \quad \beta > b_{n+1-j}.$$

Let S^* denote an optimal strategy for this situation, i.e., $M(S^*) = f_n(i,j)$. There are two possibilities for the first question of S^x.

(i) Suppose for some k the first question is "Is $\alpha \gtrless b_k$?". We can assume without loss of generality that $1 \leqslant k < i$ since no information can be gained if $k \geqslant i$. If the answer is "$\alpha < b_k$" then we are faced with the knowledge

$$\alpha < \beta, \quad b_1 < \ldots < b_n, \quad \alpha < b_k, \quad \beta > b_{n+1-j}.$$

By definition this situation requires $f_n(k,j)$ additional questions in order to determine A. On the other hand, if the answer is "$\alpha > b_k$", then we know

$$\alpha < \beta, \quad b_1 < \ldots < b_n, \quad \alpha > b_k, \quad \beta > b_{n+1-j}.$$

Note that now we can deduce $\beta > \alpha > b_k$ so that the elements $b_1, b_2, \ldots, b_k$ must just be the integers $1, 2, \ldots, k$. Hence, we are faced with the reduced situation

$$\alpha < \beta, \quad b_{k+1} < \ldots < b_n, \quad \alpha < b_i, \quad \beta > b_{n+1-j}.$$

Again, by definition, this requires $f_{n-k}(i-k,j)$ additional questions to determine A, where we make the convention that $f_m(x,y) = f_m(m+1,y)$ if $x \geqslant m+1$ with a similar convention holding if $y \geqslant m+1$.

Thus, we can write

$$f_n(i,j) \leqslant 1 + \max\big(f_n(k,j),\ f_{n-k}(i-k,j)\big).$$

(ii) Suppose for some k the first question of S^* is "Is $\beta \gtrless b_{n+1-k}$?". We can assume $1 \leqslant k < j$. Arguing as before we can conclude

$$f_n(i,j) \leqslant 1 + \max\big(f_n(i,k),\ f_{n-k}(i,j-k)\big).$$

Since in both (i) and (ii) the choice of k was arbitrary and since the first question of S^* must be either of type (i) or of type (ii) then we have the following recurrence:

$$f_n(i,j) = 1 + \min\left\{ \min_{1 \leqslant k < i} \max\big(f_n(k,j),\ f_{n-k}(i-k,j)\big), \quad \min_{1 \leqslant k < j} \max\big(f_n(i,k),\ f_{n-k}(i,j-k)\big) \right\} \tag{1}$$

for $n \geqslant 1$, $1 \leqslant i$, $j \leqslant n+1$, $(i,j) \neq (1,1)$, where $f_n(1,1) = 0$, for $n \geqslant 0$. It is this rather formidable looking recurrence which we shall solve.

4. The Structure of $f_n(i, j)$

We first make some preliminary remarks. It is easily checked that $f_n(i,j) = f_n(j,i)$. Further, if the value of α is determined by questions which never contain β, then since $\alpha < b_i$, we know that $h(1, i-1) = 1 + [\log_2(i-1)]$ questions are required (where we take $\log_2(0) \equiv -1$). Thus, by determining α and β independently we see that

$$f_n(i,j) \leqslant 2 + [\log_2(i-1)] + [\log_2(j-1)] \equiv R(i,j). \tag{2}$$

On the other hand, there are exactly $\left[ij - \binom{i+j-n-1}{2}\right]$ choices of A which satisfy the required conditions (where the binomial coefficient $\binom{x}{2}$ is taken to be 0 for $x < 2$). A standard argument of information theory shows

$$f_n(i,j) \geqslant \log_2\left[ij - \binom{1+j-n-1}{2}\right]. \tag{3}$$

A straightforward calculation yields

$$R(i,j) - \log_2\left[ij - \binom{1+j-n-1}{2}\right] < 3 \tag{4}$$

from which it follows, since $f_n(i,j)$ is an integer,

$$f_n(i,j) = R(i,j), \quad R(i,j) - 1 \quad \text{or} \quad R(i,j) - 2. \tag{5}$$

We shall call these values *regular*, *special* and *extraspecial* and denote them by R, S and X, respectively.†

As examples, we list the values of $f_n(i,j)$, $0 \leqslant n \leqslant 7$, in the form of square arrays in Table I. For a fixed value of n, the variables i, j each range over $1 \leqslant i, j \leqslant n+1 \equiv N$. The values of $f_n(1, 1)$ and $f_n(1, n+1)$ are the upper left-hand and upper right-hand entries respectively. Thus $f_3(3, 3) = 4$, $f_6(3, 5) = 5$, etc.

Notice that all the entries in the tables for $N = 1, 2, 4$ and 8 are *regular*. This is not accidental.

Let us define certain subregions of the Nth array (the array which gives the values of $f_{N-1}(i,j)$). Write $N = 2^k + t$, $0 < t \leqslant 2^k$.

For $0 \leqslant r \leqslant k-1$, define the *rth region* of the array to be the set of coordinate values (i,j) for which $2^r < i \leqslant 2^{r+1}$, $2^k < j \leqslant N$. For $r = k$, the kth (or *critical*) *region* is defined by $2^k < i, j \leqslant N$.

† Thus, the knowledge that $\alpha < \beta$ can result in a savings of at most *two* questions.

TABLE I

$N = 1$

0

$N = 2$

0	1
1	2

$N = 3$

0	1	2
1	2	3
2	3	3

$N = 4$

0	1	2	2
1	2	3	3
2	3	4	4
2	3	4	4

$N = 5$

0	1	2	2	3
1	2	3	3	4
2	3	4	4	4
2	3	4	4	4
3	4	4	4	5

$N = 6$

0	1	2	2	3	3
1	2	3	3	4	4
2	3	4	4	5	5
2	3	4	4	5	5
3	4	5	5	5	5
3	4	5	5	5	5

$N = 7$

0	1	2	2	3	3	3
1	2	3	3	4	4	4
2	3	4	4	5	5	5
2	3	4	4	5	5	5
3	4	5	5	5	5	5
3	4	5	5	5	5	5
3	4	5	5	5	5	6

$N = 8$

0	1	2	2	3	3	3	3
1	2	3	3	4	4	4	4
2	3	4	4	5	5	5	5
2	3	4	4	5	5	5	5
3	4	5	5	6	6	6	6
3	4	5	5	6	6	6	6
3	4	5	5	6	6	6	6
3	4	5	5	6	6	6	6

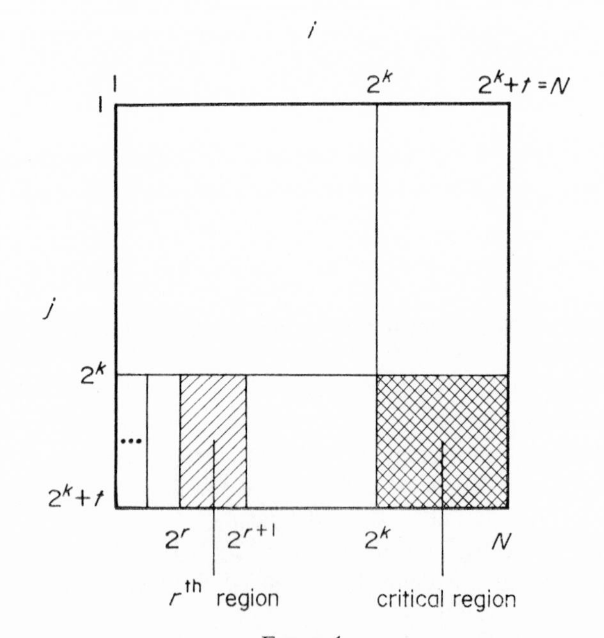

FIGURE 1

We next state a sequence of facts, whose proofs, as we have mentioned, will not be given here.

1. All entries $f_n(i,j)$, $1 \leqslant i, j \leqslant 2^k$, are *regular*, i.e., $f_n(i,j) = R(i,j)$.

2. All entries $f_n(1,j) = f_n(i,1)$, $1 \leqslant i, j \leqslant N$, are *regular*, i.e., $f_n(1,j) = 1 + [\log_2(j-1)]$.

3. The only *extraspecial* values occur in the critical region.

4. In general, the structure of the values of $f_n(i,j)$ in the rth region (and its symmetrical counterpart $f_n(j,i)$) is as follows, for $0 \leqslant r < k$:

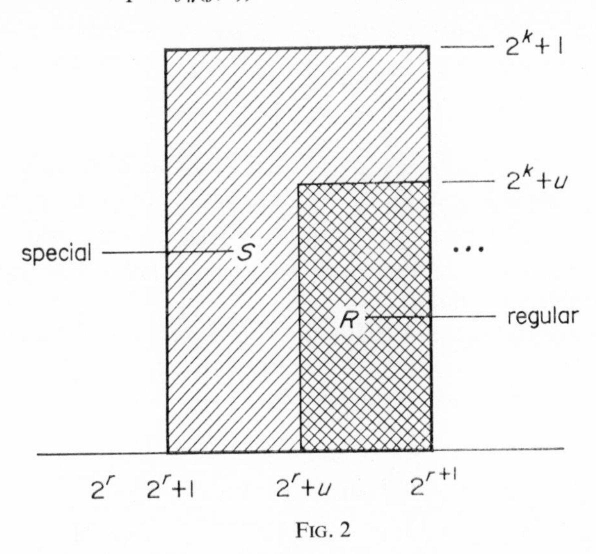

FIG. 2

There is a "strip" of *special* values of some uniform width u (possibly 0) which borders a rectangle of *regular* values as shown in Fig. 2. More precisely, for some $u \geqslant 0$, the values $f_n(i,j)$, $2^r + u < i \leqslant 2^{r+1}$, $2^k + u < j \leqslant N$, are regular. The remaining values in the rth region are special. For $r = k$, the same behaviour occurs in the critical region, both for a rectangle† of regular values bordered by a strip of special values, as well as a rectangle† of special values bordered by a strip of extraspecial values. The structure of the values in the critical region directly determines the structure of the values at these coordinates for larger values of N.

5. *Define* $F(N)$ by

$$F(N) = \begin{cases} X & \text{if all values in critical region are extraspecial,} \\ S & \text{if all values in critical region are special,} \\ u & \text{if there is a strip of } S(X) \text{ values of width } u \text{ which borders a} \\ & \text{rectangle of } R(S) \text{ values.} \end{cases}$$

† Which is in this region, in fact, a square.

We give a recursive definition for F from which the values of $f_n(i,j)$ can be immediately deduced. As usual, $[x]$ denotes the greatest integer $\leqslant x$. Let $F(1) = 0$, $F(2) = 0$, $F(3) = S$, $F(4) = 0$. For $N = n + 1 = 2^k + t$, $0 < t \leqslant 2^k$, $k \geqslant 2$,

$$F(N) = \begin{cases} X & \text{for } 0 < t \leqslant [\frac{3}{7}\cdot 2^{k-1}], \\ F(t) & \text{for } [\frac{3}{7}\cdot 2^{k-1}] < t \leqslant 2^{k-2}, \\ S & \text{for } 2^{k-2} < t \leqslant [\frac{5}{7}\cdot 2^{k}], \\ F(t - 2^{k-2}) + 2^{k-1} & \text{for } [\frac{5}{7}\cdot 2^{k}] < t \leqslant 3\cdot 2^{k-2}, \\ 2^{k-1} & \text{for } 3\cdot 2^{k-2} < t \leqslant [\frac{6}{7}\cdot 2^{k}], \\ F(t) & \text{for } [\frac{6}{7}\cdot 2^{k}] < t \leqslant 2^{k}. \end{cases} \tag{6}$$

Of course, the values of $F(N)$ for $[\frac{3}{7}\cdot 2^{k-1}] < t \leqslant 2^{k-2}$ correspond to the width of a strip of X values bordering a rectangle of S values. It is granted that this particular expression for $F(N)$ might not be the first thing that one would guess. It is true, however, that with sufficient patience one can indeed inductively establish (6) (and (7)).

Of course, in order to establish (6) one also must know $F^{(r)}(N)$, the width of the strip of S values which borders the rectangle of R values in the rth region, $1 \leqslant r < k$. These are recursively given by

$$F^{(r)}(N) = \begin{cases} S & \text{for } 0 < t \leqslant [\frac{5}{7}\cdot 2^{r}], \\ F(t - 2^{r-2}) + 2^{r-1} & \text{for } [\frac{5}{7}\cdot 2^{r}] < t \leqslant 3\cdot 2^{r-2}, \\ 2^{r-1} & \text{for } 3\cdot 2^{r-2} < t \leqslant [\frac{6}{7}\cdot 2^{r}], \\ F(t) & \text{for } [\frac{6}{7}\cdot 2^{r}] < t \leqslant 2^{r}, \\ 0 & \text{for } 2^{r} < t \leqslant 2^{k}. \end{cases} \tag{7}$$

6. It follows from (6) that $f_n(n+1, n+1)$ increases by 1 as N goes from $2^k + [\frac{3}{7}\cdot 2^{k-1}]$ to $2^k + [\frac{3}{7}\cdot 2^{k-1}] + 1$ and as N goes from $2^k + [\frac{5}{7}\cdot 2^k]$ to $2^k + [\frac{5}{7}\cdot 2^k] + 1$. Hence, the least value of n for which $h(2,n) = f_n(n+1, n+1) = t$ is exactly $[\frac{12}{7}\cdot 2^{(t-2)/2}]$ if t is even and $[\frac{17}{7}\cdot 2^{(t-3)/2}]$ if t is odd. Asymptotically, as n increases from 2^k to 2^{k+1}, the first 3/14 of the values of $h(2,n)$ are extraspecial, the next 1/2 are special and the final 2/7 are regular.

5. Concluding Remarks

The behaviour of $h(2,n)$ has recently been determined independently by Hwang and Lin (to appear). They did this without determining the general values of $f_n(i,j)$, by a combination of various bounds on $h(2,n)$ together with an ingenious explicit optimal strategy.

The determination of $h(3, n)$ would appear to be possible (but somewhat more difficult) using the techniques of this paper. The exact value of $h(m, n)$ for general m, however, would seem to require new ideas.

The author wishes to express his thanks to J. Evnine, F. K. Hwang, and S. Lin for stimulating discussions on this subject.

Reference

F. K. Hwang and S. Lin. An optimal algorithm for merging a linearly ordered set with an ordered set with two elements. *Acta Informatica.* To be published.

Spectra of Determinant Values in (0, 1) Matrices

N. METROPOLIS*

University of California
Los Alamos Scientific Laboratory
Los Alamos, New Mexico, U.S.A.

1. Introduction

As part of a broad program in computer development, a small nucleus at the Laboratory is concerned with problems in number theory and combinatorial analysis. The interest extends to a high level programming language that is convenient for such studies; moreover, attention is also given to questions of logical design of a computer so that the characteristic computational features of such studies may be implemented in an efficient manner. Clearly, there is a definite correlation between the efficiency of execution and the level of ambition of a computer problem; this seems of particular relevance to applications in number theory and combinatorics.

Two separate studies are described; each has utilized the MADCAP programming language, due to Wells (1970); the calculations were performed on MANIAC II, the computer designed and built at the Laboratory. P. R. Stein and M. B. Wells participated in the first part and W. A. Beyer, C. J. Everett, and J. R. Neergaard in the second.

2. Spectra of Determinant Values

A natural consequence of earlier work on a class of (0,1) matrices with vanishing determinants (Metropolis and Stein, 1967) and on permanents of cyclic (0,1) matrices (Metropolis, Stein and Stein, 1969) was an interest in the complete spectra of determinant magnitudes of $n \times n$ (0,1) matrices *with distinct ordered rows*. The number of such matrices is, of course, given by the binomial coefficient $\binom{2^n}{n}$. For $n = 6$, direct enumeration is not practical;

* This work was done under the auspices of the U.S. Atomic Energy Commission.

however, using branch merging techniques, it has been possible to find the spectra up to $n = 7$, with $n = 8$ being feasible on the most advanced computer.

The enumeration may be described in terms of a search tree that systematically accounts for all possibilities; the aim is to combine at each level all "equivalent" branches of the tree. The idea of branch merging here is to seek, for each k, a set of $k \times n$ inequivalent matrices, with $1 \leqslant k \leqslant n$. Two matrices are equivalent if they can be made equal by (i) row and/or column

TABLE I

Spectrum of determinant magnitudes for (0, 1) matrices

$n = 6$		$n = 7$	
M	N	M	N
0	34 923 518	0	40 885 781 314
1	25 666 809	1	26 883 246 720
2	10 746 288	2	16 511 989 560
3	2 135 343	3	4 650 079 360
4	1 163 064	4	3 511 706 880
5	176 701	5	744 944 448
6	129 360	6	833 612 648
7	17 885	7	161 359 296
8	13 930	8	208 846 176
9	1 470	9	57 084 608
		10	42 833 560
		11	9 880 640
		12	17 749 760
		13	2 437 120
		14	2 432 640
		15	806 400
		16	759 360
		17	80 640
		18	135 240
		19	0
		20	26 880
		21	0
		22	0
		23	0
		24	1 920
		25	0
		26	0
		27	0
		28	0
		29	0
		30	0
		31	0
		32	30

M is the determinant magnitude; N is the number of $n \times n$ of matrices (with distinct rows) having that magnitude.

interchanges, or (ii) as Williamson (1946) has shown, adding modulo 2 the elements of row i to the corresponding members of row $j \neq i$. It is advantageous to select as representative that matrix with the least number of 1's from certain equivalent matrices treated under Williamson's transformation. A generalized row- and column-sum has also been used as a preliminary check. Finally, it is efficient to restrict k to $k_{max} < n$ and consider all remaining possibilities, i.e., eliminating further tests for equivalent types. Empirically, it was found that $k_{max} = n - 1$ optimized computing time for small n. For $n = 6$, the time was approximately fifteen minutes; $n = 7$ required fifteen hours of early morning (stand-by) time.

For $n < 7$, all magnitudes of determinant values from zero to the corresponding maxima occur; however, as shown in Table I, gaps exist in the case $n = 7$. It is perhaps interesting to note that there is considerable structure (lack of monotonicity) in the distribution even where no gaps exist. An explanation of the gaps remains an open question; it appears complicated but not profound.

3. Square Roots of Integers in Various Bases

D. H. Lehmer has frequently remarked that the requirements of a number theorist with respect to computer arithmetic differ considerably from those of other users. The former is usually concerned with integer manipulation free from round-off considerations and other approximations. On the other hand, his interest is often in very large integers requiring multi-word arithmetic operations. An interest in the structure of arithmetic processors for dealing efficiently with such operations was in part a motivation of the present study.

Secondly, most statistical studies of the distribution of digits in the square root expansion of small integers have been confined to base 2 (or some integral power of 2) and base 10. Schmidt (1960) has proved that there exist numbers that may be normal in one base but not in another. Table II gives a summary of the radicals studied in various bases.

A variety of statistical tests have been applied.

(1) For expansions in base 2, the distribution of runs or "clusters" of 1's was examined, i.e., of sequences of 1's preceded and terminated by 0. Von Mises (1964) has shown that the probably of obtaining r runs, each of length m, in a sequence of n symmetric Bernoulli trials is given by a Poisson distribution. An examination of all binary expansions (cf. Table II) showed good agreement with that expected from a random sequence (Beyer *et al.*, 1969).

(2) The binary sequences were also studied to determine the intervals over which one digit would dominate the other; this is sometimes referred to as the "lead" test. The most extreme case found was in the case of the expansion

of $\sqrt{14}$: beginning with the digit in position 144, digit 1 leads all the way to 88062, having an excess there of 582. Feller (1968) has derived an arc sine law for the corresponding probability that is as large as 0·025. The next smallest probability was that for $\sqrt{3}$: digit 1 leads all the way after position 658; probability is 0·055. These illustrations show that the probability of such long leads is greater than one might expect intuitively.

TABLE II

Number of digits of $\sqrt{n}$ in base b investigated

n	b	number of digits
2, 3, 5, 6, 7 10, 11, 13, 14, 15	2	~88,000
2, 3, 5	3	~55,000
2, 3, 5	5	~37,000
2, 3, 5	6	~33,000
2, 3, 5	7	~31,000
2, 3, 5	10	~25,000

For the generalized serial test, the table has been extended to include expansions in $b = 11$, 13, 14, 15 for all n.

(3) χ^2 test for frequencies. The χ^2 values of the cumulative frequency distributions have been examined for all cases in Table II. In addition, it was trivial to include bases 4, 8, 16. A rather small probability level of ~0·008 for $(\sqrt{10})_{16}$ is noted in the region of position 20,000. For $(\sqrt{5})_{10}$ a level near 0·999 is maintained for the interval between 16500 and 17000. Other cases were undistinguished.

(4) Generalized serial test. Good and Gover (1967) have applied this test for randomness to the first 10,000 binary digits of $\sqrt{2}$. In the present study, it has been extended to all n values shown in Table II; moreover, these integers serve both as radicands and as bases.

Let a sequence of N digits in base b be given and let the sequence be circularized, i.e., the last digit is considered as being followed by the first. Let n_I be the number of occurrences of the v-plet $I = (i_1, i_2, \dots i_v)$. Define

$$\psi_v^2 = \frac{b^v}{N} \sum_I \left(n_I - \frac{N}{b^v}\right)^2, \qquad \psi_0^2 = \psi_{-1}^2 = 0.$$

The quantity of interest is the second difference $\Delta^2 \psi_v^2 = \psi_v^2 - 2\psi_{v-1}^2 + \psi_{v-2}^2$; it has the property of being asymptotically of chi-square form, with $b^v - 2b^{v-1} + b^{v-2}$ degrees of freedom.

For $b = 2$, $v = 1, 2, \ldots, 10$ has been used; for larger b, the set of v gradually decreases to only $v = 1, 2$. The sequences consisted of blocks of either 5,000 or 10,000, in addition to the full sets. Of the many sequences tested, eleven have levels between 0·0002 and 0·006, the smallest being the block between 10,001–20,000 in $(\sqrt{3})_3$.

4. Side Remarks

(a) As a consequence of the present study, the peripheral question arose of generalizing Newton's method for the rth root of integer N. It turns out that there is a class of functions $g_K(w)$, $K \geqslant 2$, for which the recursive sequences $w_{n+1} = g_K(w_n)$ converge to $N^{1/r}$ with relative error $e_{n+1} \cong C(K)\, e_n{}^K$ (Everett and Metropolis, 1969). Newton's method results when $K = 2$. For $r = 2$: if $w_1 = x_1/y_1$, where x_1, y_1 is the first positive solution of Pell's equation $x^2 - Ny^2 = 1$, then $w_{n+1} = x_{n+1}/y_{n+1}$ is the $K^n p$th or $2K^n p$th convergent of the continued fraction for $\sqrt{N}$, its period p being even or odd.

(b) The implications on the logical structure of arithmetic processors has been another peripheral consequence of our present study. Too often, the designer is unaware of the needs of the users, or else has a restricted point of view. It has been possible to develop a unified concept in which one recognizes two basic types of computer numbers—(i) fully representable, and (ii) imprecise quantities with error specification. The general form for the two types is very similar, the distinction being achieved with a tag bit. It would not be appropriate to discuss here the basic arithmetic manipulations; suffice it to say that in general the output of two input operands is fully representable provided both inputs are of that type, otherwise it is of the second type. The detection and implementation would be fully automatic and would not imply additional programming chores. The simplification for programming languages and compilers could be of some importance. The number theorist who uses computers has often called attention to some of their peculiarities and awkwardness.

References

Beyer, W. A., Metropolis, N. and Neergaard, J. R. (1970). Statistical study of digits of some square roots of integers in various bases. *Math. Comp.* **24,** 455–471.

Everett, C. J. and Metropolis, N. (1969). Approximation to the r-th root of N. *Studies in Appl. Math.* To be published.

Good, I. J. and Gover, T. N., (1967). The generalized serial test and the binary expansion of 2. *J. Roy. Stat. Soc. A,* **130,** 102–107.

Metropolis, N. and Stein, P. R. (1967). On a class of (0, 1) matrices with vanishing determinants. J. *Comb. Theory.* **3,** 191–198.

Metropolis, N., Stein, M. L. and Stein, P. R. (1969). "Permanents of cyclic (0, 1) matrices". *J. Comb. Theory.* **7,** 291–321.

Schmidt, W. (1960). On normal numbers. *Pacific J. Math.* **10,** 661–672.
von Mises, R. (1964). "Mathematical Theory of Probability and Statistics". Academic Press, London and New York.
Wells, M. B. (1971). "Elements of Combinatorial Computing". Pergamon Press, Oxford.
Williamson, J. (1946). Determinants whose elements are 0 and 1. *Monthly, Amer. Math. Soc.* **53,** 427–434.

On the Nonexistence of Certain Perfect Codes

J. H. VAN LINT

Technological University, Eindhoven, Netherlands

1. Introduction

Let R^n be the vector space of dimension n over the field $GF(q)$. In coding theory the field is called the alphabet and the vectors of R^n are called words. The Hamming distance of two words is defined to be the number of places in which the words differ; Hamming weight of a word is the number of non-zero elements. With this distance function R^n is a metric space. If $\mathbf{x} \in R^n$ and e is a nonnegative integer the sphere $B(\mathbf{x}, e)$ is the set $\{\mathbf{y} \in R^n: d(\mathbf{x}, \mathbf{y}) \leqslant e\}$, where d is the distance. A subset V of R^n is an *e-error-correcting code* if for each pair of different codewords $\mathbf{x} \in V$, $\mathbf{y} \in V$ the spheres $B(\mathbf{x}, e)$ and $B(\mathbf{y}, e)$ are disjoint. An e-error-correcting code V is called perfect if $R^n = U_{\mathbf{x} \in V} B(\mathbf{x}, e)$. Hamming (1950) and Golay (1949) discovered essentially all the perfect codes that are known today, although very many mathematicians have worked very hard to find more. About a year ago my name was added to the long list of persons who have rediscovered one of the two known non-trivial perfect codes with $e > 1$. Recovering from the disappointment I decided to prove the non-existence of a number of perfect codes and indeed succeeded to do so for $q = 2$ and a number of small odd values of e. Of course these results turned out to be known (*cf.* e.g. Shapiro and Slotnick (1959), Leont'ev (1965)). The next stages of the research used the computer and produced a number of results which I have not seen elsewhere. The next sections will be on these computational results. It would be very useful to find out exactly what is known on the nonexistence of perfect codes and to have the results together in one place. (See Berlekamp, 1968.)

2. A simple computer-search

If V is a perfect e-error-correcting code with word-length n over the field $GF(q)$ where $q = p^\alpha$ then the volume of a sphere $B(\mathbf{x}, e)$ must divide the total number of elements of R^n. The number of elements of R^n is q^n. Hence:

$$\sum_{i=0}^{e} \binom{n}{i} (q-1)^i = p^\beta. \tag{2.1}$$

Since

$$\sum_{i=0}^{n} \binom{n}{i} (q-1)^i = q^n$$

we find by subtraction

$$q^n - p^\beta \equiv 0 \pmod{q-1},$$

which implies $p^\beta = q^k$. Therefore a necessary condition for the existence of a perfect e-error-correcting code over $GF(q)$ with word-length n is

$$\sum_{i=0}^{e} \binom{n}{i} (q-1)^i = q^k. \tag{2.2}$$

In the special case $q = 2$, (2.2) takes the form

$$\sum_{i=0}^{e} \binom{n}{i} = 2^k. \tag{2.3}$$

A number of solutions of (2.3) is obvious, e.g., $e = k = n$ (code consists of one word); $n = 2e + 1$ (code consisting of the all-zero and all-one words); $n = 2^k - 1$ and $e = 1$ (corresponding to the Hamming codes, *cf.* Peterson (1961), Golay (1949)). Two more solutions of (2.3) are known, namely $n = 23$, $e = 3$ (corresponding to the Golay (23,12) code) and $n = 90$, $e = 2$ (there is no perfect code with these parameters). For $q > 2$ the equation (2.2) has one known non-trivial solution $n = 11, e = 2, q = 3$ (corresponding to the Golay (11,6) code). Using known theorems on diophantine equations E. L. Cohen (1964) showed that for $e = 2$ the equation (2.2) has no other solutions for $q \leqslant 6$ and using a computer he showed that for $e = 2$, $3 \leqslant q$ (odd) $\leqslant 125$ and $3 \leqslant k \leqslant 40000$ there are no new solutions to (2.2). E. L. Alter (1968) proved that there is no perfect code with $q = 7$, $e = 2$.

If e is odd the polynomial in n on the left-hand side of (2.3) is divisible by $(n + 1)$. Hence by elementary methods all solutions, for fixed e, can be found (Shapiro and Slotnick (1959)). As far as I know this has been done for $e < 20$. By computer search M. H. McAndrew (1965) showed that there are no new perfect binary codes with $e \leqslant 20$ and $n \leqslant 2^{70}$.

As a result of our research I can announce:

Computer result: For $e \leqslant 10^3$, $q \leqslant 10^2$, $n \leqslant 10^3$ *there are no new solutions to the equation* (2.2). (*Only prime powers q are considered.*)

The program was written by B. van Bree, J. Jansen and H. Willemsen who first wrote new multi-length arithmetic subroutines to keep computing time within reasonable bounds. In the program

$$\text{pol}\,(n, e) := e! \sum_{i=0}^{e} \binom{n}{i} (q-1)^i$$

was generated recursively. The polynomial $\text{pol}\,(n, e) - e!q^k$ has one zero $n > e$. This zero was determined using the Newton–Raphson method. Afterwards a neighborhood of this zero was checked to see if an integer solution of (2.2) was present. Computing time was 7800 sec. on the *EL-X8* at the Technological University Eindhoven.

3. The case q = 2, e = 4

A more interesting application of computer methods will now be treated. The condition (2.2) was found by considering volumes only. By considering possible weight distributions of code vectors in a perfect code S. P. Lloyd (1957) found the following theorem:

THEOREM (Lloyd): *If there exists a perfect e-error-correcting code with word-length $n+1$ over $GF(2)$ then*

$$\phi_e(n, x) := \sum_{i=0}^{e} (-1)^i \binom{n-x}{i} \binom{x}{e-i}$$

has e distinct integer zeros $<n$. Here

$$\binom{x}{i} = \frac{x(x-1)\cdots(x-i+1)}{i!}.$$

Recently Baker and Davenport (1969) continued an investigation of the author, (van Lint, 1968) on a set of diophantine equations and completely solved the problem. It is possible to use their method to attack the case $q = 2$, $e = 4$ as will be shown below.

The polynomial $\phi_4(n, x)$ in Lloyd's theorem can be factored. We find

$$\phi_4(n, x) = \frac{2}{3}\left(x^2 - nx + \frac{n^2 - 3n + 4 + \sqrt{6n^2 - 18n + 16}}{4}\right) \cdot \left(x^2 - nx + \frac{n^2 - 3n + 4 - \sqrt{6n^2 - 18n + 16}}{4}\right). \tag{3.1}$$

Hence a necessary condition for the existence of a binary perfect 4-error-correcting code is that

$$n \pm (3n - 4 \pm \sqrt{6n^2 - 18n + 16})^{\frac{1}{2}}$$

is an even integer for all four possible choices of the sign $\pm$. This leads to the following set of diophantine equations:

$$\begin{aligned} 6n^2 - 18n + 16 &= x^2, \\ 3n - 4 - x &= y^2, \\ 3n - 4 + x &= z^2. \end{aligned} \tag{3.2}$$

Combination of these yields, with $yz = 3t$,

$$(n-1)^2 - 1 = 3t^2, \tag{3.3}$$

$$3(2n-3)^2 + 5 = 2x^2. \tag{3.4}$$

The solution of (3.3) is

$$n - 1 = \tfrac{1}{2}(2 + \sqrt{3})^m + \tfrac{1}{2}(2 - \sqrt{3})^m, \qquad m = 0, 1, \ldots. \tag{3.5}$$

The equation (3.4) has two sequences of solutions:

$$2n - 3 = (\tfrac{1}{2} + \tfrac{1}{3}\sqrt{6})(5 + 2\sqrt{6})^k + (\tfrac{1}{2} - \tfrac{1}{3}\sqrt{6})(5 - 2\sqrt{6})^k, \tag{3.6}$$

$$2n - 3 = (\tfrac{3}{2} + \tfrac{2}{3}\sqrt{6})(5 + 2\sqrt{6})^k + (\tfrac{3}{2} - \tfrac{2}{3}\sqrt{6})(5 - 2\sqrt{6})^k. \tag{3.6a}$$

The method of Baker and Davenport enables us to find all pairs m, k such that the values of n given by (3.5) and (3.6) or (3.6a) respectively, are equal. There are 3 intervals which need separate treatment, two of them computational.

First, large n are excluded by Baker's (1969) effective form of a theorem of Gelfand.

THEOREM (A. Baker): *Suppose that $l \geqslant 2$ and that $\alpha_1, \ldots, \alpha_l$ are non-zero algebraic numbers, whose degrees do not exceed d and whose heights do not exceed A, where $d \geqslant 4$ and $A \geqslant 4$. If the rational integers $b_1, \ldots, b_l$ satisfy*

$$0 < |b_1 \log \alpha_1 + \ldots + b_l \log \alpha_l| < e^{-\delta H}$$

where $0 < \delta \leqslant 1$ and

$$H = \max(|b_1|, \ldots, |b_l|),$$

then

$$H < (4^{l^2} \delta^{-1} d^{2l} \log A)^{(2l+1)^2}.$$

If n satisfies (3.5) and (3.6) then

$$|m \log(2+\sqrt{3}) - k \log(5+2\sqrt{6}) - \log(\tfrac{1}{2}+\tfrac{1}{3}\sqrt{6})| < (3.73)^{-m}. \quad (3.7)$$

For (3.5) and (3.6a) the last logarithm is to be replaced by $\log(\frac{3}{2}+\frac{2}{3}\sqrt{6})$. We can apply the theorem of Baker with $\alpha_1 = 2+\sqrt{3}$, $\alpha_2 = 5+2\sqrt{6}$ and $\alpha_3 = \frac{1}{2}+\frac{1}{3}\sqrt{6}$ or $\frac{3}{2}+\frac{2}{3}\sqrt{6}$, taking $\delta = 1$, $d = 4$, $A = 36$. The result is that $k < m < 10^{470}$.

A lower bound for n is found by a sieve-method used in van Lint (1968). (As far as practical use of perfect codes is concerned this lower bound is high enough.) The sequences of values for n given by (3.5) and (3.6) or (3.6a) are considered mod p for consecutive primes p. These sequences are periodic, with period $P(p)$, and the assumption that the sequences have an element in common leads to congruence conditions modulo $P(p)$ on the indices m and k. Combination of these conditions (Chinese remainder theorem) then gives congruence conditions, modulo a very large number, on m and k. This was done on the *EL-X*8 for $p \leqslant 43$. In 30 sec. computing time it was found that the sequences (3.5) and (3.6) or (3.6a) do not give the same values for n with $n < 10^{400000}$ except $n = 2$, $n = 3$ and $n = 8$.

For the remaining interval Baker and Davenport give the following lemma:

LEMMA (Baker and Davenport): *Let $K > 6$, M, p, q be positive integers satisfying*

$$1 \leqslant q \leqslant Km, \quad |\theta q - p| < 2(KM)^{-1}, \quad \|q\beta\| \geqslant 3K^{-1}$$

where θ and β are irrationals and $\|z\|$ denotes the distance of a real number z from the nearest integer. Then the inequality

$$|m\theta - k + \beta| < c^{-m}$$

has no solution in the range

$$\frac{\log K^2 M}{\log c} < m < M.$$

From (3.7) we find

$$\left| m \frac{\log(2+\sqrt{3})}{\log(5+2\sqrt{6})} - k - \frac{\log(\frac{1}{2}+\frac{1}{3}\sqrt{6})}{\log(5+2\sqrt{6})} \right| < 2(2+\sqrt{3})^{-m} \qquad (3.8)$$

and a corresponding inequality with $\frac{3}{2}+\frac{2}{3}\sqrt{6}$ instead of $\frac{1}{2}+\frac{1}{3}\sqrt{6}$.

To exclude the remaining possible values of k we make suitable choices for

K and M and find values of p and q satisfying the conditions of the lemma by computing

$$\theta = \frac{\log(2 + \sqrt{3})}{\log(5 + 2\sqrt{6})}$$

to 1000 decimal places and then taking for p/q a suitable convergent to the continued fraction of this approximation. We omit the details since the method is fully described in the Baker and Davenport (1969) paper. This part of the computation was done at the Atlas Computer Laboratory by Mr. W. F. Lunnon. For this assistance the author expresses his gratitude. Concerning the program Mr. Lunnon gave the following details. The logarithms are done by Thiele's expansion to 1175 decimal places (100-length) and checked by reversing the expressions. The continued fraction is done by straightforward iteration stopping when q exceeds a given bound. The computation used the *ABC* multilength system on ATLAS I at Chilton and took two hours.

Combining the results of all these computations we have

THEOREM: *There are no non-trivial binary perfect* 4-*error-correcting codes.*

References

Alter, R. (1968). On the Nonexistence of Close-Packed Double Hamming-Error-Correcting Codes on $q = 7$ Symbols. *J. of Comp. and Syst. Sci.* **2,** 169–176.

Baker, A. and Davenport, H. (1968). The Equations $3x^2-2=y^2$ and $8x^2-7=z^2$. *Quart. J. of Math.* **20,** 129–137.

Berlekamp, E. R. (1968). Algebraic Coding Theory. McGraw Hill, New York.

Cohen, E. L. (1964). A Note on Perfect Double Error-Correcting Codes on q Symbols. *Inf. and Control.* **7,** 381–384.

Golay, M. J. E. (1949). Notes on Digital Coding. *Proc. I.R.E.* **37,** 657.

Hamming, R. W. (1950). Error-Detecting and Error-Correcting Codes. *B.S.T.J.* **29,** 147–160.

Leont'ev, V. K. (1965). On the Existence of Densely Packed Codes (Russian). *Problemy Kibernet.* **15,** 252–257.

Lloyd, S. P. (1957). Binary Block Coding. *B.S.T.J.* **36,** 517–535.

McAndrew, M. H. (1965). An Algorithm for Solving a Polynomic Congruence, and its Application to Error-Correcting-Codes. *Math. of Comp.* **19,** 68–72.

Peterson, W. W. (1961). "Error-Correcting Codes". MIT Press, Cambridge, Massacheussets.

Shapiro, H. S. and Slotnick (1959). On the Mathematical Theory of Error-Correcting Codes. *IBM J. of Res. and Dev.* **3,** 25–34.

van Lint, J. H. (1968). "On a Set of Diophantine Equations". Report 68–WSK–03 of the Technological University, Eindhoven.

A Class of Theorems in Additive Number Theory which Lend Themselves to Computer Proof

STEFAN A. BURR

Bell Telephone Laboratories Incorporated, Whippany, New Jersey, U.S.A.

1. Introduction

Let T be an arbitrary sequence of integers. Define $P(T)$ to be the set of all integers which are representable as a sum of distinct terms of T. ("Distinct" means having distinct indices; if a value occurs twice in T, it may be used twice in a representation.) We say T is complete if $P(T)$ contains all sufficiently large integers. An obvious example of a complete sequence is the powers of two (starting with 2^0), which generates all integers. We can modify this example to give one where the completeness starts only at a large value:

$$1001, 2, 4, 8, 16, \ldots, 2^n, \ldots ;$$

clearly 999 is not representable, but every number from 1000 on is.

2. A Completeness Algorithm

An interesting feature of the study of completeness of sequences is that it is possible to use a computer to prove sequences complete. Probably the most convenient method for doing this in general is an unpublished one given by R. L. Graham, embodied in the following result. Let $T = \{t_1, t_2, \ldots\}$ be a sequence for which a k and an α exist such that $1 \leqslant \alpha < 2$ and

$$t_i < t_{i+1} \leqslant \alpha t_i \quad \text{for} \quad i \geqslant k.$$

Suppose that an X and Y exist such that

$$t_k \leqslant X \leqslant \left(1 - \frac{\alpha}{2}\right) Y,$$

and such that for every $X \leqslant x \leqslant Y$, $x \in P(T)$. Then $P(T)$ contains every $x \geqslant X$.

The proof of this is very simple. Clearly, for any $y \geqslant t_k$, integral or not, there exists an n such that

$$t_{n-1} < y \leqslant t_n \leqslant \alpha t_{n-1} < \alpha y;$$

thus

$$y \leqslant t_n < \alpha y.$$

In particular there is an n such that

$$\tfrac{1}{2} Y + 1 \leqslant t_n < \alpha(\tfrac{1}{2} Y + 1).$$

Consider the number $Y + 1 - t_n = t$. We wish to show that $X \leqslant t \leqslant Y$. Clearly $t \leqslant Y$, and

$$Y + 1 - t_n > Y + 1 - \alpha(\tfrac{1}{2} Y + 1) = \left(1 - \frac{\alpha}{2}\right) Y + 1 - \alpha$$

$$\geqslant X + 1 - \alpha > X - 1.$$

But both ends of the inequality are integers, so

$$t = Y + 1 - t_n \geqslant X.$$

Since t is in the interval $[X, Y]$ we have the representation

$$t = t_{i_1} + \ldots + t_{i_r};$$

but none of these t_i can equal t_n, since

$$t = Y + 1 - t_n \leqslant Y + 1 - (\tfrac{1}{2} Y + 1) = \tfrac{1}{2} Y < t_n.$$

Hence we have

$$Y + 1 = t + t_n = t_{i_1} + \ldots + t_{i_r} + t_n,$$

and $Y + 1 \in P(T)$. Continuing the argument by induction, we have the desired result.

To use a computer to prove a sequence complete by this method, we first estimate α by hand (this is usually trivial), and then generate a segment of $P(T)$ on the computer and check for the existence of a suitable interval $[X, Y]$. If successful this also yields the last number not in $P(T)$; this we call the threshold of completeness. We do not discuss computer implementations of these algorithms here; suffice it to say that it is often possible to generate $P(T)$ into the millions (and to check for suitable $[X, Y]$) in less than a minute.

The completeness theorems proved in this manner are of interest in themselves (especially since they may lead to conjectures of more general results), but there is considerable theoretical interest in the method, independent of the results obtained. The form of proof is to check a possibly very large number of cases to get a suitable initial condition: $[X, Y] \subset P(T)$, followed by a simple induction. In other words, the overall proof is an induction in which the induction step is easy, but in which the initial condition is hard to check. This cooperation between man and machine is quite necessary; neither can do without the other.

3. An Incompleteness Algorithm

In this paper we will discuss another example of man–machine cooperation to produce proofs. The method is completely unlike the previous one; for instance this method in effect shows sequences not complete, and it can be applied to only a very restricted class of sequences. But the most striking difference is the different nature of the cooperation; the proofs generated are again in the form of an induction, but this time the initial condition is easy and the induction step is the hard part that the computer checks.

We are going to be concerned with what we will call recurring sequences, which we define to be integer sequences T which satisfy a linear recursion relation of the form

$$a_0 t_n + a_1 t_{n+1} + \dots + a_{k-1} t_{n+k-1} = t_{n+k}, \tag{1}$$

where the a_i are fixed integers. A common example is the sequence of Fibonacci numbers, which satisfy

$$t_n + t_{n+1} = t_{n+2}, \quad \text{with} \quad t_0 = 0, \qquad t_1 = 1.$$

We call k the order of the equation or the sequence. It is clear that an initial condition of k consecutive terms is sufficient to determine the sequence from that point on. (Our sequences do not necessarily start at $n = 0$.)

In studying the completeness properties of recurring sequences it appears especially fruitful to consider not completeness as such but a stronger property we will call essential completeness. We define a sequence to be essentially complete if it remains complete when any finite set of terms is removed from it. Clearly, necessary conditions for this are that a sequence have an infinite number of positive terms and that there exist no prime p such that p divides all but a finite number of terms. A few results are known for recurring sequences. For instance, Roth and Szekeres (1954) and Graham (1964a) have shown that any integer-valued polynomial satisfying the above two conditions is essentially complete. Also, the Fibonacci sequence, defined above, is not essentially complete (Brown, 1961). On the other hand the sequence defined by

$$t_n = F_n - (-1)^n,$$

where $\{F_n\}$ is the Fibonacci sequence, is essentially complete (Graham, 1964b). The proof of this last result can be generalized to other sequences; but we will be concerned with generalizations of the preceding result. We will give an algorithm by which a computer can prove certain recurring sequences not essentially complete. It also has some application to other sequences than recurring sequences.

We will state the method in the form of a theorem. Since this statement is somewhat complex, we make a few preliminary definitions. We say that two sequences T_1 and T_2 are equivalent if one is eventually a shifting of the other. We say that a sequence $T = \{t_n\}$ is definite if $t_n = 0$, $t_n \to \infty$, or $t_n \to -\infty$ as $n \to \infty$; otherwise indefinite. Let T and S be a pair of fixed sequences. (In the theorem, S will be nearly the sum sequence of T.) Let $U = \{u_n\}$ and $V = \{v_n\}$ be sequences. If for some k, unrestricted in sign, $v_n = u_n - t_{n+k}$ eventually, we say that V is a derived sequence of U. (We use the word "eventually" in this definition because it is convenient to ignore the starting point of a sequence.) Consider simultaneously the sequence $W = \{w_n\}$ defined by $w_n = s_{n+k} - u_n$. If $v_n \to \infty$ and $w_n \to \infty$ as $n \to \infty$, we say that V is a permissible derived sequence of U. If for every choice of k, the sequences V and W resulting from U are definite, we say U is unambiguous. We may now state our result.

THEOREM 1. *Let $T = \{t_n\}$ be an increasing sequence of integers, and let $S = \{s_n\}$ differ from its sum sequence $\left\{\sum_{i=1}^{n} t_i\right\}$ by a bounded amount. (S is not restricted to integer values.) Suppose there exists a finite set $\mathcal{T} = \{T_1, \ldots, T_N\}$ of increasing sequences which are unambiguous (relative to T and S as above), and such that every permissible derived sequence of any member of $\mathcal{T}$ is equivalent to some member of $\mathcal{T}$. Then T is not essentially complete.*

Proof. We will show, by the method of descent, that, for some m, if the first m terms are removed from T, then no positive number of the form $u_j - 1$, where u_j is an element of a member of $\mathcal{T}$, is in $P(T)$.

We must first choose a suitable value of m. Let $U = \{u_n\}$ be a member of $\mathcal{T}$. Then those k for which the corresponding derived sequences are permissible form a finite interval of values $k_0, \ldots, k_l$. Then, for sufficiently large j, $u_j \geqslant s_{j+k_0-1}$ and $u_j \leqslant t_{j+k_l+1}$, since no derived sequence of U is indefinite. Furthermore, for sufficiently large j, $t_{j+k_l} < u_j - 1$ for the same reason.

Since T and S are increasing, the same inequalities hold with k_0 replaced by any smaller number in the first, k_l by any larger in the second, and k_l by any smaller in the third. Thus we see that $t_{j+k} < u_j < s_{j+k}$ implies that the corresponding derived sequence is permissible, provided j is sufficiently large. With the same proviso, $t_n \neq u_j - 1$ for any n. We will call such a u_j normal.

We now choose m sufficiently large that if u_j is any term of a sequence in $\mathscr{T}$ for which $u_j > t_m$, then u_j is normal. In addition we choose m large enough to ensure that

$$\sum_{i=m+1}^{n} t_i < s_n - 1$$

for every $n > m$.

Let $u_j - 1$, with u_j an element of a sequence in $\mathscr{T}$, be the smallest number of that form which is a sum of distinct elements $t_i > t_m$ of T. Let t_n be the largest such element; then

$$\sum_{i=m+1}^{n} t_i \geqslant u_j - 1.$$

Thus $s_n > u_j$. But clearly $u_j > t_n$, so that (since u_j is normal) the derived sequence $V = \{\ldots, u_j - t_n, u_{j+1} - t_{n+1}, \ldots\}$, with $k = n - j$, is permissible. We have seen that $t_n \neq u_j - 1$; hence $u_j - 1 - t_n = v_i - 1$, with v_j a term of some sequence $T_K \in \mathscr{T}$. But this is a contradiction; so the theorem is proved.

We now consider how this theorem can be applied to recurring sequences. In fact, as we shall later see, it may only be applicable to two very special subsets of these sequences. Let T be a recurring sequence satisfying equation (1). Clearly, any shifting of T also satisfies (1), as does any linear combination of sequences satisfying (1). Thus all derived sequences of T, their derived sequences in turn, and so on, satisfy (1) eventually. The algorithm will consist of taking $T_1 = T$ and generating permissible derived sequences until the process terminates or machine capacity is exhausted. We will assume that the characteristic polynomial of T is irreducible and of degree at least 2. Hence t_n has the form

$$c_1\alpha_1{}^n + \ldots + c_k\alpha_k{}^n, \tag{2}$$

where $\alpha_1, \ldots, \alpha_k$ are the roots of the characteristic polynomial. The sum sequence therefore has the form

$$C_1\alpha_1{}^n + \ldots + C_k\alpha_k{}^n + C.$$

In fact, $C_i = c_i \cdot \alpha_i(\alpha_i - 1)$ and $C = \sum_{i=1}^{k} c_i/(\alpha_i - 1)$. We will take S to be the sequence defined by

$$s_n = C_1\alpha_1{}^n + \ldots + C_k\alpha_k{}^n;$$

thus S has the desired property. It has the valuable additional property of satisfying the original difference equation (1). Furthermore, s_n assumes only rational values. To see this, we note that the sum sequence assumes only integral values, so we must show that C is rational. But $\{s_n\}$ satisfies (1), so we have

$$a_0(t_0 - C) + a_1(t_0 + t_1 - C) + \dots + a_{k-1}(t_0 + \cdots + t_{k-1} - C)$$
$$= t_0 + \dots + t_k - C.$$

Hence, C is a rational number. (The coefficient of C in the above expression cannot be zero, since then 1 would be a root of the characteristic polynomial.) For computational purposes it is convenient to multiply both T and S by the denominator of C, so as to deal only with integers.

To implement Theorem 1 we need means of identifying and comparing sequences. A straightforward way of identifying a sequence satisfying (1) is by specifying k consecutive terms, since this uniquely determines such a sequence. However, a one-to-one relationship is very desirable, although not absolutely necessary. An obvious way of approaching this end is to specify that the last term be greater than or equal to some fixed number B, and that the preceding $k - 1$ terms be less than B. We call this a semi-canonical representation. This is still not enough to guarantee uniqueness, so that some additional criterion is needed.

Two simple special criteria come to mind, sufficient for most cases of interest. First, suppose that all the coefficients a_i of (1) are nonnegative. Then if all the terms appearing in a semi-canonical representation are positive, it is easy to see that the representation is the unique one with that property. Second, suppose that all the sums $a_0 + \dots + a_{k-1}, a_1 + \dots + a_{k-1}, \dots, a_{k-1}$ are at least 1. (The first is then at least 2.) Then a simple application of Abel summation shows that if k consecutive terms of a sequence satisfying (1) are positive and increasing, so are all later terms. Thus, if either of the two conditions on the a_i are satisfied, a semi-canonical representation becomes truly canonical if we specify that the k terms be positive and increasing. Of course, a given sequence need not have such a canonical representation, but this is no great harm in a tentative algorithm. Needless to say, if the algorithm fails to find such a representation for a sequence, it should terminate immediately and indicate failure.

This also gives us a means of comparing sequences, since if a sequence has a canonical representation it must approach infinity, and if it has the negative of a canonical representation it must approach minus infinity. If it has neither and is not identically zero it should be treated as if it were indefinite and the algorithm terminated.

4. Generalizations of the Algorithm

As we shall see shortly, Theorem 1 is not very widely applicable in practice. We will give a generalization of this theorem which extends its range considerably, namely to repetitions of sequences. Let $T = \{t_n\}$ be a recurring sequence. Set

$$T^{(1)} = t_1, t_2, t_3, \ldots = T,$$

$$T^{(2)} = t_1, t_1, t_2, t_2, t_3, t_3, \ldots,$$

$$T^{(3)} = t_1, t_1, t_1, t_2, t_2, t_2, \ldots,$$

$$\ldots\ldots\ldots\ldots\ldots\ldots\ldots\ldots\ldots\ldots\ldots$$

Clearly each $T^{(i)}$ is a recurring sequence.

Theorem 1 does not apply directly to these sequences, primarily because some derived sequence will be indefinite. However, we may easily generalize Theorem 1 to this case.

THEOREM 2. *In the statement of Theorem* 1, *let S be replaced by a sequence differing from the ith multiple of the sum sequence by a bounded amount. Then if the conditions of the theorem hold for this S, then $T^{(i)}$ is not essentially complete.*

The proof of this theorem differs only trivially from that of Theorem 1; its implementation as an algorithm is the same as that of Theorem 1, except that S is multiplied by i.

As stated the algorithms deal with a sequence specified by a difference equation and a set of initial conditions; but we shall soon see that the initial conditions are unimportant in the most important case. We first state, as a lemma, a result proved by Cohoon and Burr (1970).

LEMMA 1. *Let S be a sequence of rationals satisfying a difference equation whose characteristic equation is irreducible, so that we have*

$$t_n = c_1\alpha_1{}^n + \ldots + c_k\alpha_k{}^n,$$

where the α_i are the roots of the characteristic equation. Then no c_i is zero unless the sequence is identically zero.

Let us define a smooth sequence to be a recurring sequence whose characteristic equation is irreducible and which has a unique root of largest magnitude, which is positive. From the above lemma it follows that a smooth sequence, the largest root of whose characteristic equation is α, is asymptotic to $c\alpha^n$ for some c. Hence any smooth sequence is definite. Moreover, the sequence is uniquely determined by c. We now show that Theorem 1 does not depend in any very essential way on the initial conditions, if the sequence is smooth.

THEOREM 3. *Let T be a smooth sequence which satisfies the hypotheses of Theorem 2, and suppose that every sequence of $\mathcal{T}$ is ultimately an integral linear combination of shiftings of T. Then if T' is a recurring sequence satisfying the same difference equation, $T^{(i)}$ is not essentially complete.*

Proof. We mention in passing that the above assumption about the sequences of $\mathcal{T}$ is for convenience and can be omitted. (In any case, if $\mathcal{T}$ is generated as a succession of sequences derived ultimately from T, the assumption is justified.)

Let the difference equation in question be (1), as usual. Then it is clear from Lemma 1 that any recurring sequence satisfying (1) is ultimately nonpositive or it is ultimately positive and increasing. We need only consider the latter case. Let the coefficient of α_i^n in the expression for t_n be c_i, and let $\gamma_i c_i$ be the coefficient in the corresponding expression for t_n', the general term of T'. The coefficient of α_i^n in S is $c_i\alpha_i/(\alpha_i - 1)$. If we let S' be the sequence defined by $s_n' = \gamma_1\alpha_1/(\alpha_1 - 1) + \ldots + \gamma_k\alpha_k/(\alpha_k - 1)$, then S' differs from the sum sequence of T' by a constant. Let $\mathcal{T}'$ be the set of sequences T_j' which are the same linear combinations of shiftings of T' that the sequences in $\mathcal{T}$ are of T.

Thus all the linear relationships among T, S, and members T_j of $\mathcal{T}$ are preserved when primes are affixed. Moreover, the coefficient d_i of α_i^n in T_j is mapped into $\gamma_i d_i$ in T_j'; and we have seen that the same is true for T and S. Hence, letting α_1 be the root largest in magnitude, we see that all the inequality relations among T, S, and the T_j are preserved when primes are affixed. Thus, T', S', and $\mathcal{T}'$ satisfy the hypotheses of Theorem 2, and the theorem is proved. This result is interesting in that it means that one finite sequence of integers represents not just an infinite sequence, but an infinite number of infinite sequences.

We see that this algorithm is as described before; we prove a sequence not essentially complete by induction. Starting the induction is easy, since once we know how much the induction step requires we can always remove enough terms of the sequence to start the induction. On the other hand, the induction step is difficult to check; but the computer does this part of the work.

5. Results of the Algorithm

The algorithm is of course interesting in itself; but we also seek results. The algorithm has been tried on a considerable number of sequences and so far it has been successful on sequences satisfying a considerable number of different difference equations, including cases involving repetitions. However, as we shall see, these all belong so far to a special class for which a different proof may be given. Before going into more detail let us eliminate certain sequences from consideration. First, we will only consider smooth sequences or repetitions thereof, since other sequences will in general lead to an indefinite derived sequence. Moreover, we can easily see by a counting argument that if the characteristic equation of a recurring sequence T has a root greater than 2^i the sequence $T^{(i)}$ is not even complete. But we can show considerably more:

THEOREM 4. *Let $i \geqslant 1$ be given and let T be a smooth sequence whose characteristic equation has a root greater than* $\max(i, (1+\sqrt{5})/2)$. *Then $T^{(i)}$ is not essentially complete.*

In view of Theorem 3 this is an immediate consequence of:

THEOREM 5. *Let $i \geqslant 1$ be given and let $T = \{t_n\}$ be an increasing sequence of positive integers. If $i = 1$ suppose that $t_n \geqslant t_{n-1} + t_{n-2} + 1$ for large n; if $i \geqslant 2$ suppose that $t_n - (i-1)(t_1 + \ldots + t_{n-1}) \to \infty$ and $n \to \infty$. Then $T^{(i)}$ is not essentially complete.*

Proof. Let $S = \{s_n\}$ be defined by $s_n = i \cdot (t_1 + \ldots + t_n)$. If $i = 1$, let $\mathscr{T}$ consist of the single sequence $U = \{u_n\}$ defined by $u_n = t_n + t_{n-2} + t_{n-4} + \ldots$, terminating at t_1 or t_2. We have

$$\begin{aligned} u_n - s_{n-1} &= (t_n + t_{n-2} + t_{n-4} + \ldots) - (t_{n-1} + t_{n-2} + \ldots + t_1) \\ &\geqslant (t_{n-1} + t_{n-2} + \ldots + t_1 + (n-1)/2) \\ &\quad - (t_{n-1} + t_{n-2} + \ldots + t_1) \\ &\to \infty \quad \text{as} \quad n \to \infty. \end{aligned}$$

On the other hand,

$$\begin{aligned} t_{n+1} - u_n &\geqslant (t_n + t_{n-1} + 1) - (t_n + t_{n-2} + \cdots) \\ &= t_{n-1} - u_{n-2} + 1 \\ &\geqslant t_{n-3} - u_{n-4} + 2 \\ &\ldots\ldots\ldots\ldots\ldots\ldots\ldots\ldots \\ &\geqslant (n-1)/2 \to \infty \quad \text{as} \quad n \to \infty. \end{aligned}$$

Thus the only permissible derived sequence of $\{u_n\}$ is $\{u_n - t_n\} = \{u_{n-2}\}$. Thus, by Theorem 1, T is not essentially complete. This result is basically due to J. H. Folkman, see Graham (1964c).

Now suppose $i \geqslant 2$. Let $\mathscr{T}$ consist of the single sequence $U = \{u_n\}$ defined by $u_n = t_1 + \ldots + t_n$. We have

$$\begin{aligned} u_n - s_{n-1} &= (t_1 + \ldots + t_n) - i(t_i + \ldots + t_{n-1}) \\ &= t_n - (i-1)(t_1 + \ldots + t_{n-1}) \\ &\to \infty \quad \text{as} \quad n \to \infty. \end{aligned}$$

On the other hand,

$$t_{n+1} - u_n = t_{n+1} - (t_1 + \dots + t_n) \to \infty \quad \text{as} \quad n \to \infty.$$

Thus the only permissible derived sequence of $\{u_n\}$ is $\{u_n - t_n\} = \{u_{n-1}\}$. Thus, by Theorem 2, $T^{(i)}$ is not essentially complete.

A considerable number of sequences of appropriate type have been examined by a program written for the General Electric 635. So far, the only sequences for which the algorithm has been successful have been those whose characteristic equation is that of a so-called P–V number, or replications of such a sequence. We will call such sequences P–V sequences. P–V numbers are defined to be algebraic integers whose other conjugates are all less than one in absolute value. They derive their name from the fact that they were first studied by Pisot (1938) and Vijayaraghavan (1941).

Unfortunately for the usefulness of the algorithm as a generator of new theorems, there is a direct way of proving that P–V sequences and their replications not essentially complete. The basic idea appears in Cassels (1960), page 123; the author is grateful to Professor Cassels for pointing out the idea and its application to the present case. We will give a proof of the result for the sake of completeness. As usual, put $\|x\| = \min_n |x - n|$, where n ranges over all positive and negative integers.

LEMMA 2 (Cassels). *Let $T = \{t_n\}$ be a sequence of integers, and suppose that for some real α, $\sum_n \|\alpha t_n\| < \infty$. Then T is not essentially complete.*

Proof. Let T' be a sequence formed by removing a finite number of terms of T, and for which $\sum_{t \in T'} \|\alpha t\| < \frac{1}{4}$; this is clearly possible. There are infinitely many integers b for which $\|\alpha b\| \geqslant \frac{1}{4}$; but clearly none of these can be in $P(T')$. This proves the result.

We may now prove

THEOREM 6 (Cassels). *Let T be a P–V sequence with corresponding P–V number $\alpha > 1$. Then $T^{(i)}$ is not essentially complete for any i.*

Proof. The theorem is trivial if α is a (rational) integer, so let α be irrational. Let t_n have the representation (2) of Section 3 with $\alpha_1 = \alpha$. Then we have

$$\begin{aligned}\|\alpha t_n\| &\leqslant |\alpha t_n - t_{n+1}| \\ &= |\alpha \cdot c_1 \alpha^n + \alpha \cdot c_2 \alpha_2{}^n + \cdots + \alpha \cdot c_k \alpha_k{}^n - c_1 \alpha^{n+1} - \dots - c_k \alpha_k{}^{n+1}| \\ &= |(\alpha - \alpha_2)\, c_2 \alpha_2{}^n + \dots + (\alpha - \alpha_k)\, c_k \alpha_k{}^n|.\end{aligned}$$

But $\alpha_2, \ldots, \alpha_k$ are all less than 1 in absolute value; therefore $\Sigma_n \|\alpha t_n\| < \infty$. This completes the proof, since the same inequality holds for $T^{(i)}$.

Although the fact that the theorems thus far proved on the computer may be proved in another way is disappointing, it is nevertheless interesting that such fairly intricate computer proofs are possible at all. Note further that Theorem 6 does not say that the algorithm must terminate in these cases. Moreover, it is not ruled out that other sequences may be found to which the algorithm may be applied. Finally, the set $\mathscr{T}$, with the relations among its members is of interest in itself.

The results so far proved by the program are summarized in Table I. The first column is just an index; the next two columns give the order k and the coefficients $a_0, \ldots, a_{k-1}$ of the difference equation of a sequence T. The fourth column gives the value of the largest root α. The other columns, labeled 1, 2, ..., represent $T^{(1)}$, $T^{(2)}$, ...; the occurrence of an asterisk or a number means that the sequence in question is not essentially complete. An asterisk means that the sequence is covered by Theorem 4. A number means that the program was successful in generating a proof by means of Theorem 2; the number itself is N, the number of sequences in $\mathscr{T}$ as generated.

Sequences numbered 2–4, 5–7, and 8–10 represent three infinite families of P–V numbers approaching $(1 + \sqrt{5})/2$ from below. These families, plus the isolated cases numbered 1 and 11 represent the only P–V numbers $\leqslant (1 + \sqrt{5})/2$; this was proved by Dufresnoy and Pisot (1955a). Hence, in view of Theorem 4, we see that Theorem 1 alone can only be applied to a small number of P–V sequences. On the other hand, Dufresnoy and Pisot (1955b) have shown that the set of accumulation points of the set of P–V numbers has in turn an accumulation point at 2. Thus we can expect to find quite a large number of P–V sequences to which Theorem 2 can be applied, even with $i = 2$.

A point of interest is that a few of the P–V numbers appearing in the table are powers of other P–V numbers in the table, and therefore the corresponding sequences are subsequences of other sequences. Thus sequences 13 and 19 are subsequences of sequences 5 and 8 respectively, and sequences 21 and 23 are subsequences of sequence 1.

It is interesting to consider, for fixed T, the N corresponding to $T^{(1)}$, $T^{(2)}$, ...; rather surprisingly, N appears to grow more slowly than an exponential. This is especially interesting because within a family of sequences (say sequences 2–4), the growth of N as k is increased seems very rapid.

It should be noted that the visible distinction between P–V sequences and others goes beyond simple success or failure of the algorithm; the distinction is clear in the detailed behaviour of the algorithm. One measure of how well the algorithm is doing is its backlog of sequences as it proceeds. This is the difference between the number of sequences added to $\mathscr{T}$ and the number of

TABLE I

	k	(a_j)	α	1	2	3	4	5	6	7	8	9	10	11	12
1.	2	1, 1	1·618	1	3	10	15	28	45	55	78	91	120	153	171
2.	3	1, 0, 1	1·466	2	43	190	473	866	1612	2636					
3.	5	1, 0, 1, 0, 1	1·570	21	2097										
4.	7	1, 0, 1, 0, 1, 0, 1	1·610	208											
5.	3	1, 1, 0	1·325	14	129	522	1242	2405							
6.	5	1, 1, 1, 0, 0	1·534	77											
7.	7	1, 1, 1, 1, 1, 1, 0	1·590	1296											
8.	4	1, 0, 0, 1	1·380	48	2051										
9.	5	1, 0, −1, 1, 1	1·443	167											
10.	6	1, 0, −1, 0, 1, 1	1·502	668											
11.	6	1, −1, 1, 0, −1, 2	1·562	587											
12.	3	1, −1, 2	1·755	*	18	74	173	320	563	935	1434	2071			
13.	5	1, −1, 1, −1, 2	1·674	*	810										
14.	3	1, 1, 1	1·839	*	4	39	97	227	377	603	848	1355	1810		
15.	5	1, 0, 0, 1, 1	1·705	*	594										
16.	3	−1, 1, 2	2·247	*	*	19	43	81	179	275	464	643	962	1250	1785
17.	4	−1, 0, 2, 1	1·905	*	61	473	1841								
18.	6	−1, 0, 1, 1, 1	1·778	*	1756										
19.	2	1, 2	2·414	*	*	2	6	12	12	20	30	42	56	72	
20.	2	1, 3	3·303	*	*	*	3	3	9	9	18	18	30	30	30
21.	2	1, 4	4·236	*	*	*	*	4	4	4	12	12	12	24	24
22.	2	2, 3	3·562	*	*	*	4	5	13	14	25	26	28	45	63
23.	2	−1, 3	2·618	*	*	4	7	11	16	22	29	37	46	56	67
24.	2	−1, 4	3·732	*	*	*	3	7	7	13	13	21	21	31	31

sequences whose derived sequences have been computed. With a P–V sequence the backlog at first rises at a moderate rate, levels off, and then declines more or less linearly until it becomes zero. With other sequences, the backlog rises rapidly from the start and shows no sign of leveling off. In fact, the rise may accelerate. Furthermore, the sequences generated soon become visibly irregular, often forcing the algorithm to terminate after a small number of steps. This occurs even when the sequence is one believed not to be essentially complete.

The algorithm has not been successful on any sequence not either one related to a P–V sequence or one covered by Theorem 4. It may very well be true that it will be successful only on these sequences. However, the existence of Theorem 4 suggests that exceptions might exist, analogous in some way to the sequences of Theorem 4. Another possible source of exceptions might be sequences whose characteristic equation is that of a number in the class (T), first studied by Salem (1945). These are algebraic integers whose other conjugates are less than or equal to one in absolute value, with equality occurring for at least one conjugate. So far, no such sequences have been tested.

There is other evidence than that presented here for the conjecture that the algorithm will be successful on any P–V sequence. R. L. Graham has given an ingenious proof (unpublished) of this fact for the sequence $F^{(i)}$, where F is the Fibonacci sequence. Unfortunately, this result does not appear to generalize easily. Also, it is easy to see that the same is true if $T = \{k^n\}$, where k is an integer $\geqslant 2$. Since any integer $\geqslant 2$ is a P–V number, this lends additional weight to the conjecture. (Generally, 1 is excluded from the set of P–V numbers; if we choose not to do so, $\{1^n\}$ becomes a trivial exception to the conjecture.)

The completeness algorithm described in Section 2 is of value in studying the behaviour of sequences, not necessarily with irreducible characteristic equations, under successive removal of the early terms. For some sequences the resulting thresholds appear to grow exponentially, that is, as a power of the first term remaining. Thus these sequences appear to be essentially complete. For others the thresholds appear to grow hyper-exponentially. These include some, such as $\{F_n + 1\}$, which are known to be essentially complete. Finally, some sequences display thresholds which grow moderately and suddenly blow up, suggesting that the sequences are not essentially complete. These are not limited to P–V sequences. Thus the completeness properties of recurring sequences display a considerable amount of interesting structure. It is to be hoped that computer algorithms discussed here may lead to the discovery and proof of theorems of a more general character.

6. Conclusion

We briefly discuss here a generalized algorithm and also some possible improvements of the algorithm. The generalization that we discuss is the extension to sequences formed by combining two or more different sequences having the same difference equation. Thus, we may consider the sequence 1, 1, 2, 3, 3, 4, 5, 7, 8, 11, 13, 18, 21, ... formed by joining together the Fibonacci sequence with its associated sequence. Of course, the result of Cassels applies to these sequences too, as can be seen immediately. The generalized algorithm is somewhat tedious to state, and we shall not do so. However, it should be relatively clear to the reader what to do; the principal change involves generalizing the definition of a permissible derived sequence. Although no computer program has been written implementing this generalization, a hand calculation has shown the above sequence not to be essentially complete; $\mathcal{T}$ in this case had 13 members. Thus the scope of the generalization is nontrivial.

We will not discuss such improvements in the implementation of the results as packed representations of sequences, arranging $\mathcal{T}$ so as to simplify searching it, and so on, although these can be very effective. We will instead consider a mathematical improvement of Theorems 1 and 2 which can substantially reduce the size of $\mathcal{T}$. A rigorous statement of the result would be very complicated, so we will merely indicate its direction. Basically, we consider the genesis of a sequence in $\mathcal{T}$, as well as the sequence itself. Thus, suppose a sequence $\{v_n\}$ occurs in $\mathcal{T}$ only as a result of being the derived sequence $\{u_n - t_{n+k}\}$ of $\{u_n\}$ in $\mathcal{T}$. Then it is not necessary to consider the derived sequence $\{v_n - t_{n+k+1}\}$ of $\{v_n\}$, since we may regard a derived sequence as being the result of subtracting the largest member of T occurring in the representation of u_n. There are substantial complications of this approach. For instance, regard $\mathcal{T}$ as an ordered graph under the relation of being a permissible derived sequence. Then we must ensure that every loop in the graph contains at least one sequence all of whose permissible derived sequences are in $\mathcal{T}$.

A computer program has been written by the author implementing this improvement in the case of Theorem 1 only. The improvement is very substantial; thus the fourth, seventh and tenth entries in the above table gave only 12, 15, and 20 sequences, respectively. Also, a few sequences of higher order than have been given here were handled successfully. In fact, this program was written first; the results led the author to recognize the possibility of using the simpler results of the present work. The reasons for implementing and presenting the results that we have considered here were basically two. First, more efficient result is almost impossible to state both rigorously and intelligibly, and second, any general result concerning applicability to P–V sequences will probably follow the lines of the present results, not the complicated and inelegant lines of the other. Moreover, the more efficient algorithm has so far been no more successful on non-P–V sequences than the present one.

The author wishes to thank Mr. L. P. Horowitz for writing the program implementing Theorem 2, and for performing a number of tedious hand calculations.

References

Brown, J. L. (1961). On complete Sequences of Integers. *Amer. Math. Monthly* **68,** 557–560.

Cassels, J. W. S. (1960). On the Representation of Integers as the Sums of Distinct Summands Taken from a Fixed Set. *Acta Sci. Math.* **21,** 111–124.

Cohoon, D. K. and Burr, S. A. (1970). A Theorem on Difference Equations and its Generalization to Operator Equations. *J. Math. Anal. and Appl.* **30,** 718–729.

Dufresnoy, J. and Pisot, C. (1955a). Etude de Certaines Fonctions Méromorphiques Bornées sur la Cercle Unite. Application à un Ensemble Fermé d'Entiers Algebriques. *Ann. Sci. Ec. Norm. Sup.* **72,** 69–92.

Dufresnoy, J. and Pisot, C. (1955b). Sur les Eléments d'Accumulation d'un Ensemble Fermé d'Entiers Algébriques. *Bull. Sci. Math.* (2) **79,** 54–64.

Graham, R. L. (1964a). Complete Sequences of Polynomial Values. *Duke Math. J.* **31,** 275–286.

Graham, R. L. (1964b). A Property of Fibonacci Numbers. *Fibonacci Quarterly* **2,** 1–10.

Graham, R. L. (1964c). On a Conjecture of Erdös in Additive Number Theory. *Acta Arithmetica* **10,** 63–70.

Pisot, C. (1938). La Répartition Modulo 1 et les Nombres Algébriques. *Annali di Pisa* (2) **7,** 205–248.

Roth, K. F. and Szekeres, G. (1954). Some Asymptotic Formulae in the Theory of Partitions. *Quarterly J. Math.* (2) **5,** 241–259.

Salem, R. (1945). Power Series with Integral Coefficients. *Duke Math. J.* **12,** 153–172.

Vijayaraghavan, T. (1941). On the Fractional Parts of the Powers of a Number (II). *Proc. Cambridge Phil. Soc.* **37,** 349–357.

Difference Bases

Three Problems in Additive Number Theory

J. C. P. MILLER

Mathematics Laboratory, University of Cambridge, England

1. Introduction

In this paper we compare three problems concerning the set of sums of a finite number of consecutive elements chosen from an ordered set of integers. One of these leads to *Perfect Difference Sets*, where there is a satisfactory and virtually complete mathematical theory. The second problem is that of *Restricted Difference Sets*, where there is a body of solutions with a mathematical theory, intermingled with many less easily explained sporadic solutions of the problem; this problem turns out to be rather suitable for a large scale computer search. The third problem is that of *Unrestricted Difference Sets*, where there is less theory and no obvious rapid methods of search for particular solutions, involving very great expenditure of time for relatively meagre results.

2. Restricted Difference Bases

We start by stating the problem for which computers have been used extensively.

I. Consider a ruler of length 24 inches, marked originally in inches. The ruler is worn so that only $k + 1$ division marks remain (including the ends), yet it remains possible to measure all exact inch lengths from 1 to 24. What is the least value of k possible? How many distinct solutions are there for this value of k?

The answer here is that $k = 8$, and that there are 472 distinct solutions (i.e. not counting reversal of ends as providing a distinction).

Sample solutions are with marks at

$$k = 8, n = 24 \qquad \left.\begin{matrix} 0 & 1 & 13 & 14 & 16 & 18 & 20 & 22 & 24 \\ 0 & 1 & 3 & 7 & 11 & 15 & 19 & 23 & 24 \end{matrix}\right\} \qquad (2.11)$$

or at

Note that the largest gap is as much as 12 in the first case, and as little as 4 in the second; these are extreme values.

For a ruler of length 36 inches, on the other hand, $k = 9$ and there is only *one* solution

$$k = 9,\ n = 36 \qquad 0\ 1\ 3\ 6\ 13\ 20\ 27\ 31\ 35\ 36 \tag{2.12}$$

We can restate the problem, a little more generally, in two ways:

(a) Consider a set of $k + 1$ distinct integers

$$0 = a_0 < a_1 < a_2 < a_3 < \ldots < a_k = n \tag{2.2}$$

such that the set of differences $a_r - a_s$, $r > s$, includes all integers from 1 to n. For given n, what is the least possible value of k, and how many solutions are there having this value of k?

(b) Consider an ordered set of k positive integers

$$b_1, b_2, \ldots, b_k \tag{2.3}$$

such that the set of sums of one or more consecutive members of the set includes all integers from 1 to n. For given n, what is the least possible value of k, and how many solutions are there for this value of k?

Clearly

$$a_i = \sum_{r=1}^{i} b_r \tag{2.4}$$

The formulation (a) corresponds to the examples (2.11, 2.12) already given; with formulation (b) they become

$$k = 8,\ n = 24 \qquad \left.\begin{matrix} 1\ 12\ 1\ 2\ 2\ 2\ 2\ 2 \\ 1\ \ 2\ 4\ 4\ 4\ 4\ 4\ 1 \end{matrix}\right\} \tag{2.51}$$

$$k = 9,\ n = 36 \qquad 1\ 2\ 3\ 7\ 7\ 7\ 4\ 4\ 1 \tag{2.52}$$

The latter formulation, with examples as illustrated immediately above, is, on the whole, more illuminating at a glance, and will be adopted in almost all cases in the rest of this paper. One may call the set of sums (2.1) a *Difference Basis*, and the set of differences $b_r = a_r - a_{r-1}$, (2.3), a *Difference Set*. The sets discussed in this paragraph are called *Restricted*, because the *span n* of the integers represented by the differences or sums is equal to the *range*, i.e. the largest integer in the basis, or the sum of all integers in the difference set.

3. Difference Bases

We shall now generalise the problem in two ways.

II. *Difference Cycles.* Consider an ordered *cycle* of k positive integers

$$b_1, b_2, \ldots, b_k \tag{3.1}$$

and the set of sums of consecutive integers

$$\sum_{i}^{[i+s]} b_r \tag{3.2}$$

where $0 \leqslant s < k - 1$, and $[i + s]$ is reduced modulo k if $i + s > k$. What is the least value of k for which a set b_i exists such that the set of sums (3.2) includes all integers $1, 2, \ldots, n - 1$?

The set (3.1) will be called a *Difference Cycle* or a *Cyclic Difference Set.* As an example we may take

$$k = 6, n = 31 \qquad 1\ 3\ 2\ 7\ 8\ 10 \tag{3.3}$$

where, e.g. 11 is given by $10 + 1$, $22 = 8 + 10 + 1 + 3$ and so on.

III. An *Unrestricted Difference Set* is the solution to a problem stated in the words of I(a) or I(b) except that

$$a_k = \sum_{1}^{k} b_i \geqslant n \text{ replaces } a_k = \sum_{1}^{k} b_i = n \tag{3.4}$$

Thus our ruler may exceed in length the maximum n of exact inch lengths to be measured, i.e. the range may exceed the span.

An example is the set

$$k = 7, n = 24 \qquad 8\ 10\ 1\ 3\ 2\ 7\ 8 \tag{3.5}$$

which gives the span 1(1)24, and also 31 and 39, the range. It will be noted that this gives a smaller value of k than one of the best restricted sets such as (2.51).

4. Perfect Difference Cycles

We consider first the case most amenable to mathematical treatment. This concerns cyclic difference sets, II.

Singer (1938) uses finite field theory to show that:

A sufficient condition that there exists a set of k integers

$$a_0, a_1, \ldots, a_{k-1}$$

having the property that the $k^2 - k$ *differences* $a_i - a_j$, $i \neq j$, *are congruent modulo* $k^2 - k + 1 = n$ *to the integers*

$$1, 2, 3, \ldots, k^2 - k$$

in some order is that $k - 1$ *be the power of a prime,* p^α.

We consider, rather than the a_i, the ordered cycle of differences

$$b_1, b_2, \ldots, b_k$$

such that $b_{i+1} = a_{i+1} - a_i$, $i = 0\,(1)\,k-1$, $a_k = k^2 - k + 1$. The example (3.3) gives such a cycle. Such a cycle, where each difference occurs an equal number λ (here $\lambda = 1$) of times, is called a *perfect difference cycle.*

Given one such cycle, others may be readily constructed by use of multipliers, for the differences of the set $Ma_0, Ma_1, \ldots, Ma_{k-1}$, also give a complete set of residues, mod $k^2 - k + 1$, if $(M, k^2 - k + 1) = 1$, and the set of $a_i' = Ma_i \pmod{k^2 - k + 1}$ are all distinct, and need not be (though may be) a rearrangement of the original set. Thus with $n = 13$, $(a_i) = (0\,1\,3\,9)$, we find $(a_i') = (2a_i) = (0\,2\,6\,18) = (0\,2\,5\,6) \pmod{13}$. The corresponding b_i-cycles are 1 2 6 4 and 1 3 2 7. Singer also considers the number of distinct cycles, and gives evidence in support of $N = \phi(n)/3\alpha$, where reversal is counted as distinct.

Singer (1938, p. 384) gives one cycle for each $p^\alpha \leqslant 16$. O'Beirne (1965) gives likewise one set of a_i for each $p^\alpha \leqslant 19$. Leech (1965, p. 162) gives a set of a_i for $p^\alpha = 32$.

A substantial theory of *Perfect Difference Sets*, with many references, is described in Ryser (1963), and in Selmer (1966). These generalise the idea to include cases where each residue mod n is included the same number $\lambda > 1$ of times. Selmer in Table 5, p. 206, gives all cycles with $\lambda < \frac{1}{2}n$, $n \leqslant 40$ and Hall 1956 gives 46 sets with $n \leqslant 50$.

The Singer cycles, called PLANAR by Ryser (1969), all have $\lambda = 1$. They appear all to be obtainable, for any n, from one another by use of multipliers. However, for $n = 31$, $\lambda = 7$ there exist two cycles not so related. There seems scope for further study here.

Singer's condition, that $k - 1 = p^\alpha$ for a perfect cycle has been shown necessary as well as sufficient for $n \leqslant 1600$ (see Ryser, 1963).

5. General Difference Cycles

It is also of interest to study cycles when n is not of the form $p^{2\alpha} + p^\alpha + 1$, which shall have minimum k, and cover all residues at least once. The use of a multiplier M still gives a complete set of residues, provided $(M, n) = 1$, but solutions are generally more numerous than for perfect sets (where these exist) with given number of elements $k + 1$, and are not all obtainable from one another by use of multipliers.

Enumerations have been carried out up to $n = 21$, and counts are listed below (including perfect cycles). In this N is the number of distinct solutions (excluding reversal), and T is the number of types that remain distinct under use of multipliers.

$k+1$	n	N	T	$k+1$	n	N	T	$k+1$	n	N	T
1	1	1	1	4	8	3	2	5	14	25	9
2	2	1	1		9	7	3		15	38	11
	3	1	1		10	4	2		16	16	5
					11	5	1		17	24	3
3	4	1	1		12	2	1		18	9	3
	5	2	1		13	2	1		19	9	1
	6	1	1						20	NIL	
	7	1	1						21	1	1

It is of interest that no cycle of 5 elements exists for $n = 20$, although there is one for $n = 21$. It is simple, and seems worthwhile, to give a proof of the existence of this gap. There are just 20 differences (as for $n = 21$) to cover 19 residues. Thus one and only one must appear twice; this is clearly 10. One half-cycle is either undivided, or divided into two parts (since there are 5 parts in all).

If a half cycle is undivided, by representing 11, 12, 14 in turn once each only, we are led uniquely to the b_i

$$3 \quad 1 \quad 10 \quad 2 \quad 4$$

which does not give 5. This half-cycle must therefore be divided if a solution exists; it is one of (1, 9) (2, 8) (3, 7) or (4, 6). The available divisions (unordered) on the other half are (1, 2, 7) (1, 3, 6) (1, 4, 5) and (2, 3, 5), It is not difficult to see that no pair, one from each set, can be fitted without repetition.

Incidentally, the impossibility of 1-part and 4-part half-cycles together proves that no restricted difference set with 4 elements can be made for $n = 10$, for the latter would imply the former.

6. Outline of Studies

Restricted differences bases form the main study of this paper. These have been studied by Leech (1956) who gives references to some earlier work by Haselgrove and Leech (1957) and by Wichmann (1962).

Much of the earlier work is concerned with obtaining bounds for $[l(n)]^2/n$, where $l(n)$ is the least value of $k + 1$. Leech (1956) showed that $[l(n)]^2/n \geqslant 2{\cdot}434$, whilst Wichmann (1963) constructed bases for which $l(n) \leqslant \sqrt{n3} + 3$.

Wichmann (1963) also exhibits a few solutions with strong patterns of integers in the sets, and his general, very effective basis is highly patterned. He also points out that bases for 76 and 78 can be achieved with ordered sets of 14 elements, but that he did not know of a similar set for $n = 77$.

These circumstances, and access to the tables prepared and referred to by Leech (1956), and, in particular, a knowledge of the method by which he prepared them, led me to program a machine search for solutions, in order to extend the tables of complete results.

TABLE I. Numbers of restricted difference sets.

k	n	N	Range max b_i		k	n	N	Range max b_i		k	n	N	Range max b_i	
2	3	1	2		8	24	472	12	4	11	44	2446	21	6
						25	230	12	4		45	1057	22	6
3	4	2	2			26	83	12	5		46	342	22	6
	5	2	3	2		27	28	13	5		47	119	21	6
	6	1	3			28	6	11	6		48	34	20	7
4	7	6	4	2		29	3	12	6		49	11	21	7
	8	4	4	3							50	2	20	7
	9	2	4	3	9	30	1018	14	5					
						31	445	15	5	12	51	8159	25	6
5	10	19	5	3		32	152	15	5		52	3175	25	6
	11	15	5	3		33	60	16	6		53	1143	25	6
	12	7	6	3		34	10	14	6		54	418	26	7
	13	3	6	4		35	5	15	6		55	165	25	7
						36	1	7			56	54	24	7
6	14	65	7	3							57	12	25	7
	15	40	7	3	10	37	1339	18	5		58	6	24	9
	16	16	8	4		38	487	18	6					
	17	6	7	4		39	181	19	6	13	59	15999	28	6
						40	50	17	6		60	6126	29	7
7	18	250	9	3		41	18	18	6		61	2480	29	7
	19	163	9	4		42	2	16	7		62	903	30	7
	20	75	10	4		43	1	7			63	334	29	7
	21	33	10	4							64	119	28	7
	22	9	9	5							65	43	29	8
	23	2	9	6							66	3	28	11
											67	6	11	9
											68	2	11	10

$n = 69$ needs $k = 14$.

Over the last few years, I have organised a series of more and more rapid programs on EDSAC II and TITAN (the last is about 250,000 times faster than the first), with the following results:

The coverage of complete results now extends to $n = 68$.
The gap at $n = 77$ has been filled in—but a gap still remains at $n = 99$ to replace it.
Wichmann's upper limit has been replaced by $l(n) \leqslant [\sqrt{3n}] + 2$, a fairly trivial advance.

I have also devised a condensed proof that Wichmann's set is a basis. This method of proof can be used in other, less efficient, but still useful, bases.

A major result of the search, besides the full lists of solutions which are much too long to give *in toto*, is the count of numbers $N(n)$ of solutions for given n and least k. These are listed in Table I, together with the range of $\max b_i$ amongst these solutions.

7. Patterns

A glance at Table II shows immediately a number of patterned solutions. These may be grouped into sets, e.g.

n		$n = 6k - 20$
16	1 2 6 4 1 2	
22	1 2 6 6 4 1 2	
28	1 2 6 6 6 4 1 2	
34	1 2 6 6 6 6 4 1 2	

and solutions for $n = 40, 46, 52$ can be forecast. There is also one for $n = 58$, but k is too large with this pattern and its usefulness lapses. This pattern has a *rate of increase* 6, and can be written in the form $1\ 2\ 6^{k-5}\ 4\ 1\ 2$, *the index denoting repetition* as in the theory of partitions.

Consider now several patterns with rates of increase 7, 11:

$$1\ 2\ 3\ 7^{k-6}\ 4\quad 4\ 1 \qquad \text{with } n = 7k - 27$$
$$1\ 2\ 3\ 3\quad 7^{k-6}\ 4\ 1 \qquad n = 7k - 28$$
$$1\ 1\ 3\ 7^{k-6}\ 2\quad 4\quad 2 \qquad n = 7k - 29$$

$$1\ 1\ 3\ 5\ 5\quad 11^{k-10}\ 6\quad 6\ 6\ 1\ 1 \qquad n = 11k - 75$$
$$1\ 1\ 3\ 5\ 5\quad 5\quad 11^{k-6}\quad 6\ 6\ 1\ 1 \qquad n = 11k - 76$$

The first one in each set was generalised by Wichmann (1963) to give a two-parameter pattern:

$$1^r \quad (r+1) \quad (2r+1)^r \quad (4r+3)^s \quad (2r+2)^{r+1} \quad 1^r$$

with

$$k = 4r + s + 2, \quad n = qk - 3(2r+1)^2, \quad q = 4r + 3$$

TABLE II. Complete lists of difference sets for selected n.

k	n	
3	4	1 1 2
		1 2 1
	5	1 1 3
		1 2 2
	6	1 3 2
4	7	1 1 1 4
		1 1 2 3
		1 1 3 2
		1 3 1 2
		1 2 2 2
		1 2 3 1
	8	1 1 3 3
		1 3 2 2
		1 4 1 2
		1 2 4 1
	9	1 1 4 3
		1 3 3 2
5	11	1 1 1 4 4
		1 1 4 2 3
		1 1 3 3 3
		1 1 5 1 3
		1 1 3 4 2
		1 1 4 3 2
		1 1 5 3 1
		1 3 1 4 2
		1 3 2 3 2
		1 3 3 2 2
		1 2 1 5 2
		1 5 1 2 2
		1 4 3 1 2
		1 2 2 5 1
		1 2 3 4 1
	12	1 1 1 5 4
		1 1 4 3 3
		1 3 1 5 2
		1 3 3 3 2
		1 6 1 2 2
		1 2 2 6 1
		1 2 4 4 1
	13	1 1 4 4 3
		1 3 1 6 2
		1 5 3 2 2
6	16	1 1 1 5 4 4
		1 1 4 3 4 3
		1 1 4 4 3 3
		1 1 5 5 3 1
		1 3 1 3 6 2
		1 3 3 5 2 2
		1 2 1 5 5 2
		1 2 6 4 1 2
		1 5 3 3 2 2
		1 8 1 2 2 2
		1 4 1 7 1 2
		1 2 2 2 8 1
		1 2 2 4 6 1
		1 2 4 4 4 1
		1 2 6 2 4 1
		1 2 7 1 4 1
	17	1 1 1 5 5 4
		1 1 4 4 4 3
		1 1 6 4 2 3
		1 1 6 4 3 2
		1 3 6 2 3 2
		1 7 3 2 2 2
7	22	1 1 1 5 5 5 4
		1 1 8 5 3 3 1
		1 3 1 7 2 6 2
		1 2 6 6 4 1 2
		1 5 1 3 8 2 2
		1 5 3 7 2 2 2
		1 6 1 9 1 2 2
		1 2 2 9 1 6 1
		1 2 3 7 4 4 1
	23	1 1 9 4 3 3 2
		1 3 6 6 2 3 2
8	28	1 1 11 5 3 3 3 1
		1 2 6 6 6 4 1 2
		1 8 1 11 1 2 2 2
		1 6 4 9 3 2 1 2
		1 2 2 2 11 1 8 1
		1 2 3 3 7 7 4 1
	29	1 1 12 4 3 3 3 2
		1 3 6 6 6 2 3 2
		1 2 3 7 7 4 4 1

TABLE II. (*continued*)

k	n	
9	34	1 1 1 12 5 4 4 3 3
		1 1 7 1 4 11 3 3 3
		1 1 3 7 7 7 2 4 2
		1 1 14 5 3 3 3 3 1
		1 2 6 6 6 6 4 1 2
		1 5 3 9 3 7 2 2 2
		1 10 1 13 1 2 2 2 2
		1 2 2 2 2 13 1 10 1
		1 2 2 9 2 10 1 6 1
		1 2 3 3 3 7 10 4 1
	35	1 1 15 4 3 3 3 3 2
		1 3 1 11 2 7 2 6 2
		1 3 6 6 6 6 2 3 2
		1 2 3 3 7 7 7 4 1
		1 1 3 3 10 4 7 4 1
	36	1 2 3 7 7 7 4 4 1
10	41	1 1 1 12 6 4 5 4 4 3
		1 1 1 17 5 1 4 4 4 3
		1 1 1 11 5 7 5 4 4 2
		1 1 1 17 5 3 4 3 4 2
		1 1 4 8 1 8 8 3 4 3
		1 1 3 7 7 7 7 2 4 2
		1 1 18 4 3 3 3 3 3 2
		1 3 1 3 10 6 6 2 7 2
		1 3 2 5 8 8 8 1 3 2
		1 3 6 6 6 6 6 2 3 2
		1 3 6 4 4 11 5 2 3 2
		1 5 1 7 3 9 9 2 2 2
		1 5 8 8 8 2 2 3 2 2
		1 7 1 3 5 5 13 2 2 2
		1 8 1 10 3 11 1 2 2 2
		1 6 4 9 4 9 3 2 1 2
		1 2 1 3 8 8 8 5 4 1
		1 2 3 3 3 7 7 10 4 1
	42	1 1 1 16 5 4 4 4 3 3
		1 2 3 3 7 7 7 7 4 1
	43	1 2 3 7 7 7 7 4 4 1
11	49	1 1 1 21 5 1 4 4 4 4 3
		1 1 2 8 2 10 7 9 5 1 3
		1 1 4 2 9 9 9 3 7 3 1
		1 3 2 5 8 8 8 8 1 3 2
		1 5 8 8 8 8 2 2 3 2 2
		1 4 3 4 9 9 9 5 1 2 2
		1 10 1 12 3 13 1 2 2 2 2
		1 2 1 3 8 8 8 8 5 4 1
		1 2 3 3 7 7 7 7 7 4 1
		1 2 3 3 7 10 4 7 7 4 1
		1 2 3 3 10 4 10 4 7 4 1

k	n	
11	50	1 1 1 20 5 4 4 4 4 3 3
		1 2 3 7 7 7 7 7 4 4 1
12	57	1 1 1 25 5 1 4 4 4 4 4 3
		1 1 5 5 2 17 4 4 9 3 3 3
		1 1 5 7 2 10 10 10 3 4 1 3
		1 1 3 5 5 11 11 6 6 6 1 1
		1 3 1 3 4 10 10 10 6 3 4 2
		1 3 2 5 8 8 8 8 8 1 3 2
		1 5 1 7 1 7 3 9 17 2 2 2
		1 5 8 8 8 8 8 2 2 3 2 2
		1 12 1 14 3 15 1 2 2 2 2 2
		1 2 1 2 3 3 10 10 10 7 7 1
		1 2 1 3 8 8 8 8 8 5 4 1
		1 2 3 7 7 7 7 7 7 4 4 1
	58	1 1 1 24 5 4 4 4 4 4 3 3
		1 1 6 7 1 10 10 10 3 4 2 3
		1 1 4 2 9 9 9 9 3 7 3 1
		1 4 3 4 9 9 9 9 5 1 2 2
		1 2 3 11 3 7 8 10 4 4 4 1
		1 2 3 11 7 3 11 7 4 4 4 1
13	66	1 1 1 28 5 4 4 4 4 4 4 3 3
		1 2 5 7 11 7 11 6 4 6 3 1 2
		1 2 3 15 7 3 11 7 4 4 4 4 1
	67	1 1 5 7 2 10 10 10 10 3 4 1 3
		1 1 4 2 9 9 9 9 9 3 7 3 1
		1 1 3 5 5 5 11 11 11 6 6 1 1
		1 3 1 3 4 10 10 10 10 6 3 4 2
		1 4 3 4 9 9 9 9 9 5 1 2 2
		1 2 1 2 3 3 10 10 10 10 7 7 1
	68	1 1 6 7 1 10 10 10 10 3 4 2 3
		1 1 3 5 5 11 11 11 6 6 6 1 1

$n = 69$ needs $k = 14$

The second will also generalise to give a pattern with

$$1^r \quad (r+1) \quad (2r+1)^{r+1} \quad (4r+3)^s \quad (2r+2)^r \quad 1^r$$

with the same

$$k = 4r + s + 2,$$

but

$$n = qk - 3(2r+1)^2 - 1.$$

However, the third pattern with increase 7 does not obviously generalise, and appears to have only a centre pattern.

Another kind of pattern is shown by

n	
34	1 1 1 12 5 4 4 3 3
42	1 1 1 16 5 4 4 4 3 3
50	1 1 1 20 5 4 4 4 4 3 3
58	1 1 1 24 5 4 4 4 4 4 3 3

with the rate of increase 8 and $n = 8k - 38$; another with the same rate and $n = 8k - 39$ is

n	
41	1 8 1 10 3 11 1 2 2 2
49	1 10 1 12 3 13 1 2 2 2 2
57	1 12 1 14 3 15 1 2 2 2 2 2
65	1 14 1 16 3 17 1 2 2 2 2 2 2

and others can be found.

One use for full tables of solutions is in finding such patterns. There are, however, other sporadic solutions that show no obvious pattern at all. For example

$$n = 22 \qquad 1\ 3\ 1\ 7\ 2\ 6\ 2$$

seems difficult to fit into any group. Another example is

$$n = 58 \qquad 1\ 2\ 3\ 11\ 3\ 7\ 8\ 10\ 4\ 4\ 1$$

which retains the end pattern of Wichmann's solution, but has rather lost it in the middle. On the other hand the pair

58	1 2 3 11 7 3 11 7 4 4 4 1
66	1 2 3 15 7 3 11 7 4 4 4 4 1

will generalise with rate of increase 8, $n = 8k - 38$.

8. Construction of Solutions

The identification of patterns enables us to construct solutions for $n > 68$, where complete sets have not yet been sought. For this purpose we need other two parameter solutions besides Wichmann's solutions to cover intermediate cases. Several have been constructed and the more useful ones are listed below. None gives a better ratio n/k^2 than Wichmann's best, but they do fill in some gaps.

One of the simplest and most obvious will be used for illustration; it is given by

$$1^{r-2}\ r^{s-t}\ (r-1)^t \qquad \text{(A)}$$

which has

$$k = r + s - 2, \qquad n = rs + r - t - 2.$$

This is a three-parameter solution, but the parameter t merely provides solutions for s successive values of n, $t = 1(1)s$; these have the same k, and the best ratio n/k^2 is given by $t = 1$. This continuous coverage immediately below the best n is conspicuously absent with Wichmann's basis. With $t = 1$ we readily find that the best value of s, measured by the ratio n/k^2, is given by $s = r - 1$, $k = 2r - 3$, $n = r^2 - 3$. We find also that $(r, s) = (\lambda, \lambda)$, and $(r, s) = (\lambda + 1, \lambda - 1)$ give the same values $k = 2\lambda - 2$, $n = \lambda^2 + \lambda - 2$. This pattern (A) provides a solution for minimum k for $n = 1(1)22$, 24(1)27, 30(1)33, 37, 38, 39, 44, 45, 46, 51, 52, 53, 59, 60, 61, 69 and no more.

Wichmann's bases have peak values for $s = 2r-2(1)2r+4$, and have $s = 2r + 1$ giving the best value of n/k^2, and are given in Section 7; they are repeated here for comparison with others that follow.

$$\left.\begin{array}{llllllll} 1^r & (r+1) & (2r+1)^r & (4r+3)^s & (2r+2)^{r+1} & 1^r & & n = N \\ 1^r & (r+1) & (2r+1)^{r+1} & (4r+3)^s & (2r+2)^r & 1^r & & n = N - 1 \end{array}\right\} \quad \text{(B)}$$

In each case $k = 4r + s + 2$

$$\left.\begin{aligned} N &= qk - 3(2r+1)^2 \qquad q = 4r + 3 \\ &= 4r(r+s) + 8r + 3s + 3 \end{aligned}\right\} \quad (8.1)$$

These give only a consecutive pair of values of n, thus leaving gaps to be filled otherwise when $r \geqslant 1$. Wichmann (1963) fills these by adding up to 4 extra integers at one (either) end, each not exceeding $r + 1$. We can however do a little better than this by seeking other bases, still general, but having smaller values of n for the same k. The following have been found:

(i) with $s = 2s' + 1$, odd

$$\left.\begin{array}{ll}
1^r(r+1)(2r+1)^r(4r+3)^{s'}(2r+2)(4r+3)^{s'}(2r+2)^{r+1}1^r & n = N-2r-1 \\
1^r(r+1)(2r+1)^r(4r+3)^{s'}(2r+1)(4r+3)^{s'}(2r+2)^{r+1}1^r & n = N-2r-2 \\
1^r(r+1)(2r+1)^{r+1}(4r+3)^{s'}(2r+2)(4r+3)^{s'}(2r+2)^r1^r & n = N-2r-2 \\
1^r(r+1)(2r+1)^{r+1}(4r+3)^{s'}(2r+1)(4r+3)^{s'}(2r+2)^r1^r & n = N-2r-3 \\
\text{(ii) with s} = 2s' + 2\text{, even} & \\
1^r(r+1)(2r+1)^r(4r+3)^{s'+1}(2r+2)(4r+3)^{s'}(2r+2)^{r+1}1^r & n = N-2r-1 \\
1^r(r+1)(2r+1)^r(4r+3)^{s'+1}(2r+1)(4r+3)^{s'}(2r+2)^{r+1}1^r & n = N-2r-2 \\
1^r(r+1)(2r+1)^{r+1}(4r+3)^{s'}(2r+2)(4r+3)^{s'+1}(2r+2)^r1^r & n = N-2r-2 \\
1^r(r+1)(2r+1)^{r+1}(4r+3)^{s'}(2r+1)(4r+3)^{s'+1}(2r+2)^r1^r & n = N-2r-3
\end{array}\right\} \quad \text{(C)}$$

Here k, N are given by (8.1).

A proof of Wichmann's basis in very condensed form is given in Appendix A. Similar arguments may be used to establish other bases. In practice I have been content to use the families quoted to suggest solutions (this is the difficult problem) which have been verified individually— it is usually easy to establish one-parameter sets by induction.

All these solutions have rate of increase $q = 4r + 3$, i.e. 3, 7, 11, 15, 19, etc. It is also of interest to construct solutions with intermediate rates of increase for which, as with Wichmann's, $\lim k^2/n = 3$. I have found

$$\left.\begin{array}{lllllll}
1^r & (2r+1)^{r+1} & (4r+1)^s & (2r)^r & (r+1) & 1^{r-1} & n = N \\
1^r & (2r+1)^r & (4r+1)^s & (2r)^{r+1} & (r+1) & 1^{r-1} & n = N-1 \\
1^{r-1} & (2r+1)^{r+1} & (4r+1)^s & (2r)^r & r & 1^r & n = N-1 \\
1^{r-1} & (2r+1)^r & (4r+1)^s & (2r)^{r+1} & r & 1^r & n = N-2
\end{array}\right\} \quad \text{(D)}$$

In these $k = 4r + s + 1$ and, with $q = 4r + 1$, the rate of increase, we have

$$N = qk - 2r(6r + 1) = 4r(r + s) + 6r + s + 1$$

As with (B) for $q = 4r + 3$, to produce (C), it is permissible here to reduce a "middle" element $(4r + 1)$ to $2r$ or $2r + 1$; there are sixteen cases, since s even and s odd differ, we shall not give these in detail here, but use the label (E) where appropriate.

Finally for increase $4r$ we have

$$\left.\begin{array}{llllllll}
 & 1^2 \; 2^{r-2} & (2r+1)^r & (4r)^s & (2r-1^r & 1 & 2^{r-1} & n = N \\
\text{and} & 1^2 \; 2^{r-2} & (2r+1)^{r-1} & (4r)^s & (2r-1)^{r+1} & 1 & 2^{r-1} & n = N-2
\end{array}\right\} \quad \text{(F)}$$

with $k = 4r + s - 1$, $N = qk - 12r^2 + 8k - 3 = 4r(r + s) + 4r - 3$. This is not very good, and has not been examined further.

Consider now the Wichmann solutions with increase 7. Examination of the tables exhibits a two-parameter family

$$1\ 2\ 3^t\ 7^{s-t}\ (3t + 1)\ 4\ 1 \qquad t = 1(1)s - 1$$

thus filling the gap between Wichmann's solutions (B) and the others (C). I have not been able to find a similar family for increase 11.

Likewise two solutions at the end of Section 7 foreshadow an increase of 8 from a variation of the peak Wichmann solution at $n = 50$. Not enough evidence is yet available to guess how this might be extended.

Next we consider the solutions labelled $12x$ and $12z$ in Tables III and IV. These are

$$k = 14\ n = 71 \quad 1\ 1\ 1\ 1\ 1\ 19\ 8\ 7\ 6\ 6\ 6\ 5\ 5\ 4$$
$$n = 73 \quad 1\ 1\ 1\ 1\ 1\ 21\ 8\ 7\ 6\ 6\ 6\ 5\ 5\ 4$$

with a pattern of the differences on the right ranging from $2r$ to r; the rate of increase is 12, one extra 6 and an increase of 6 in the large difference at each increase.

A similar pattern occurs with increase 10

$$k = 13\ n = 65 \quad 1\ 1\ 1\ 1\ 19\ 7\ 6\ 6\ 5\ 5\ 5\ 4\ 4$$

It is not identical in form, but of the same type.

This suggests the possibility of trying larger versions of the same type of pattern which has led to two effective patterns with increase 14. A third is due to John Leech. The best is

$$k = 17\ n = 106 \quad 1\ 1\ 1\ 1\ 1\ 1\ 34\ 9\ 8\ 7\ 7\ 7\ 6\ 6\ 6\ 5\ 5$$

Others are listed in Table IV, and it seems clear that extensions to higher increase are possible, though the progress is not regular but seems to have sporadic variations. This possibliity merits further study.

We also use in Table IV cases where a best Wichmann solutions 15B has one further difference added at the end—this can be up to $r + 1$, but only a unit is actually needed, for $n = 154$, and a 4 for $n = 142$ with present knowledge. We indicate this by $15B+$. Some other cases of $11B+$, $15B+$ are included for greater completeness. The same kind of addition can also be made in other cases, it has been used with $12z$ and 14β.

9. Solutions for $69 \leqslant n \leqslant 168$.

In Tables III and IV we give solutions constructed for $69 \leqslant n \leqslant 168$. These are indicated in Table III by the rate of increase (from 7 to 19) associated with

TABLE III. Rates of increase of solutions known for $69 \leqslant n \leqslant 168$

n	$k = 14$	n	$k = 15$
69	7*a*, 8*A*, 9*A*, 10*k*, 11*B*+, 12*F*	80	8*d*, 9*h*, 10*l*, 11*B*+, 13*D*
70	7*B*, 8*b*, 9*g*, 10*l*, 11*t*	81	8*e*, 9*D*, 10*m*, 11*t*, 12*F*, 13*D*
71	7*B*, 8*c*, 9*h*, 10*m*, 11*u*, 12*x*	82	8*f*, 9*D*, 10*n*, 11*u*
72	8*d*, 9*D*, 10*n*, 11*C*, 12*y*	83	9*D*, 10*v*, 11*C*, 12*x*
73	8*e*, 9*D*, 10*v*, 11*C*, 12*z*	84	10*p*, 11*C*, 12*y*
74	8*f*, 9*D*, 10*p*, 11*C*	85	9*j*, 10*q*, 11*C*, 12*z*
75	10*q*, 11*v*	86	11*v*
76	9*j*	87	10*r*
77	10*r*	88	10*s*
78	10*s*, 11*B*	89	11*B*
79	11*B*	90	11*B*

n	$k = 16$	n	$k = 17$
91	9*D*, 10*M*, 11*b*+, 12*F*	102	10*n*, 11*b*+, 12*w*
92	9*D*, 10*n*, 11*t*, 13*D*, 15*B*	103	9*j*, 10*v*, 11*t*, 12*F*, 14*α*
93	10*v*, 11*u*, 12*F*, 13*D*, 15*B*	104	10*p*, 11*u*
94	9*j*, 10*p*, 11*C*, 13*D*	105	10*q*, 11*C*, 12*F*, 13*D*
95	10*q*, 11*C*, 12*x*	106	11*C*, 13*D*, 14*β*
96	11*C*, 12*y*	107	10*r*, 11*C*, 12*x*, 13*D*, 15*B*
97	10*r*, 11*v*, 12*z*	108	10*s*, 11*v*, 12*y*, 15*B*
98	10*s*	109	12*z*
99	—	110	—
100	11*B*	111	11*B*
101	11*B*	112	11*B*

n	$k = 18$	n	$k = 19$
113	10*v*, 11*B*+, 12*z*+, 13*E*, 14*β*+	124	10*p*, 11*B*+, 12*z*+, 13*t*, 14*β*+, 15*B*+
114	10*p*, 11*t*, 12*w*, 13*E*, 15*C*	125	10*q*, 11*t*, 12*z*+, 13*E*, 14*β*+, 15*B*+, 16*F*
115	10*q*, 11*u*, 12*F*, 15*C*	126	11*u*, 12*w*, 13*E*, 14*β*+, 15*B*+
116	11*C*, 14*γ*, 15*C*	127	10*r*, 11*C*, 12*F*, 13*t*, 14*β*+, 15*B*+
117	10*r*, 11*C*, 12*F*, 14*α*	128	10*s*, 11*C*
118	10*s*, 11*C*, 13*D*	129	11*C*, 12*F*, 15*C*
119	11*v*, 12*x*, 13*D*	130	11*v*, 14*γ*, 15*C*
120	12*y*, 13*D*, 14*β*	131	12*x*, 13*D*, 14*α*, 15*C*
121	12*z*	132	12*y*, 13*D*
122	11*B*, 15*B*	133	11*B*, 12*z*, 13*D*
123	11*B*, 15*B*	134	11*B*, 14*β*
		135	—
		136	—
		137	15*B*
		138	15*B*

TABLE III. (*continued*)

n	$k = 20$	n	$k = 21$
139	11C, 12F, 13t, 14β+, 15B+, 16F, 17D	154	14β+, 15B+
140	11C, 13E, 14β+, 15B+, 17D	155	11B, 12x, 14β+, 15B+, 16F 17D, 19B
141	11v, 12F, 14β+, 15B+, 16F	156	11B, 12y, 15B+, 17D, 19B
142	15B+	157	12z, 13D, 15B+, 16F, 17D
143	12x	158	13D, 14γ
144	11B, 12y, 13D, 14γ, 15C	159	13D, 14α, 15C
145	11B, 12z, 13D, 14α, 15C	160	15C
146	13D, 15C	161	15C
147	—	162	14β
148	14β	163	—
149	—	164	—
150	—	165	—
151	—	166	—
152	15B	167	15B
153	15B	168	15B

TABLE IV. Solution patterns referred to in Table III.

n																
$7k - 29 =$	69	1	1	3	7	7	7	7	7	7	7	7	2	4	2	a
$7k - 28 =$	70	1	2	3	3	7	7	7	7	7	7	7	7	4	1	B
$7k - 26 =$	71	1	2	3	7	7	7	7	7	7	7	7	4	4	1	B
$8k - 43 =$	69	1	1	1	1	1	1	8	8	8	8	8	8	8	7	A
$8k - 42 =$	70	1	1	1	24	1	1	4	4	4	4	4	4	4	3	b
		1	1	2	6	8	8	8	8	8	8	5	2	4	1	
$8k - 41 =$	71	1	1	1	3	8	8	8	8	8	8	8	2	5	2	c
		1	1	2	1	8	8	8	8	8	8	8	6	1	3	
$8k - 40 =$	72	1	1	4	8	8	8	8	8	8	8	1	2	4	3	d
		1	2	2	2	2	2	2	2	19	1	19	1	16	1	
$8k - 39 =$	73	1	2	1	3	8	8	8	8	8	8	8	5	4	1	e
$8k - 38 =$	74	1	1	1	32	5	4	4	4	4	4	4	4	3	3	f
$9k - 57 =$	69	1	1	1	1	1	1	1	9	9	9	9	9	9	8	A
$9k - 56 =$	70	1	2	1	2	9	9	9	9	9	9	4	3	1	2	g
$9k - 55 =$	71	1	1	2	6	9	9	9	9	9	5	6	1	1	3	h
$9k - 54 =$	72	1	1	2	3	4	4	9	9	9	9	9	5	5	1	D
$9k - 53 =$	73	1	1	2	4	4	5	9	9	9	9	5	5	5	1	D
		1	1	5	5	9	9	9	9	9	4	4	4	3	1	
$9k - 52 =$	74	1	1	5	5	5	9	9	9	9	9	4	4	3	1	D
$9k - 50 =$	76	1	4	3	4	9	9	9	9	9	9	5	1	2	2	j
		1	1	4	2	9	9	9	9	9	9	3	7	2	1	

TABLE IV. (*continued*)

n		
$10k - 71 = 69$	1 1 1 1 1 25 7 4 5 5 5 5 2 6	k
$10k - 70 = 70$	1 2 1 2 3 3 10 10 3 10 10 7 7 1	l
	1 2 1 2 3 3 3 10 10 7 10 10 7 1	
$10k - 69 = 71$	1 1 1 1 30 4 4 3 1 5 5 5 5 5	m
	1 1 1 1 1 1 1 9 10 10 10 8 9 8	
$10k - 68 = 72$	1 1 5 1 10 10 10 10 10 3 2 4 3 2	n
$10k - 67 = 73$	1 2 1 2 3 3 3 10 10 10 10 10 7 1	o
$10k - 66 = 74$	1 2 1 2 3 3 10 10 7 10 10 7 7 1	p
$10k - 65 = 75$	1 1 1 1 30 3 6 5 5 5 5 5 3 4	q
$10k - 73 = 77$	1 3 1 3 4 10 10 10 10 10 6 3 4 2	r
	1 1 5 7 2 10 10 10 10 10 3 4 1 3	
	1 2 1 2 3 3 10 10 10 10 10 7 7 1	
$10k - 62 = 78$	1 1 6 7 1 10 10 10 10 10 3 4 2 3	s
$11k - 85 = 69$	1 1 3 5 5 11 11 11 6 6 6 1 1 1	$B+$
	1 1 1 3 5 5 11 11 11 6 6 6 1 1	
$11k - 84 = 70$	1 1 1 1 5 7 2 9 11 11 8 3 6 4	t
$11k - 83 = 71$	1 1 1 2 2 3 11 11 11 11 4 8 1 4	u
	1 1 1 3 3 7 11 11 11 8 4 5 3 2	
$11k - 82 = 72$	1 1 3 5 5 5 11 5 11 11 6 6 1 1	C
$11k - 81 = 73$	1 1 3 5 5 11 11 5 11 6 6 6 1 1	C
	1 1 3 5 5 5 11 6 11 11 6 6 1 1	
$11k - 80 = 74$	1 1 3 5 5 11 11 6 11 6 6 6 1 1	C
$11k - 79 = 75$	1 1 6 7 3 11 11 11 11 4 4 1 3 1	v
$11k - 76 = 78$	1 1 3 5 5 5 11 11 11 11 6 6 1 1	B
$11k - 75 = 79$	1 1 3 5 5 11 11 11 11 6 6 6 1 1	B
$12k - 104 = 124$	1 1 1 1 1 45 8 7 6 6 6 6 6 6 6 5 5 4 3	$z+$
$12k - 103 = 113$	1 1 1 1 1 39 8 7 6 6 6 6 6 6 5 5 4 4	$z+$
$12k - 102 = 102$	1 1 1 1 1 33 8 7 6 6 6 6 6 5 5 4 5	$z+$, w
	1 1 1 1 1 42 1 1 7 6 6 6 6 6 6 5 5	
$12k - 101 = 91$	1 1 2 7 7 12 12 12 12 5 5 5 5 1 2	F
$12k - 99 = 69$	1 1 2 7 7 7 12 12 5 5 5 1 2 2	F
$12k - 97 = 71$	1 1 1 1 1 19 8 7 6 6 6 5 5 4	x
$12k - 96 = 72$	1 3 1 5 1 4 12 12 12 6 2 5 6 2	y
$12k - 95 = 73$	1 1 1 1 1 21 8 7 6 6 6 5 5 4	z
$13k - 123 = 124$	1 1 1 3 6 6 6 6 13 13 6 13 13 13 7 7 7 1 1	E
$13k - 122 = 125$	1 1 1 3 6 6 6 6 13 13 7 13 13 13 7 7 7 1 1	E
	1 1 1 3 6 6 6 13 13 13 6 13 13 7 7 7 7 1 1	
	1 1 4 6 6 6 6 13 13 6 13 13 13 7 7 7 1 1 1	
$13k - 121 = 126$	1 1 1 3 6 6 6 13 13 13 7 13 13 7 7 7 7 1 1	E
	1 1 4 6 6 6 6 13 13 7 13 13 13 7 7 7 1 1 1	
	1 1 4 6 6 6 13 13 13 6 13 13 7 7 7 7 1 1 1	
$13k - 120 = 127$	1 1 4 6 6 6 13 13 13 7 13 13 7 7 7 7 1 1 1	E
$13k - 116 = 131$	1 1 1 3 6 6 6 6 13 13 13 13 13 13 7 7 7 1 1	D

TABLE IV (*continued*)

n		
$13k - 115 = 132$	1 1 1 3 6 6 6 13 13 13 13 13 13 7 7 7 7 1 1	D
	1 1 4 6 6 6 6 13 13 13 13 13 13 7 7 7 1 1 1	
$13k - 114 = 133$	1 1 4 6 6 6 13 13 13 13 13 13 7 7 7 7 1 1 1	D
$14k - 142 = 124$	1 1 1 1 1 1 41 9 8 7 7 7 7 6 6 6 5 5 4	$\beta+$
$14k - 141 = 125$	1 1 1 1 1 1 41 9 8 7 7 7 7 6 6 6 5 5 5	$\beta+$
$14k - 140 = 126$	1 1 1 1 1 1 41 9 8 7 7 7 7 6 6 6 5 5 6	$\beta+$
$14k - 139 = 127$	1 1 1 1 1 1 41 9 8 7 7 7 7 6 6 6 5 5 7	$\beta+$
$14k - 136 = 102$	1 1 1 1 1 1 35 6 3 8 7 7 7 7 6 6 4	γ
$14k - 135 = 103$	1 1 1 1 1 1 31 9 8 7 7 7 6 6 6 5 5	α
$14k - 132 = 106$	1 1 1 1 1 1 34 9 8 7 7 7 6 6 6 5 5	β
$15k - 161 = 124$	1 1 1 4 7 7 7 15 15 15 15 8 8 8 8 1 1 1 1	B+
	1 1 1 1 4 7 7 7 15 15 15 15 8 8 8 8 1 1 1	
$15k - 160 = 125$	1 1 1 4 7 7 7 15 15 15 15 8 8 8 8 1 1 1 2	B+
	2 1 1 1 4 7 7 7 15 15 15 15 8 8 8 8 1 1 1	
$15k - 159 = 126$	1 1 1 4 7 7 7 15 15 15 15 8 8 8 8 1 1 1 3	B+
	3 1 1 1 1 4 7 7 7 15 15 15 15 8 8 8 8 1 1	
$15k - 158 = 127$	1 1 1 4 7 7 7 15 15 15 15 8 8 8 8 1 1 1 4	B+
	4 1 1 1 4 7 7 7 15 15 15 15 8 8 8 8 1 1 1	
$15k - 156 = 114$	1 1 1 4 7 7 7 7 15 7 15 15 8 8 8 1 1 1	C
$15k - 155 = 115$	1 1 1 4 7 7 7 15 15 7 15 8 8 8 8 1 1 1	C
	1 1 1 4 7 7 7 7 15 8 15 15 8 8 8 1 1 1	
$15k - 154 = 116$	1 1 1 4 7 7 7 15 15 8 15 8 8 8 8 8 1 1	C
$15k - 148 = 122$	1 1 1 4 7 7 7 7 15 15 15 15 8 8 8 1 1 1	B
$15k - 147 = 123$	1 1 1 4 7 7 7 15 15 15 15 8 8 8 8 1 1 1	B
$16k - 181 = 139$	1 1 2 2 9 9 9 16 16 16 7 7 7 1 2 2 2	F
$16k - 179 = 141$	1 1 2 2 9 9 9 9 16 16 16 16 7 7 7 7 1 2 2 2	F
$17k - 202 = 138$	1 1 1 9 9 9 9 17 17 17 8 8 8 8 8 4 1 1 1 1	D
$17k - 201 = 139$	1 1 1 9 9 9 9 9 17 17 17 8 8 8 8 4 1 1 1 1	D
	1 1 1 1 9 9 9 9 17 17 17 8 8 8 8 8 5 1 1 1	
$17k - 200 = 140$	1 1 1 1 9 9 9 9 9 17 17 17 8 8 8 8 5 1 1 1	D
$19k - 244 = 155$	1 1 1 1 5 9 9 9 9 9 19 19 19 10 10 10 10 1 1 1 1	B
$19k - 243 = 156$	1 1 1 1 5 9 9 9 9 19 19 19 10 10 10 10 10 1 1 1 1	B

the solution, together with an identifying letter corresponding to a full solution given in Table IV; this in turn is usually the solution for the smallest relative value of n that occurs in Table III. Capital letters refer to general solutions mentioned in Section 8, and cover several values of n; lower case letters refer to solutions with one parameter, listed separately in Table IV.

Table IV is arranged in groups according to rate of increase. The relative rarity of solutions with rates of increase 14, 16 is notable, and is due to lack of such solutions in the search area $n \leqslant 68$. Likewise other solutions with increase $\geqslant 12$ are, in general, incomplete, for the same reason.

10. Discussion

We can now survey our results, and see what can be added to Wichmann's conclusions.

(i) The gap at $n = 77$ is now filled. Three solutions $10r$ were found for $n = 47$, ($n = 37$ has too many solutions, and the rate of increase is too ill-defined, for these to have been identified at this stage).

The first gap now appears at $n = 99$ and the next at $n = 110$; it is difficult to forecast whether those may be expected to be filled. Table II, at $n = 66$, shows that a relative scarcity of solutions can occur, and it seems probable that eventual failure *will* occur. However the solutions give hope for extra solutions in other cases, which might help to fill the gaps. Also the solutions for increase 12 may not be all known yet—identification is still difficult near $n = 68$. Gaps for $n \geqslant 135$ have less significance and indicate possibly considerable lack of knowledge.

(ii) The solution $12z$ came to light rather late, from an examination of the 2480 solutions from $n = 61$. Its form was very suggestive and a short search resulted in 14α, β, the two best solutions in Table IV with increase 14. These solutions do not form an obvious *family* of two-parameter solutions, but the *type* would seem to be fairly general. The best one given 14β seems highly effective, nothing as good for increase 16 has been found, while the best solution for increase 10 does not quite make the tables. Note that the first case given for 14β, $n = 106$, is the first that occurs; there is *not* one for $n = 92$ with one 7 fewer and 27 in place of 34.

(iii) Wichmann compares $l(n) = \max(k + 1)$ with $\sqrt{3n}$, finding that $l(n) \leqslant [\sqrt{3n}] + 3$ for every n, where [ ] indicates integral part. In fact his construction gives $l(n) \leqslant [\sqrt{3n}] + 3$ because starting with one of his best solutions, which give $l(n) \leqslant [\sqrt{3n}]$, he adds in turn, 4 extra differences of not more than $n + 1$ at one end. However before the fourth is needed, at an increase $3n + 4$ in n, the value of $[\sqrt{3n}]$ has always increased by a unit, as he remarks. The increase occurs, in fact, when $3(n_0 + \delta n)$ first becomes a square, where $n_0 = 4r(r + s) + 8r + 3s + 3$.

Now in Wichmann's solution $s = 2r + 1 + \varepsilon$ where $\varepsilon = -2, -1, 0, 1, 2$ for each particular value of r in turn. We then have

$$k = 6r + \varepsilon + 3$$

$$3n_0 = (6r + \varepsilon + \tfrac{9}{2})^2 - \varepsilon^2 - \tfrac{9}{4}$$

whence

$$\sqrt{3n_0} \doteqdot k + \tfrac{3}{2} - \tfrac{1}{2} \cdot \frac{4q^2 + 9}{4k + 6}$$

The increase occurs when $3n$ just exceeds $(6r + \varepsilon + 5)^2$.

But

$$(6r + \varepsilon + 5)^2 - 3n_0 = 6r + \varepsilon^2 + \varepsilon + 7,$$

so

$$\delta n = n - n_0 \geqslant 2r + \frac{\varepsilon^2 + \varepsilon + 7}{3}$$

Hence

$$\delta n = 2r + 3 \text{ for } \varepsilon = -1, 0, 1$$

$$\delta n = 2r + 5 \text{ for } \varepsilon = -2, +2$$

while the final, fourth difference is added in Wichmann's argument when $8n = 3r + 4 > 2r + 5$ when $r > 1$.

Further, the solutions (C) of Section 8, show that when $8n = 2n + 1$, the two additional differences at one end of the basis can be replaced by a single difference $2n + 1$ in the middle of the basis, with a further pair of differences of sum $2n + 2$ at one end, if needed.

Combining these two arguments we find eventually that $l(n) \leqslant [\sqrt{3n}] + 2$ for all n.

It is also of interest to examine the difference $3n - k^2$ for $n \leqslant 168$. This is positive or zero throughout the whole range. Only for $n = 147$ can it be equal to zero, though this is perhaps unlikely. This suggests $k \leqslant [\sqrt{3n}]$ as an aim that might perhaps be achieved, though not if we ever reach a stage when the gap, for given k, between qB and qC, $(q = 4r + 3)$, is complete or large. This gap contains $2r - 1$ values of n.

Finally, we may note that patterns (C) and (D) strictly alternate in taking the lead for successive values of k, these provide a firm second rung in the ladder of established solutions.

11. Unrestricted Difference Sets

For an outline of previous work we refer to Leech (1956), which gives references and discusses and extends the work of Rédei and Rényi. Leech successfully constructed some interesting bases by combining simple unrestricted bases with Singer's (1938) perfect difference bases to give larger bases. He has also made a systematic search for the simpler bases, exhaustively to $k = 7$, $n = 18$, with some beyond. Such a search seems much harder for unrestricted than for restricted bases. Almost all the theory of constructing the larger bases depends on the use of different cycles, usually perfect difference cycles, though the more general difference cycles of Section 5 can also be used.

We have not extended the search for unrestricted bases, but, now that complete results for restricted bases to $n \leqslant 68$ are available, it seems worth while to extend the record of Leech (1956) beyond $k = 11$.

The method used is that of "opening-up" judiciously one of the perfect differences cycles; this was used by Leech (1956), and mentioned by O'Beirne (1965).

Firstly, we note that every unrestricted set in which the range is less than twice the span can be joined up at the ends to give a general difference cycle, though not necessarily with minimum k.

Conversely we can cut a cycle at one of its points to give e.g.

$$b_{i+1}, b_{i+2}, \ldots, b_k, b_1, b_2, \ldots b_i$$

This will be an unrestricted basis with $n = b_i + b_{i+1} - 1$, for $b_i + b_{i+1}$ is now unrepresented, and every split sum exceeds this, because it contains both b_i and b_{i+1}.

We can also open up *with overlap*, e.g.

$$b_{i+1}, b_{i+2}, \ldots b_k, b_1, \ldots b_i, b_{i+1}, \ldots b_{i+j-1}.$$

Then the first sum missing is $b_i + b_{i+1} \ldots + b_{i+j} = \mu$ so that the span is $n = b_i + b_{i+1}, \ldots + b_{i+j} - 1 = \mu - 1$. The number of elements is $k + j = K$

We simply need to list cycles and seek large μ for given K.

Consider as an example the cycle given by Singer for $q^\alpha = 3^2$, $k = 10$, $n = 91$. From this cycle five others can be obtained by using multipliers. The six cycles are:

$$n = 91 \quad k = 10 \qquad 1\ 2\ 6\ 18\ 22\ 7\ 5\ 16\ 4\ 10 \tag{11.1}$$

$$1\ 3\ 9\ 11\ 6\ 8\ 2\ 5\ 28\ 18 \tag{11.2}$$

$$1\ 4\ 2\ 20\ 8\ 9\ 23\ 10\ 3\ 11 \tag{11.3}$$

$$1\ 4\ 3\ 10\ 2\ 9\ 14\ 16\ 6\ 26 \tag{11.4}$$

$$1\ 5\ 4\ 13\ 3\ 8\ 7\ 12\ 2\ 26 \tag{11.5}$$

$$1\ 6\ 9\ 11\ 29\ 4\ 8\ 2\ 3\ 18 \tag{11.6}$$

The maximum sum of consecutive pairs is 46 in (11.2). If we split this pair we get the unrestricted set

$$n = 45 \quad k = 10 \qquad 18\ 1\ 3\ 9\ 1\ 6\ 8\ 2\ 5\ 28$$

which compares with $n = 43$, $k = 10$ for restricted sets. The best sum of three consecutive elements is 51 also in (11.2) which does not help us. The best sum of four consecutive elements, however, is 62 in (11.4) which gives, with an overlap of two

$$n = 61 \quad k = 12 \qquad 16\ 6\ 26\ 1\ 4\ 3\ 10\ 2\ 9\ 14\ 16\ 6$$

which compares with $n = 58$, $k = 12$ for restricted sets.

The Table V below gives known solutions that surpass the restricted bases in span—the best for n to 68, and the best known, the Wichmann bases, for greater n up to the limit of Table III.

An outstanding problem is to find, or disprove the existence of, a basis with $k = 8$, $n = 30$.

TABLE V. Unrestricted difference bases

k	n	R	Bases
6	18	24	6 3 1 7 5 2
		25	8 1 3 6 5 2
		31	14 1 3 6 2 5
		31	13 1 2 5 4 6
7	24	39	8 10 1 3 2 7 8
9	37	73	16 1 11 8 6 4 3 2 22
		64	7 15 5 1 3 8 2 16 7
10	45	91	18 1 3 9 11 6 8 2 5 28
11	51	83	2 8 14 1 4 7 6 3 28 2 8
12	61	113	16 6 26 1 4 3 10 2 9 14 16 6
	60	133	35 1 8 10 5 7 21 4 2 11 3 26
	59	108	8 9 23 10 3 11 1 4 2 20 8 9
13	70	127	14 16 6 26 1 4 3 10 2 9 14 16 6
	69	131	8 9 23 10 3 11 1 4 2 20 8 9 23
15	93	175	20 9 13 35 1 4 7 3 16 2 6 17 20 9 13
	91	210	27 42 3 4 11 2 19 12 10 16 8 1 5 23 27
17	113	206	24 6 22 10 11 18 2 5 8 1 3 23 24 6 22 10 11
18	127	247	11 8 10 35 31 1 4 20 2 12 3 6 7 33 11 8 10 35

These are originally due to John Leech.

It may be of interest to note that John Leech's basis with $k = 131$, $N = 6539$ has a span exceeding the Wichmann span, $N = 5850$, by as much as 689. This is based on a more spohisticated construction, which he describes (1956), and is the best of many he found, the smallest has $n = 199$.

12. What I have tried to do in this account is to show how the use of a computer can provide a certain amount of material for discussion and thought in problems of this nature. The results in mathematical terms have not added a great deal to what is known of these problems, but the search itself has been interesting and has helped to provide material for student problems!

Many problems still remain, it is hoped to continue machine searches— in a more selective way for the restricted difference sets, but a comprehensive attack on the general difference cycles, and unrestricted difference sets for small n would also be of interest.

My thanks are due to John Leech, for useful criticism of this account, and for various additions to the results quoted.

Appendix A.

Condensed Proof of Wichmann's Difference Basis

Indices indicated repeated differences, which are ordered.

Thus

$$1\ 2\ 3\ 7\ 7\ 7\ 7\ 4\ 4\ 1$$

becomes

$$1\ 2\ 3\ 7^4\ 4^2\ 1.$$

Proof is arranged so that for $q = 4r + 3$, $n = mq + t$, each fixed value of t is treated separately as m varies. We use standard *moves*, or *operations*, and *exchange moves*, repetitively, so that condensed description is possible. The number of repetitions, each involving unit increase in m, that is, an increase q in n, is indicated in parentheses; a zero increase sometimes occurs on exchange. A complete line covers fixed t, and indicates representations for $t + mq$, $q = 0(1)r + s - 1$ or $r + s$; moves are listed from left to right in the table.

We use $r = 2a$ and $r = 2b - 1$ to separate even and odd cases.

MOVES

M Add $(2r + 1)^2$ and 1

N Add $(2r + 2)^2$ and remove 1

P Add $(4r + 3)$

Q Add $(2r + 1)$ and $(2r + 2)$

$S = S(0)$ Shift or slide, without change of m, from one end to other of a group $(2r + 2)^{r+1}$, $(4r + 3)^s$, or $(1^r(r + 1))(2r + 1)^r$.

X Change ends, e.g.

$$1^t\,(r + 1)\,(2r + 1)^u = (2r + 2)^u\,1^{t-u+r+1}$$

Particularly

$X(0)$ $$1^{t-1}\,(r + 1)\,(2r + 1)^r = (2r + 2)^r\,1^t$$

$X(1)$ $\quad 1^{t-a-1} \quad (2a+1) \quad (4a+1)^{2a} \quad (8a+3)^s$ to
$\quad (4a+1) \quad (8a+3)^s \quad (4a+2)^{2a+1} \quad 1^{t-a}$, with increase $q = 8a+3$.

PROOFS

Basis $1^{2a} \quad (2a+1) \quad (4a+1)^{2a} \quad (8a+3)^s \quad (4a+2)^{2a+1} \quad 1^{2a}$

$N = (2a+s+1)(8a+3)+2a \qquad k = s+8a+2$

Range.in t	Start	Moves
$(0, a)$	1^t on Right	$N(t)SP(s)Q(2a+1-2t)M(t)$
$(a+1, 2a)$	1^t on Right	$N(a)X(0)P(s)X(1)M(a)$
$(2a+1, 3a+1)$	$1^{t-2a-1}(2a+1)$	$M(a)P(s)N(a)$
$(3a+2, 4a+1)$	$1^{t-2a-1}(2a+1)$	$M(4a+1-t)SP(s)Q(2t-6a-2)N(4a+1-t)$
$(4a+2, 5a+2)$	$(4a+2)1^{t-2a-2}$	$N(t-4a-2)SP(s)Q(10a+4-2t)M(t-4a-2)$
$(5a+3, 6a+2)$	$(4a+2)1^{t-2a-2}$	$N(a)P(s)M(a)$
$(6a+3, 7a+2)$	$1^{t-6a-2}(2a+1)(4a+1)$	$M(a-1)X(1)P(s)X(0)N(a)$
$(7a+3, 8a+2)$	$1^{t-6a-2}(2a+1)(4a+1)$	$M(8a+2-t)SP(s)Q(2t-14a-5)N(8a-t-3)$

Basis $1^{2b-1} \quad (2b) \quad (2b-1)^{2b-1} \quad (8b-1)^s \quad (4b)^{2b} \quad 1^{2b-1}$

$N = (2b+s)(8s+1)+2b-1 \qquad k = s+8b-2$

Range in t	Start	Moves
$(0, b)$	1^t on Right	$N(t)SP(s)Q(2b-2t)M(t)$
$(b+1, 2b-1)$	1^t on Right	$N(b)P(s)M(b)$
$(2b, 3b-1)$	$1^{t-2b}(2b)$	$M(b-1)X(1)P(s)X(0)N(b-1)$
$(3b, 4b-1)$	$1^{t-2b}(2b)$	$M(4b-1-t)SP(s)Q(2t-6b-1)N(4b-1-t)$
$(4b, 5b-1)$	$(4b)1^{t-4b}$	$N(t-4b)SP(s)Q(10b-2t-1)M(t-4b)$
$(5b, 6b-1)$	$(4b)1^{t-4b}$	$N(b-1)X(0)P(s)X(1)M(b-1)$
$(6b, 7b-1)$	$1^{t-6b+1}(2b)(4b-1)$	$M(b-1)P(s)N(b)$
$(7b, 8b-2)$	$1^{t-6b+1}(2b)(4b-1)$	$M(8b-2-t)SP(s)Q(2t-14b+2)N(8b-t-1)$

References

Hall, M. (1956). A survey of difference sets. *Proc. Amer. Math. Soc.* **7,** 975–986.

Haselgrove, C. B. and Leech, J. (1957). Note on restricted difference bases. *J. London Math. Soc.* **32,** 228–231.

Leech, J. (1956). On the representation of 1, 2, ... , n by differences. *J. London Math. Soc.* **31,** 160–169.

O'Beirne, T. H. (1965). "Puzzles and Paradoxes", Oxford University Press, London.
Ryser, H. J. (1963). "Combinatorial Mathematics". Wiley, New York.
Selmer, E. S. (1966). "Linear Recurrence Relations over Finite Fields". Dept. Maths., University of Bergen, Norway.
Singer, J. (1938). A theorem in finite projective geometry and some applications to number theory. *Trans. Amer. Math. Soc.* **43,** 377–385.
Wichmann, B. (1962). A note on restricted difference bases. *J. London Math. Soc.* **38,** 465–466.

A Natural Generalization of Steiner Triple Systems

N. S. MENDELSOHN

University of Manitoba, Winnipeg, Canada

1. Introduction and Summary

In the Atlas Symposium No. 1, Knuth (1970) showed how a computer could be used for solving word problems in universal algebras. This led the present author to a series of studies on the use of groupoids in combinatorial design. Computer computations for a number of small values of the parameters led to a series of number theoretic models of appropriate groupoids. This paper describes a successful end result for generalizations of Steiner Triple Systems although the heuristic use of the computer in leading to the final result is not described.

Let S be a set of v elements. Let T be a collection of b subsets of S, each of which contains three elements arranged cyclically, and such that any ordered pair of elements of S appears in exactly one cyclic triplet (note the cyclic triplet $\{a, b, c\}$ contains the ordered pairs ab, bc, ca but not ba, cb, ac). When such a configuration exists we will refer to it as a generalized triple system. Where ambiguity is impossible we will often refer to a generalized triple system simply as a system. If we ignore the cyclic order of the triples a generalized triple system is a *B. I. B. D.* with parameters

$$v, \qquad k = 3, \qquad \lambda = 2, \qquad b = \frac{v(v-1)}{3}, \qquad r = v - 1.$$

It is shown that not every *B. I. B. D.* with these parameters is a generalized triple system. In the case where $v \equiv 1$ or 3 mod 6, one can always construct such a generalized triple system by simply taking a Steiner triple system for the same value of v and using each triple twice, once in each of the two possible cyclic orders. It is also shown that not every generalized triple system of odd order can be separated into two Steiner triple systems. A construction is given for a generalized triple system for any $v \not\equiv 2$ mod 3 with the exception of $v = 6$, a value for which the system does not exist.

Next we introduce the notions of simplicity and purity. A generalized triple system is called simple if no proper subset of its triples forms a system. Such a system is called pure if no triple appears more than once when cyclic order is ignored. If p is a prime congruent to 1 mod 3 a system with $v = p$ is constructed which is pure and simple. For p an odd prime which is congruent to 2 mod 3 the corresponding pure and simple system is constructed with $v = p^2$.

Sometimes generalized triple systems can be decomposed into Steiner triple systems. It is shown here that the pure and simple system of order p for $p \equiv 1 \bmod 3$ which has been constructed, can be separated into two $\dfrac{(p-1)}{6}$ collections of triples

$$A_1, A_2, \ldots, A_{(p-1)/6}; \qquad A_1{}^*, A_2{}^*, \ldots, A_{(p-1)/6}$$

with the following properties. Each block A_i or $A_i{}^*$ contains p triples which, in fact, are all the translates of a single triple i.e. a block A_j or $A_j{}^*$ will consist of triples $\{a+i, b+i, c+i\}$ for $i = 0, 1, 2, \ldots, p-1$. Furthermore, each of the $2^{(p-1)/6}$ sets of $\dfrac{p(p-1)}{6}$ triples $C_1 \cup C_2 \ldots \cup C_{(p-1)/6}$ where $C_i = A_i$ or $C_i = A_i{}^*$ is a Steiner triple system.

Finally, we show that with respect to the automorphism group, two extremes may occur; systems whose automorphism group is doubly transitive on points and systems having only the identity automorphism.

2. Algebraic Representation of Generalized Triple Systems

Let S be a set of v elements for which the generalized triple system exists. Let $\{a, b, c\}$ be any such triple. Since the ordered pair $\langle a, b\rangle$ uniquely determines c we can introduce a binary operator (denoted simply by concatenation) such that any *distinct* pair a, b uniquely determines an element c such that $ab = c$, where $c \neq a$, $c \neq b$. Now the same triple is uniquely determined by the ordered pairs $\langle b, c\rangle$ and $\langle c, a\rangle$. Hence, $ab = c$ implies $bc = a$ and $ca = b$. Substituting for c we obtain $b(ab) = a$ and $(ab)a = b$. Leaving aside for a moment the definition of a^2 suppose now we have any groupoid S with a single binary relation $b(ab) = a$ for all a, b in S. It follows that $b = (ab)\big(b(ab)\big) = (ab)\,a$. The further condition that a, b and ab are distinct when $a \neq b$ will now yield the following. Suppose $ab = ac$ then $(ab)\,a = (ac)\,a$ or $b = c$. Hence we have a left cancellation law and in the same way we have a right cancellation law. Suppose now that $aa = x$. Then $a = a(aa) = ax$. But if $x \neq a$ then $ax \neq a$, a contradiction. Hence $a^2 = a$. Conversely, the identities $a^2 = a$ and $a(ba) = b$ imply that if $a \neq b$ then $ab \neq a$ and $ab \neq b$. Hence we can quote the following theorem, which has just been proved.

THEOREM 1. *Let S be a set of elements for which a generalized triple system exists. If we define a binary operation $a \cdot b = c$ if and only if $\{a, b, c\}$ is one of the cyclic triples then $\langle S, \cdot \rangle$ is a quasi-group satisfying the identities*

$$a(ba) = b \tag{1}$$

$$a^2 = a \tag{2}$$

Conversely, let $\langle S, \cdot \rangle$ be any groupoid satisfying the identities $a(ba) = b$, $a^2 = a$. Then we have that $\langle S, \cdot \rangle$ is a quasigroup and if $a \neq b$ then $ab \neq a$, $ab \neq b$ and from the elements of S a generalized triple system can be constructed by assigning to the ordered pair $\langle a, b \rangle$ the unique cyclic triple $\{a, b, a \cdot b\}$.

Suppose now we have a generalized triple system on v elements $a_1, a_2, \ldots, a_v$. Since the pair a_i, a_j appears only in the cyclic blocks $\{a_i, a_j, a_i a_j\}$ and $\{a_j, a_i, a_j a_i\}$, then ignoring cyclic order each pair of elements lies in exactly two blocks. Also an element a_i lies in exactly $v - 1$ of the blocks namely $\{a_i, a_1, a_i a_1\}$, $\{a_i, a_2, a_i a_2\}$, $\ldots$ $\{a_i, a_{i-1}, a_i a_{i-1}\}$, $\{a_i, a_{i+1}, a_i a_{i+1}\}$, $\ldots$, $\{a_i, a_v, a_i a_v\}$. Hence, ignoring cyclic order the blocks form a *B. I. B. D.* with parameters $v, k = 3, \lambda = 2, r = v - 1, b = \dfrac{v(v-1)}{3}$. Since b is an integer, $v \not\equiv 2 \bmod 3$. If $v \equiv 1$ or $v \equiv 3 \bmod 6$ one can always construct a trivial generalized triple system from the corresponding Steiner system as shown in the introduction. Generalized triple systems can be constructed in various ways as is shown below. The system can be displayed by the multiplication table of the groupoid. If the elements of S are $1, 2, 3, 4, \ldots v$, the multiplication table is a v by v latin square with the entry in the ith row, jth column being $i \cdot j$.

3. Illustrative Examples

We consider here small values of $v \not\equiv 2 \bmod 3$ and indicate various possibilities

$v = 3$; The only multiplication table is given by

$$\begin{matrix} 1 & 3 & 2 \\ 3 & 2 & 1 \\ 2 & 1 & 3 \end{matrix}$$

corresponding to the cyclic triplets $\{1, 2, 3\}$, $\{1, 3, 2\}$.

$v = 4$; The multiplication table is uniquely determined (apart from transposing the matrix) and is given by

$$\begin{matrix} 1 & 3 & 4 & 2 \\ 4 & 2 & 1 & 3 \\ 2 & 4 & 3 & 1 \\ 3 & 1 & 2 & 4 \end{matrix}$$

corresponding to the cyclic triplets $\{1, 2, 3\}, \{1, 3, 4\}, \{1, 4, 2\}, \{2, 4, 3\}$.

$v = 6$; This case is interesting. An exhaustive enumeration shows that no generalized triple system exists. However, the corresponding *B. I. B. D.* with $v = 6,\ k = 3,\ \lambda = 2,\ r = 5,\ b = 10$ exists as follows (the triples are not cyclically ordered):

$(1, 2, 3),\ (1, 2, 4),\ (1, 3, 6),\ (1, 4, 5),\ (1, 5, 6),\ (2, 3, 5),\ (2, 4, 6),\ (2, 5, 6),$
$(3, 4, 5),\ (3, 4, 6).$

That this design cannot have its triples cyclically arranged so that each ordered pair appears exactly once can be seen as follows:

$\{1, 2, 3\}$ implies $\{1, 4, 2\}$, and $\{2, 5, 3\}$. Also $\{1, 4, 2\}$ implies $\{1, 5, 4\}$; $\{2, 5, 3\}$ implies $\{3, 5, 4\}$. But we have now reached a contradiction since the ordered pair 5, 4 appears in the two cyclic triplets $\{1, 5, 4\}$ and $\{3, 5, 4\}$.

$v = 7$; Many multiplication tables are possible. In all cases the design can be separated into two Steiner triple systems. We illustrate two such non-isomorphic systems.

I

$$\begin{matrix} 1 & 3 & 2 & 5 & 6 & 7 & 4 \\ 3 & 2 & 1 & 6 & 7 & 4 & 5 \\ 2 & 1 & 3 & 7 & 4 & 5 & 6 \\ 7 & 6 & 5 & 4 & 1 & 2 & 3 \\ 4 & 7 & 6 & 3 & 5 & 1 & 2 \\ 5 & 4 & 7 & 2 & 3 & 6 & 1 \\ 6 & 5 & 4 & 1 & 2 & 3 & 7 \end{matrix}$$

II

$$\begin{matrix} 1 & 6 & 4 & 2 & 7 & 5 & 3 \\ 4 & 2 & 7 & 5 & 3 & 1 & 6 \\ 7 & 5 & 3 & 1 & 6 & 4 & 2 \\ 3 & 1 & 6 & 4 & 2 & 7 & 5 \\ 6 & 4 & 2 & 7 & 5 & 3 & 1 \\ 2 & 7 & 5 & 3 & 1 & 6 & 4 \\ 5 & 3 & 1 & 6 & 4 & 2 & 7 \end{matrix}$$

Concerning these systems we note the following.

(a) Both systems can be separated into two Steiner triple systems.

(b) If we ignore cyclic order I has repeated blocks, whereas II does not have such blocks.

(c) I and II are *not* isomorphic since I has a subsystem of order 3 while II has no proper subsystems.

It is also interesting to note that if we start with two Steiner triple systems of order 7 on the same seven elements it is not always possible to arrange the blocks in appropriate cyclic orders to form a generalized triple system. As an example: if S_1 consists of the triples $(1, 2, 4)$, $(2, 3, 5)$, $(3, 4, 6)$, $(4, 5, 7)$, $(5, 6, 1)$, $(6, 7, 2)$, $(7, 1, 3)$ and S_2 consists of the triples $(1, 2, 5)$, $(2, 3, 4)$, $(3, 5, 6)$, $(5, 4, 7)$, $(4, 6, 1)$, $(6, 7, 2)$, $(7, 1, 3)$, it is impossible to reorient these triples to form a generalized triple system.

$v = 9$; Here a new phenomenon appears. We show the existence of two distinct types of system. In the first type the system cannot be separated into two Steiner triple systems. In the second type such a separation is possible. For each type there are several non-isomorphic designs. As the second type is much the same as in the case $v = 7$ an example will not be shown. The following table illustrates a system of the first type.

1	3	4	2	6	7	8	9	5
4	2	1	3	9	8	5	6	7
2	4	3	1	7	9	6	5	8
3	1	2	4	8	5	9	7	6
9	7	8	6	5	1	3	4	2
5	8	7	9	4	6	1	2	3
6	9	5	8	2	3	7	1	4
7	6	9	5	3	2	4	8	1
8	5	6	7	1	4	2	3	9

That the system cannot be separated into two Steiner triple systems S_1 and S_2 can be seen as follows. The system contains the four triples $\{1, 2, 3\}$, $\{1, 3, 4\}$, $\{1, 4, 2\}$, $\{2, 4, 3\}$. Suppose now that $\{1, 2, 3\} \in S_1$. It follows that $\{1, 4, 2\}$, $\{1, 3, 4\}$ are in S_2. But now $\{2, 4, 3\}$ cannot lie in either of S_1 or S_2.

4. Extension of Systems

In this section it will be shown how to embed a system of order v into a larger system. The embeddings will allow us to prove the main result, namely; a generalized triple system exists for all $v \not\equiv 2 \bmod 3$ except for $v = 1$ and $v = 6$. The ideas in this section are due to R. G. Stanton. In particular, Theorems 2 and 4 are due to him.

First we note the following. Let T be a generalized triple system on the elements $1, 2, \ldots, v, v+1, v+2, \ldots u$. Suppose the triples of T which are made up from the elements of the subset $\{1, 2, \ldots, v\}$ themselves form a generalized triple system S, and let V be the remaining triples of T. We have $T = S \cup V$, and call T an extension of S. Now if S^* is any generalized triple system on the set $\{1, 2, \ldots, v\}$ then it is readily seen that $T^* = S^* \cup V$ is a generalized triple system. This is essentially because each triple of the set V has at most one of its elements in the set $\{1, 2, \ldots, v\}$. This implies that to embed a generalized triple system S in a larger system T we need pay no attention to the triples belonging to S.

In terms of the multiplication table of the quasigroup corresponding to the extension T of S its appearance is as in Fig. 1.

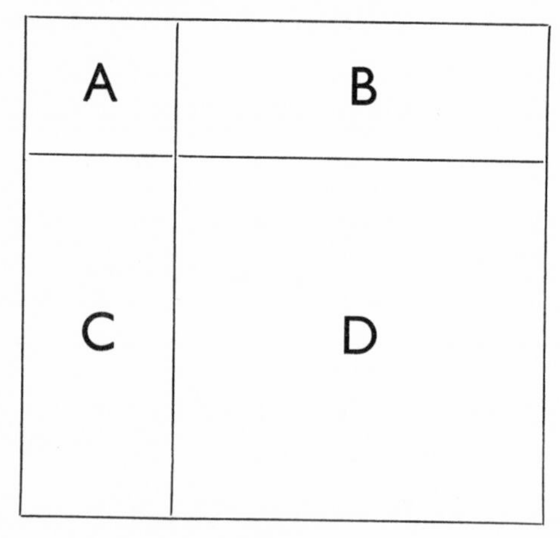

FIGURE 1

Here A is a square matrix of size $v \times v$ corresponding to the system S. B is of dimension $v \times (u-v)$, C of dimension $(u-v) \times v$, and D of dimension $(u-v) \times (u-v)$. It is readily seen that if the matrix D has all its entries filled in, then the identities $a(ba) = b$ and $(ab)a \equiv b$ completely determine the entries in the submatrices B and C. For the matrix D we will refer to the diagonals parallel to the main diagonal of D as the 1st, 2nd, ..., $(u-v)$th diagonal. In particular, the ith diagonal will contain consecutively the cells—

$$(v+1,\ v+i),\ (v+2,\ v+i+1),\ (v+3,\ v+i+2),\ \ldots,\ (u-i+1,\ u),\ (u-i+2,\ v+1),\ \ldots,\ (u, v+i-1).$$

THEOREM 2. *A generalized triple system on v elements can be extended to a system on $2v+1$ elements.*

Proof. The entries of the matrix D are taken as follows. The first diagonal contains the entries $v+1$, $v+2$, $v+3$, ..., $2v+1$. Also, for $i = 2, 3, \ldots,$ $v+1$ the entries in the ith diagonal are all taken to be i. As an illustration the following table shows an extension of a system on four elements to one on nine elements.

1	3	4	2	6	7	8	9	5
4	2	1	3	7	8	9	5	6
2	4	3	1	8	9	5	6	7
3	1	2	4	9	5	6	7	8
9	8	7	6	5	1	2	3	4
5	9	8	7	4	6	1	2	3
6	5	9	8	3	4	7	1	2
7	6	5	9	2	3	4	8	1
8	7	6	5	1	2	3	4	9

THEOREM 3. *A generalized triple system on v elements can be extended to a system on $2v+4$ elements.*

Proof. Choose the entries of D as follows. The 1st diagonal has entries $v+1$, $v+2$, ..., $2v+4$. The second diagonal contains the entries, $v+4$, $v+5$, ..., $2v+4$, $v+1$, $v+2$, $v+3$. The third diagonal contains the entries $2v+4$, $v+1$, $v+2$, ..., $2v+3$. The $(v+1)$th diagonal contains the entries $2v+3$, $2v+4$, $v+1$, $v+2$, ..., $2v+2$. There remain v diagonals. Fill any

1	3	2	7	8	9	10	4	5	6
3	2	1	9	10	4	5	6	7	8
2	1	3	10	4	5	6	7	8	9
8	6	5	4	7	10	1	9	2	3
9	7	6	3	5	8	4	1	10	2
10	8	7	2	3	6	9	5	1	4
4	9	8	5	2	3	7	10	6	1
5	10	9	1	6	2	3	8	4	7
6	4	10	8	1	7	2	3	9	5
7	5	4	6	9	1	8	2	3	10

1	3	2	9	10	4	5	6	7	8
3	2	1	10	4	5	6	7	8	9
2	1	3	7	8	9	10	4	5	6
6	5	8	4	7	10	3	9	1	2
7	6	9	2	5	8	4	3	10	1
8	7	10	1	2	6	9	5	3	4
9	8	4	5	1	2	7	10	6	3
10	9	5	3	6	1	2	8	4	7
4	10	6	8	3	7	1	2	9	5
5	4	7	6	9	3	8	1	2	10

one of these diagonals with the entry 1, a second diagonal with the entry 2, a third diagonal with the entry 3, etc. The above two tables illustrate the case for $v = 3$.

THEOREM 4. *If* $v \equiv 1$ *or* 4 mod 6 *and a generalized triple system on* v *elements exists, then the system can be extended to one on* $2v + 2$ *elements.*

Proof. The construction is identical for either case. The matrix D is filled as follows. The first diagonal has successively the elements $v + 1$, $v + 2$, ..., $2v + 2$. The second diagonal is filled with 1's; the fourth with 2's; the fifth with 3's; ..., the $v + 1$st with $(v - 1)$'s. This leaves unfilled the 3rd and $(v + 2)$nd diagonal. These are filled as follows: $v + 2$ is placed in cell $(v + 1, v + 3)$; $v + 5$ in cell $(v + 4, v + 6)$; $v + 8$ in cell $(v + 7, v + 9)$ etc.; (here we are working mod $(v + 2)$ using residues $v + 1$, $v + 2$, ..., $2v + 2$). This has filled in $v + 2$ entries, some in the third diagonal and some in the $(v + 2)$th. There remains $v + 2$ unoccupied cells. These are filled in with $2v + 2$. The proof that this construction works is a straightforward verification of the identity $a(ba) = a$ for the 3rd and $(v + 2)$nd diagonal of D. Three tables for $v = 4$, 7, 10 are given below to illustrate the construction.

				6	7	8	9	10	5
				8	9	10	5	6	7
				9	10	5	6	7	8
				10	8	9	7	5	6
10	8	7	9	5	1	**6**	2	3	4
5	9	8	10	**7**	6	1	4	2	3
6	10	9	8	3	**5**	7	1	4	2
7	5	10	6	2	3	4	8	1	**9**
8	6	5	7	4	2	3	**10**	9	1
9	7	6	5	1	4	2	3	**8**	10

THEOREM 5. *A generalized triple system exists for all* $v \not\equiv 2 \bmod 3$ *except for* $v = 1$ *and* $v = 6$.

Proof. The cases $v \equiv 0$ and $v \equiv 1 \bmod 3$ can be broken up into $v \equiv 1 \bmod 6$, $v \equiv 3 \bmod 6$, $v \equiv 4 \bmod 6$; $v \equiv 0 \bmod 12$ $v \equiv 6 \bmod 12$.

							9	10	11	12	13	14	15	16	8
							11	13	13	14	15	16	8	9	10
							12	13	14	15	16	8	9	10	11
							13	14	15	16	8	9	10	11	12
							14	15	16	8	9	10	11	12	13
							15	16	8	9	10	11	12	13	14
							16	11	12	10	14	15	13	8	9
16	14	13	12	11	10	15	8	1	**9**	2	3	4	5	6	7
8	15	14	13	12	11	16	**10**	9	1	7	2	3	4	5	6
9	16	15	14	13	12	11	6	**8**	10	1	7	2	3	4	5
10	8	16	15	14	13	9	5	6	7	11	1	**12**	2	3	4
11	9	8	16	15	14	10	4	5	6	**13**	12	1	7	2	3
12	10	9	8	16	15	14	3	4	5	6	**11**	13	1	7	2
13	11	10	9	8	16	12	2	3	4	5	6	7	14	1	**15**
14	12	11	10	9	8	13	7	2	3	4	5	6	**16**	15	1
15	13	12	11	10	9	8	1	7	2	3	4	5	6	**14**	16

For $v \equiv 1 \bmod 6$, $v \equiv 3 \bmod 6$ the trivial generalized triple system can be constructed from the corresponding Steiner triple system. The case $v = 4$ has been given in Section 3, the case $v = 12$ can be constructed from $v = 4$ using Theorem 3, the case $v = 18$ can be constructed from $v = 7$ using Theorem 3. We now have sufficient initial cases for an induction. We need only consider $v \equiv 4 \bmod 6$, $v \equiv 0 \bmod 12$, $v \equiv 6 \bmod 12$.

										12	13	14	15	16	17	18	19	20	21	22	11
										14	15	16	17	18	19	20	21	22	11	12	13
										15	16	17	18	19	20	21	22	11	12	13	14
										16	17	18	19	20	21	22	11	12	13	14	15
										17	18	19	20	21	22	11	12	14	13	15	16
										18	19	20	21	22	11	12	13	14	15	16	17
										19	20	21	22	11	12	13	14	15	16	17	18
										20	21	22	11	12	13	14	15	16	17	18	19
										21	22	11	12	13	14	15	16	17	18	19	20
										22	14	15	13	17	18	16	20	21	19	11	12
22	20	19	18	17	16	15	14	13	21	11	1	**12**	2	3	4	5	6	7	8	9	10
11	21	20	19	18	17	16	15	14	22	**13**	12	1	10	2	3	4	5	6	7	8	9
12	22	21	20	19	18	17	16	15	14	9	**11**	13	1	10	2	3	4	5	6	7	8
13	11	22	21	20	19	18	17	16	12	8	9	10	14	1	**15**	2	3	4	5	6	7
14	12	11	22	21	20	19	18	17	13	7	8	9	**16**	15	1	10	2	3	4	5	6
15	13	12	11	22	21	20	19	18	17	6	7	8	9	**14**	16	1	10	2	3	4	5
16	14	13	12	11	22	21	20	19	15	5	6	7	8	9	10	17	1	**18**	2	3	4
17	15	14	13	12	11	22	21	20	16	4	5	6	7	8	9	**19**	18	1	10	2	3
18	16	15	14	13	12	11	22	21	20	3	4	5	6	7	8	9	**17**	19	1	10	2
19	17	16	15	14	13	12	11	22	18	2	3	4	5	6	7	8	9	10	20	1	**21**
20	18	17	16	15	14	13	12	11	19	10	2	3	4	5	6	7	8	9	**22**	21	1
21	19	18	17	16	15	14	13	12	11	1	10	2	3	4	5	6	7	8	9	**20**	22

Case 1. $v \equiv 4 \mod 6$. Either $v \equiv 4 \mod 12$ or $v \equiv 10 \mod 12$. If $v = 4 + 12u$ then $v = 2(6u + 1) + 2$ and the result follows from Theorem 4. If $v = 10 + 12u$, $v = 2(6u + 4) + 2$ and again Theorem 4 yields the required result.

Case 2. $v \equiv 0 \mod 12$. Here $v = 12u = 2\{6(u - 1) + 4\} + 4$ and the theorem follows from Theorem 3.

Case 3. $v \equiv 6 \bmod 12$, $v > 6$. Here $v = 12u + 6 = 2(6u + 1) + 4$ and the theorem follows from Theorem 3.

It will be noted that Theorem 2 has not been used in the proof of Theorem 5. Nevertheless, it is very useful in connection with the notion of a *pure system* which is discussed in the next section. We note that, if a system of order v is pure then so is the system of order $2v + 1$ which is constructed using Theorem 2.

5. Purity and Simplicity

A generalized triple system is called *pure* if, when considered as a *B.I.B.D.*, all its triples are distinct. A system on v elements is called *simple* if it does not contain a sub-system on u elements $u < v$.

To discuss these properties the use of the representation of the system by a groupoid with identities $(ab)a = b$, $a^2 = a$ is appropriate. A system is pure if the corresponding groupoid is such that $ab \neq ba$ if $a \neq b$. The system is simple if any two of the elements of the groupoid generate the whole groupoid.

THEOREM 6. *If $p \equiv 1 \bmod 3$, where p is prime, there is a generalized triple system on p elements which is both pure and simple.*

Proof. Since $p \equiv 1 \bmod 3$, -3 is a quadratic residue. Hence $\lambda^2 = \lambda - 1$ has a solution in $GF(p)$. Let the elements of a system be the integers mod p and define $a \cdot b = \lambda a + (1 - \lambda)b$ where $\lambda^2 = \lambda - 1$. It is immediately verified that $(ab)a = b$ and $a^2 = a$. Now if $a \cdot b = b \cdot a$, $\lambda a + (1 - \lambda)b = \lambda b + (1 - \lambda)a$ or $(2\lambda - 1)a = (2\lambda - 1)b$. Now if $2\lambda - 1 = 0$ then from $\lambda^2 = \lambda - 1$ or $4\lambda^2 = 4\lambda - 4$ we obtain $1 = -2$ or $3 = 0$, a contradiction. Hence $2\lambda - 1 \neq 0$ or $a = b$. Hence, the system is pure.

It is now shown that the whole quasigroup is generated by two distinct elements a and b. A direct computation shows that $a(a(ab)) = 2a - b$. Hence from any two distinct elements a and b be can generate $2a - b$. Hence we can generate successively

$$\begin{aligned} &2a - b \text{ from } a \text{ and } b; \\ &4a - 3b \text{ from } 2a - b \text{ and } b; \\ &8a - 7b \text{ from } 4a - 3b \text{ and } b. \\ &\vdots \\ &2^n a - (2^n - 1)b, \text{ from } 2^{n-1} a - (2^{n-1} - 1) b \text{ and } b. \end{aligned}$$

Pur $a = b + u$, $u \neq 0$. We then have generated

$$b, b + u, b + 2u, b + 4u, b + 8u, \ldots.$$

Again from $b + 2^r u$ and $b + 2^{r+1}$ we generate

$$2(b + 2^{r+1} u) - (b + 2^r u) = b + 3(2^r u).$$

From $b + 3(2^r u)$ and $b + 2^r u$ we generate

$$2\big(b + 3(2^r u)\big) - (b + 2^r u) = b + 5(2^r u).$$

Inductively, from $b + (2k+1)2^r u$ and $b + (2k-1)\,2^r u$ we generate

$$2\{b + (2k+1)2^r u\} - \big(b + 2k - 1)\,2^r j\big) = b + (2k+3)\,2^r u.$$

In particular, for $r = 0$, we have generated

$$b, b + u, b + 3u, b + 5u, b + 7u, b + 9u, \ldots .$$

Now if $b + (2r+1)\,u = b + (2s+1)\,u$ it follows that $2r + 1 \equiv 2s + 1 \bmod p$ or $r \equiv s \bmod p$. Hence, $b + u, b + 3u, b + 5u, \ldots, b + (2p-1)\,u$ are all the elements of $GF(p)$. Hence our system is simple.

COROLLARY. *Since every element of $GF(p)$ can be represented by $b, b + u, b + 2u, \ldots, b + (p-1)\,u$ we have that b and $b + u$ generate $b + iu\ (i = 0, 1, 2, \ldots, p-1)$.*

This will be used in the next theorem.

THEOREM 7. *If $p \equiv 2 \bmod 3$, where p is prime, there is a generalized triple system on p^2 elements which is both pure and simple.*

Proof. If $p \equiv 2 \bmod 3$ and p is prime then -3 is a quadratic non-residue. Hence $x^2 - x + 1$ is irreducible over $GF(p)$ (If $p = 2$, $x^2 - x + 1$ is also irreducible). Hence we may consider the elements of $GF(p^2)$ as the set of all $u + \lambda v$ where u and v are integers $\bmod p$ and $\lambda^2 = \lambda - 1$. Now construct a quasigroup whose elements are those of $GF(p^2)$ with binary operator $a \cdot b = \lambda a + (1 - \lambda)b$ and $\lambda^2 = \lambda - 1$. As in Theorem 6, $(ab)a = b$, $a^2 = a$ and the corresponding system is pure.

Now let a and b be any two elements of the quasigroup. We show that a and b generate the whole quasigroup. The argument of Theorem 6 and the corollary show that from a and b with $a = b + u$ we can generate all the elements

$$b, b + u, b + 2u, b + 3u, \ldots, b + \ (p-1)\,u.$$

Again $a \cdot b = \lambda a + (1 - \lambda) b = \lambda(b + u) + (1 - \lambda) b = b + \lambda u$. As in the corollary to the previous theorem from b and $b + \lambda u$ we can generate

$$b, b + \lambda u, b + 2\lambda u, \ldots, b + (p - 1) \lambda u.$$

From $b + ru$ and $b + s\lambda u$ (where r and s are integers mod p) we can generate $2(b + ru) - (b + s\lambda u) = b + (2r - s\lambda) u$. As r ranges over all residues mod p so does $2r$. Hence we can generate $b + (r + s\lambda) u$ where r and s are any integers mod p. The elements $b + (r + s\lambda) u$ yield all the elements of $GF(p^2)$.

While the argument of the last paragraph is not valid for $p = 2$, the theorem is still valid as can be seen from inspecting the system with $v = 4$.

THEOREM 8. *If a pure system on v elements exists then there is a pure system on $2v + 1$ and $2v + 4$ elements.*

Proof. The constructions of Theorems 2 and 3 generate pure systems from pure systems.

6. Decompositions into Steiner Triple Systems

A generalized triple system on an odd number of elements does not necessarily separate into two Steiner triple systems. However, a question of some interest is whether one can extract and in which ways a Steiner triple system from a generalized triple system.

First let us look at a collection S of unordered triples on a set of v elements. We say that two triples are compatible if they have at most one element in common, otherwise they are incompatible. Now let S be a set of $\frac{v(v-1)}{6}$ pairwise compatible triples on a set of v elements. We assert that S is a Steiner triple system. Indeed, the condition of pairwise compatibility implies that each pair of elements lies in at most one triple. To find the average number of appearances of a pair in a triple in the set S one notes that each triple contains 3 pairs and hence in the collection S there are $\frac{v(v-1)}{2}$ pairs. The total number of pairs taken from v elements is $\frac{v(v-1)}{2}$. Hence the average number of appearances of a pair is one. Hence, each pair appears exactly once, so that S is a Steiner triple system.

In this section our triples will be taken from the elements of $GF(p)$. If $\{a, b, c\}$ is a triple the collection of p triples $\{a + i, b + i, c + i\}$ $i = 0, 1, 2, \ldots p - 1$ will be called the set of translates of $\{a, b, c\}$. Obviously the set of translates is determined by any one of its triples.

THEOREM 9. *Let* $p \equiv 1 \bmod 3$, *p prime. Let T be the generalized pure and simple triple system defined in Theorem 6, i.e. the system whose cyclic triples are given by* $\{a, b, \lambda a + (1-\lambda)b\}$ *where* $\lambda^2 = \lambda - 1 \bmod p$. *Then T can be partitioned into* $\frac{p-1}{3}$ *subsets* $A_1, A_1{}^*, A_2, A_2{}^*, \ldots, A_{(p-1)/6}, A^*_{(p-1)/6}$ *with the properties*:

(1) *Each* A_i, *or* $A_i{}^*$ *consists of the p translates of a cyclic triple.*

(2) *For any triple in* A_i *the only triples incompatible with it lie in* $A_i{}^*$.

(3) *The* $2^{(p-1)/6}$ *sets of triples* $C_1 \cup C_2 \cup \cdots \cup C_{(p-1)/6}$ *where each* $C_i = A_i$ *or* $= A_i{}^{*C_i = A_i}$ *are all Steiner triple systems.*

Proof. First, note that putting $a \cdot b = \lambda a + (1-\lambda)b$ with $\lambda^2 = \lambda - 1$, we obtain that if $a \cdot b = c$ then $(a+1)\cdot(b+1) = c+1$. Hence, the translates of any cyclic triple in T also belong to T.

Next note that any triple $\{a, b, \lambda a + (1-\lambda)b\}$ is compatible with its translates. For let $\{a+t, b+t, \lambda a + (1-\lambda)\,b + t\}$ be any translate with $t \neq 0$. Now if $b = a + t$ then $a \neq b + t$, $a \neq \lambda a + (1-\lambda)b + t$, $\lambda a + (1-\lambda)\,b \neq b + t$, $\lambda a + (1-\lambda)\,b \neq \lambda a + (1-\lambda)\,b + t$ for it is easily verified that if any of the stated inequalities were equalities then $\lambda = 0$ or $\lambda = 2$ or $\lambda = 1$ all of which contradict $\lambda^2 = \lambda - 1$.

Thirdly, we show that all of the triples in T which are incompatible with $\{a, b, \lambda a + (1-\lambda)b\}$ are in a single set of translates. In fact, the triples in T which are incompatible with $\{a, b, \lambda a + (1-\lambda)b\}$ are $\{b, a, \lambda b + (1-\lambda)a\}$ $\{a, \lambda a + (1-\lambda)\,b, (\lambda+1)\,a - \lambda b\}$ and $\{\lambda a + (1-\lambda)\,b, b, (\lambda-1)\,a + (2-\lambda)\,b\}$.

If we translate $\{b, a, \lambda b + (1-\lambda)a\}$ by $\lambda(a-b)$ we get $\{\lambda a + (1-\lambda)\,b, (\lambda+1)\,a - \lambda b, a\}$, and if we translate $\{b, a, \lambda b + (1-\lambda)\,a\}$ by $(\lambda-1)(a-b)$ we obtain $\{(\lambda-1)\,a + (2-\lambda)\,b, \lambda a + (1-\lambda)\,b, b\}$.

Fourthly, let U be the set of translates of $\{a, b, \lambda a + (1-\lambda)\,b\}$ and let U^* be the set of translates of $\{b, a, \lambda b + (1-\lambda)\,a\}$. The same argument used in the previous paragraph will show that all the triples which are incompatible with any triple in U are in U^*.

We now proceed as follows. Let $\{a, b, \lambda a + (1-\lambda)\,b\}$ be a triple of T. Let A_1 be the set of all translates of $\{a, b, \lambda a + (1-\lambda)\,b\}$ and $A_1{}^*$ be the set of all translates of $\{b, a, \lambda b + (1-\lambda)\,a\}$. This uses up $2p$ of the triples of T. Let $\{c, d, \lambda c + (1-\lambda)d\}$ be a triple not in A_1 or $A_1{}^*$ then $\{d, c, \lambda d + (1-\lambda)c\}$ is not in A_1 or $A_1{}^*$ by the result of the previous paragraph. Let the translates of these for the sets A_2 and $A_2{}^*$. Continue, obtaining

$$T = A_1 \cup A_1{}^* \cup A_2 \cup A_2{}^* \cup \ldots \cup A_{(p-1)/6} \cup A^*_{(p-1)/6}.$$

Now let $S = C_1 \cup C_2 \cup \ldots \cup C_{(p-1)/6}$ where $C_i = A_i$ or $C_i = A_i^*$.

Then S consists of $\dfrac{p(p-1)}{6}$ pairwise compatible triples. Hence S is a Steiner triple system.

The following example for $p = 13$ illustrates the decomposition. Here we take $\lambda = 4$ and each triple is of the form $\{a, b, 4a + 10b\}$. The sets A_1, A_1^*, A_2, A_2^* are listed in columns.

A_1			A_1^*			A_2			A_2^*		
0	1	10	1	0	4	0	2	7	2	0	8
1	2	11	2	1	5	1	3	8	3	1	9
2	3	12	3	2	6	2	4	9	4	2	10
3	4	0	4	3	7	3	5	10	5	3	11
4	5	1	5	4	8	4	6	11	6	4	12
5	6	2	6	5	9	5	7	12	7	5	0
6	7	3	7	6	10	6	8	0	8	6	1
7	8	4	8	7	11	7	9	1	9	7	2
8	9	5	9	8	12	8	10	2	10	8	3
9	10	6	10	9	0	9	11	3	11	9	4
10	11	7	11	10	1	10	12	4	12	10	5
11	12	8	12	11	2	11	0	5	0	11	6
12	0	9	0	12	3	12	1	6	1	12	7

It is seen immediately that the four sets $A_1 \cup A_2$, $A_1 \cup A_2^*$, $A_1^* \cup A_2$, $A_1^* \cup A_2^*$ are all Steiner triple systems.

7. Automorphisms of Generalized Triple Systems

It might appear that a generalized triple system would have a large automorphism group. This is certainly so for the system described in Theorem 6, which has an automorphism group doubly transitive on points. However, the system described in Section 3, with $v = 9$ and which has a subsystem with $v = 4$ has only the identity automorphism.

8. Acknowledgements

The author wishes to thank R. G. Stanton for the ideas of Section 4 on extensions of systems, particularly the constructions in Theorems 2 and 4. The author is also indebted to D. E. Knuth for introducing him to the idea of the use of a finite model for a universal algebra for the solution of problems in combinatorial design. Finally, R. H. Bruck (1969) has used quasigroups for the study of Steiner Triple Systems and Room Designs, although his solution in the case of Room Designs is in error.

References

Bruck, R. H. (1969). What is a loop? M.A.A. Studies in Mathematics, Vol. 2, *In* "Studies in Modern Algebra", 59–99.

Knuth, D. E. and Bendix, P. B. (1970). Simple word problems in universal algebra. *In* "Computational Problems in Abstract Algebra". Pergammon, Oxford.

Mendelsohn, N. S. (1968). An Application of Matrix Theory to a Problem in Universal Algebra, Linear Algebra and its Application, Vol. 1, 471–478.

Mendelsohn, N. S. (1969). Combinatorial Designs as Models of Universal Algebras. *In* "Recent Progress in Combinatorics". (Proceedings of the Third Waterloo Conference). Academic Press, London and New York.

Doubly Periodic Arrays

ERNST S. SELMER

Institute of Mathematics, University of Bergen, Norway

1. The diagram of Fig. 1 shows a section of the lino-tile pattern of my kitchen floor. (Disregard for now the two 6×6 squares.) Apart from some nice symmetry properties, the main feature of the pattern is that each row and each column contain *the same periodic sequence*, here of period $n = 6$, defined by the cycle

$$\text{black—white—grey—white—grey—white.} \tag{1}$$

(Grey is replaced by red in my kitchen.) If extended indefinitely in all directions, we get a *doubly periodic array.*

The sequence defined by (1) is symmetric, that is, the same read both directions. In the general, non-symmetric case, we shall agree to read the rows from left to right and the columns upwards.

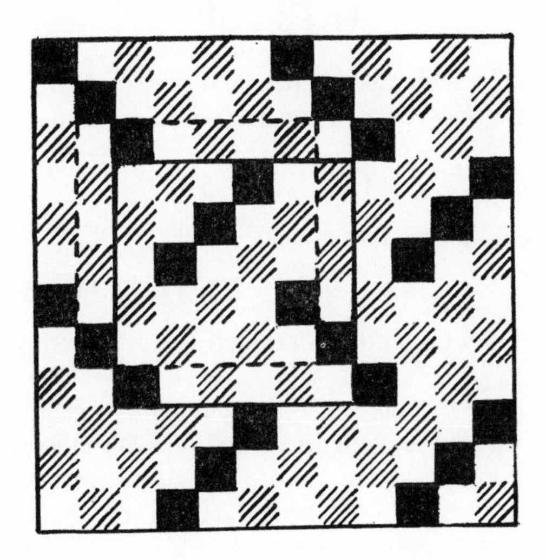

FIGURE 1

The pattern is completely characterized by any $n \times n$ $(=6 \times 6)$ square, for instance the one fully drawn. We demand that, inside any such square, the rows shall all have different starting points. In the case of (1), this means that no two black tiles shall be in the same column of a 6×6 square. It is then easily shown in general that the same property holds when rows and columns are interchanged.

Square arrays have always been popular, for instance magic squares and Latin squares. In the last few years, the above-defined doubly periodic arrays have been known to some people as "Norwegian squares". They have been studied extensively at the University of Bergen, including a few hundred hours of computing time on our IBM 360/50.

2. The fully drawn 6×6 square of Fig. 1 can be described by the translations (to the left) of rows 2–6, compared with the first (bottom) row:

$$1, 2, 5, 4, 3.$$

Because of the different starting points of the rows, this is a *permutation*

$$\begin{pmatrix} 1 & 2 & 3 & 4 & 5 \\ 1 & 2 & 5 & 4 & 3 \end{pmatrix} \tag{2}$$

of the possible translations 1–5.

The 6×6 square could, however, also have been chosen in any other position. With the dotted square of Fig. 1, we get another permutation.

$$\begin{pmatrix} 1 & 2 & 3 & 4 & 5 \\ 1 & 4 & 3 & 2 & 5 \end{pmatrix}, \tag{3}$$

which also characterizes completely the total pattern. We see that it is really 6 successive *step-lengths*, from any starting point, which determine the array, for instance

$$1 - 1 - 3 - 5 - 5 - 3. \tag{4}$$

If we "accumulate" this modulo $n = 6$:

$$1, \quad 1 + 1 = 2, \quad 2 + 3 = 5, \quad 5 + 5 \equiv 4, \quad 4 + 5 \equiv 3,$$

we get the permutation (2), and the last step $3 + 3 \equiv 0$ brings us back to the starting point.

If we begin with the second 1 of (4), accumulation gives the permutation (3), and similarly we get four other permutations. We must then consider (4) as a cycle, corresponding to the infinite periodic sequence defined by (4). We shall call it a *step-cycle*. In the general case, a step-cycle of length n gives rise to a set of d different permutations, where d is n or a proper divisor of n. I call such a set a *translation set* of permutations.

3. Given any periodic sequence of period n, in any number of symbols, we can form all possible doubly periodic arrays generated by the sequence, and the corresponding permutations on $n - 1$ symbols. It is then very simple to show that all these permutations form a group, the *solution group* of the periodic sequence. This group is of a very special form, namely *the union of disjoint translation sets*. I call such a group *translation-invariant* (whether or not it is actually the solution group of a sequence).

I can sketch only a few results about such groups. First, we notice that any periodic sequence has a trivial "staircase" solution

$$\begin{array}{ccccccc} . & . & . & . & . & . & . \\ a & b & c & d & . & . & . \\ & a & b & c & d & . & . \\ & & a & b & c & d & . \\ & & & . & . & . & . \end{array}$$

corresponding to the *identity* permutation I. This is a particular case of a *decimation* permutation: If $(\delta, n) = 1$, we get the δ-decimated sequence (cf. Selmer 1966; Ch. V,6) of the same period n by selecting every δth element of the original sequence.

Let us illustrate this for $n = 7$, $\delta = 2$, and the periodic binary sequence given by the cycle

$$1\ 0\ 1\ 0\ 0\ 0\ 1. \tag{5}$$

Picking out every second element, starting with the first one, we get the decimated cycle

$$1\ 1\ 0\ 1\ 0\ 0\ 0.$$

This is in fact (with a different starting point) the same cycle as (5), which is thus *invariant* under decimation by $\delta = 2$.

Such invariance immediately leads to a simple construction of a doubly periodic array. If we use the periodic sequence corresponding to (5) in each row, with a constant step-length of $\delta = 2$, the columns (read upwards) contain the 2-decimated sequence, which coincides with the given one. Figure 2, with a cross for 1 and a blank for 0, illustrates the resulting doubly periodic

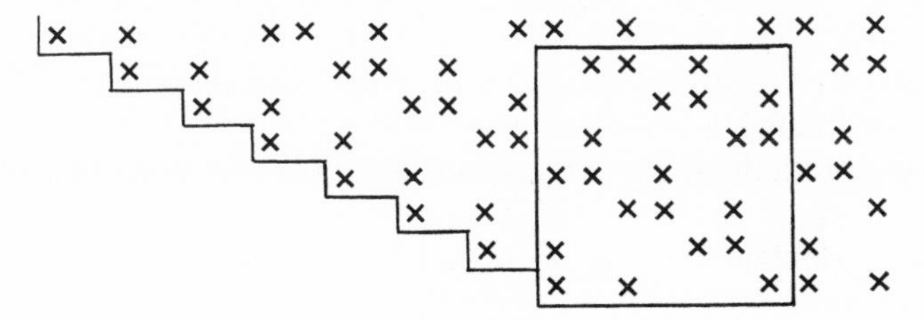

FIGURE 2

array. (It has none of the symmetry properties of Fig. 1, and would hardly be chosen for a floor tiling.)

The step-cycle of Fig. 2 contains only 2's, and yields just one permutation P_2. In the general case, P_δ is the permutation which takes t into $t\delta \pmod{n}$.

If a periodic sequence is invariant under decimation by δ, we call P_δ a trivial solution. If the sequence is *changed*, the solution group of the new sequence is simply

$$P_\delta \mathscr{G} P_\delta^{-1}, \tag{6}$$

where $\mathscr{G}$ is the solution group of the original sequence. In numerical computations, mutually decimated sequences can thus be considered *equivalent*.

In (6), the conjugate of $\mathscr{G}$ is another translation-invariant group. In general, however, transformation by an arbitrary permutation does *not* preserve translation-invariance.

4. From a translation-invariant group for given n, we can construct other groups corresponding to multiples of n. Let us for instance consider Fig. 1 as a diagram for $n = 2 \times 6 = 12$. In every row above the bottom one, we can select a black tile in such a way that no two tiles are in the same column, and consider the resulting pattern as a permutation on 11 symbols. In general, we can do the same if an $n \times n$ square is repeated m times in each direction. If we do this for all the permutations (on $n-1$ symbols) of the original translation-invariant group, and select the starting points of the rows in all possible ways in the $mn \times mn$ pattern, we get a translation-invariant group on $mn - 1$ symbols, which I call the *complete m-extension* of the original group.

We can also get extended groups (translation-invariant) for more restricted choices of the row-numberings. I cannot go into details, and only state that every translation-invariant group on $m-1$ symbols gives rise to one and only one such extension. In particular, the complete m-extension corresponds to the full symmetric group $\mathscr{S}_{m-1}$ (which is trivially translation-invariant).

When describing the doubly periodic array by a permutation, we kept the bottom row fixed, with displacement 0. In $n = 6$ symbols, the permutation (2) can be thought of as

$$\begin{pmatrix} 0 & 1 & 2 & 3 & 4 & 5 \\ 0 & 1 & 2 & 5 & 4 & 3 \end{pmatrix},$$

fixing 0. In general, it can be shown that there is a *one–one correspondence* between the following two types of groups:

(i) The translation-invariant groups on $n-1$ symbols, considered as permutation groups on n symbols fixing 0.

(ii) The permutation groups on n symbols which *contain the complete cycle*

$$(0 \ 1 \ 2 \ldots n-2 \quad n-1). \tag{7}$$

In particular, the subgroup fixing 0 of a group of type (ii) is a group of type (i).

Incidentally, if we deal with groups of type (ii), the above-mentioned extended groups correspond to *wreath products* of permutation groups on m and n symbols.

Permutation groups containing a complete cycle have been studied by Burnside and particularly by Schur; for complete references, see Wielandt (1964). Because of the cycle (7), such a group is necessarily transitive, and a famous result of Schur states that it is *either doubly transitive or imprimitive.* Using the correspondence between the groups of types (i) and (ii) above, this leads to the following

THEOREM 1. *A non-trivial translation-invariant group is either transitive or a subgroup of a (complete) group extension.*

5. Extended groups are easily constructed, and a systematic search for translation-invariant groups can therefore be restricted to the transitive ones. Such a search has been conducted on our 360/50 by S. Mossige. His methods were elaborate, and I can only sketch them here.

All permutations on $n-1$ symbols are arranged in lexicographical order, leading to a numbering (from which the permutations can be reconstructed). Each translation set is characterized by the smallest ordering number of any of its permutations. For each such set, powers and products of the permutations are constructed. We want to imbed the given set in a translation-invariant group, and must therefore add to our list the permutations of the translation sets to which the powers and products belong. Continuing this way, our list in most cases quickly "explodes", leading to the full symmetric group $\mathcal{S}_{n-1}$ (trivially translation-invariant) or the alternating group $\mathcal{A}_{n-1}$ (translation-invariant for n odd).

In order to recognize quickly such an "explosion", Mossige developed several criteria, of which a simple example is the following: The translation set of permutations corresponding to the step-cycle

$$\underbrace{1-1-\cdots-1}_{n-3}-2-(n-1)-2$$

generate the full symmetric group $\mathcal{S}_{n-1}$.

By a systematic and repeated examination, most step-cycles are thus eliminated as possible contributors to proper translation-invariant groups. For the surviving step-cycles, Mossige's program constructed the complete group-tables. In the end, we are left with *very few* transitive translation-invariant groups with $n \leqslant 10$.

6. I shall also describe briefly our computations of doubly periodic arrays. We concentrated on *binary* sequences, partly because of the ease of computation, partly because the solution groups for periodic sequences in more than two symbols are *subgroups* of solution groups in the binary case.

Incidentally, Mossige wrote his group programs in FORTRAN, but the programs for the arrays in PL/I, which has some convenient binary facilities.

We first introduce the notion of a *bigram*, which in linguistics means a pair of neighbouring letters. In a binary sequence $\{s_i\}$ of period n, we similarly distribute the n successive pairs (s_i, s_{i+1}) between the four types (0, 0), (0, 1), (1, 0) and (1, 1). It is easy to see (cf. Selmer, 1966) that when the total number of 1's in the period is given, the complete bigram statistics is known if we count the number of occurrences of one combination, say (1,1).

The bigrams thus defined are really *neighbouring* bigrams. More generally, we can define τ-*bigrams* by the pairs $(s_i, s_{\tau+i})$ of distance τ. These can be read off (vertically) if we write the sequence underneath its "τ-translate":

$$\begin{array}{l} s_0\, s_1 \ldots s_\tau\, s_{\tau+1} \ldots s_{\tau+i} \ldots s_{\tau-1} \ldots \\ \qquad\quad s_0\, s_1 \quad \ldots s_i \quad \ldots s_{n-1} \ldots . \end{array}$$

Let us consider this as one "stair" of a doubly periodic array. The n different pairs $(s_i, s_{\tau+i})$ then become *neighbouring bigrams of the vertical sequences*, which all have different starting points. The τ-bigram statistics, represented for instance by the number of pairs (1, 1), must therefore coincide with the neighbouring bigram statistics if τ shall be a possible step-length of a doubly periodic array.

This necessary condition turns out to be very strong, and often eliminates non-trivial solutions immediately. If the bigram statistics fail to give conclusive information, we may resort to trigrams and higher combinations (we actually went up to pentagrams in the worst cases). The program simply computed the complete "multigram" statistics for different combinations of step-lengths, and the results were further treated by inspection.

From the doubly periodic arrays we got by this crude method, I guessed several general rules, which were then proved. A typical example is "diluted sequences", with a cycle of the type

$$\underbrace{0\,0 \ldots 0}_{n-1}\, s_1\, \underbrace{0\,0 \ldots 0}_{n-1}\, s_2\, 0\,0 \ldots 0\, s_m.$$

More generally, all the blocks of zeros can be replaced by the same combination of digits. It turns out that the solution group in this case is completely determined as a certain *extended group*.

When these cases and a few other types are removed, we are left with very few "unpredictable" doubly periodic arrays. The search was conducted for all possible periodic binary sequences (inequivalent under decimation) with $n \leqslant 15$. Even though the cases are few, they seem to produce effective counter-examples to any conjecture concerning *sufficient* conditions for the existence of non-trivial solution groups.

7. We have already mentioned *transitivity* of translation-invariant groups in general. For solution groups of doubly periodic arrays, transitivity means that all step-lengths $\tau = 1, 2, \ldots, n-1$ must occur. This implies that the τ-bigram statistics of the cycle must be *the same* for all $\tau = 1, 2, \ldots, n-1$. I call a binary cycle with this property a *difference cycle*, since it is easily seen (cf. Selmer, 1966) that if the 1's of such a cycle occur as elements s_{d_i}, $i = 1, 2, \ldots, k$, then $\{d_1, d_2, \ldots, d_k\}$ is a *difference set* modulo n (and conversely).

This naturally raises the question: Could a binary cycle be constructed with a similar equidistribution of *trigrams* (corresponding to double transitivity), or a cycle in *more than two symbols* with the same equidistribution of bigrams? Both questions have been answered in the *negative* by McEliece (1967). For doubly periodic arrays, this leads to the following

THEOREM 2. *The solution group can be (singly) transitive only for a binary difference cycle.*

We have examined all known difference cycles up to $n \leqslant 100$. The results are rather unsystematic, with one notable exception: The *Singer* difference cycles, resulting from finite geometries, all have predictable solution groups. In particular, the Singer sets with parameters

$$(v = n, k, \lambda) = (2^r - 1, 2^{r-1} - 1, 2^{r-2} - 1)$$

give rise to binary difference cycles with transitive solution groups. These groups turn out to be imprimitive.

If these cycles are complemented (interchange of 0 and 1), we get the well known binary *maximal cycles*. A Singer cycle (not complemented) for $r = 3$ is displayed in (5). The transitive solution group, of order 24, contains the trivial permutations $P_1 = I, P_2$ (Fig. 2) and P_4 (interchange rows and columns in Fig. 2), and 21 non-trivial permutations, contained in 3 translation sets with step-cycles

$$1 - 3 - 6 - 6 - 4 - 6 - 2$$

$$2 - 6 - 5 - 5 - 1 - 5 - 4$$

$$4 - 5 - 3 - 3 - 2 - 3 - 1.$$

The second and third cycle are obtained from the first one by repeated doubling modulo 7. The corresponding arrays are displayed in Fig. 3.

As a complete non-binary example, I can mention that the solution group of the cycle (1) has order 12. In addition to the 6 permutations resulting from (4), it contains $P_1 = I$, P_5 and two permutations in each of the translation sets with step-cycles

$$1 - 3 - 1 - 3 - 1 - 3, \qquad 5 - 3 - 5 - 3 - 5 - 3.$$

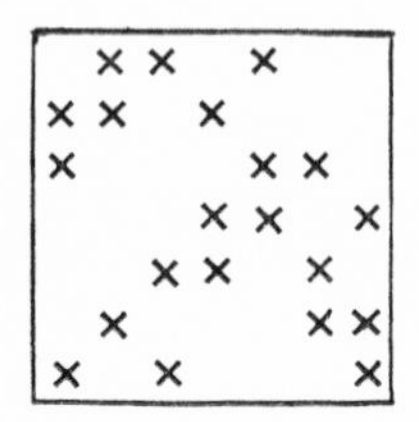
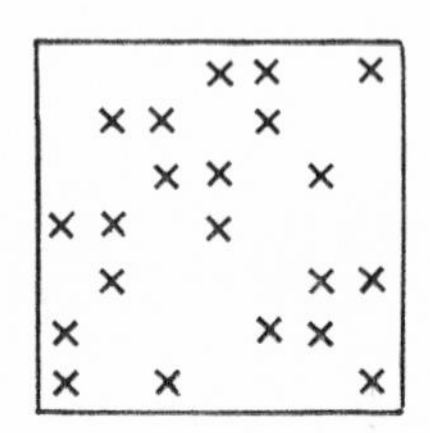
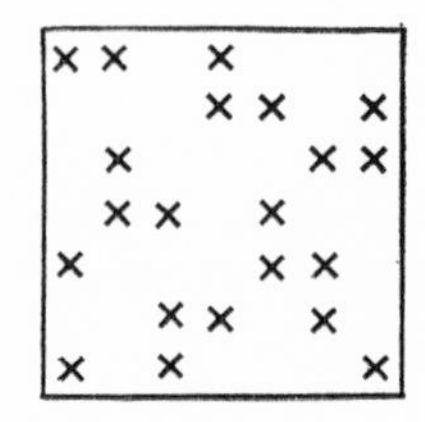

FIGURE 3

8. We finally combine the results of Theorems 1 and 2 into the following useful

THEOREM 3. *Except possibly for binary difference cycles, a non-trivial solution group is a subgroup of a (complete) group extension.*

Since extended groups occur only for composite n, this means that when the period n is a prime we only need to examine possible difference cycles.

References

McEliece, R. J. (1967). A generalization of difference sets. *Can. J. Math.* **19** (1967), 206–211.

Selmer, E. S. (1966). Linear recurrence relations over finite fields. Mimeographed lecture notes. Dept. Math., Univ. of Bergen, Norway.

Wielandt, H. (1964). "Finite permutation groups". Academic Press, London and New York.

Counting Polyominoes

W. F. LUNNON

Atlas Computer Laboratory, Chilton, Didcot, England

1. Introduction; Symmetries

A 'p-mino' is an edge-connected set of p squares ('cells') from a chessboard pattern. (Synonyms in the literature are 'polyomino' and 'animal'.) 'Fixed' p-minos are equivalent if one is a translation of the other; 'free', if in addition a reflection or rotation. (1.1) shows a pair of fixed 4-minos which both correspond to the same free 4-mino. Below them (1.2) is

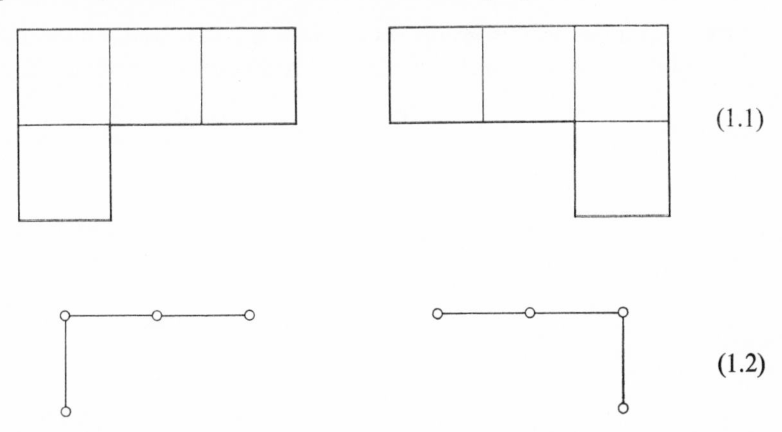
(1.1)

(1.2)

a more convenient representation which we shall normally use; we think of the blobs (centres of squares) as integer points on the Cartesian lattice. The 'bounding rectangle' (b.r.) of a p-mino is the smallest lattice rectangle containing all its cells: the b.r. of (1.2) is of size 3×2.

We shall be concerned with counting all distinct p-minos for given p. $PE(p)$ and $PX(p)$ shall be the totals of free and fixed p-minos—e.g. $PE(4) = 5$ and $PX(4) = 19$—and $PEB(p, m, n)$ and $PXB(p, m, n)$ the totals of free and fixed p-minos whose b.r. is of size $m \times n$. $PE(p)$ is of course the intuitively interesting function; the rest arise in attempting to evaluate and bound it.

In Section 9 is presented a table of these functions for $p = 1(1)18$. It includes counts of the totals of each symmetry type; we shall now describe these for completeness.

The square lattice is invariant under the usual well-known group of integer translations, rotations through $\pi/2$, etc. Translations are uninteresting because nothing can translate into itself, so we factor them out (as in the definition of a fixed p-mino). We are left with the group of symmetries of a square. Taking this to be $0 \leqslant x \leqslant a$, $0 \leqslant y \leqslant b$, $a = b$, we catalogue the group in (1.3). (We shall use a, b for $m - 1$, $n - 1$.)

Co-ordinate transform $(x, y) \rightarrow$	Description	(1.3)
(x, y)	*identity*	
$(a - x, y)$	*reflection in vertical axis*	
$(x, b - y)$	*reflection in horizontal axis*	
$(a - x, b - y)$	*π rotation*	
$(y, a - x)$	*$3\pi/2$ rotation*	
$(b - y, x)$	*$\pi/2$ rotation*	
$(b - y, a - x)$	*reflection in principal diagonal*	
(y, x)	*reflection in other diagonal.*	

The operations in this group which leave a given fixed p-mino invariant constitute a subgroup which describes its symmetry. However the same free

Type	Index	Example
I	8	
*R*2	4	
S	4	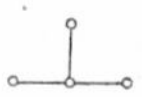
D	4	
R	2	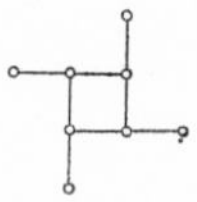
SS	2	
DD	2	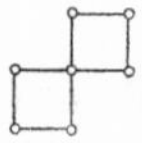
G	1	

(1.4)

p-mino, fixed in different positions, may have different but conjugate subgroups: so a symmetry type of p-mino corresponds to a conjugacy class of subgroups of the square group. These are catalogued in (1.4). The "index" = 8 divided by the order of the subgroup = number of fixed p-minos corresponding to each free p-mino of the type.

2. Theoretical Results

Below are summarised the known explicit formulae for counting p-minos. They are not very impressive, but serve a useful purpose (as we shall see) in checking computational results.

Read (1962) gives a general method of obtaining a generating function for $PXB(p, m, n)$, given n. We return to this in Section 7. He finds an explicit form when $n = 2$:

$$PXB(p, m, 2) = \sum_i \binom{p-m+1}{m-i}\binom{i}{p-m}. \qquad (2.1)$$

The free case is more complicated, as usual. Let g stand for $PXB(p, m, 2)$ and let

$$h(q, c) = PXB(q, c, 2) + 2PXB(q, c, 1)$$

(the second term being simply δ_{qc}). Then provided $2 < m < p < 2m$ (excluding a few easy special cases),

$$PEB(p, m, 2) =$$

$$\begin{array}{ll} \frac{1}{4}g & \text{if } p \text{ odd, } m \text{ even;} \\ \frac{1}{4}[g + h(\frac{1}{2}p + \frac{1}{2}, \frac{1}{2}m + \frac{1}{2}) - h(\frac{1}{2}p - 1\frac{1}{2}, \frac{1}{2}m - \frac{1}{2})] & \text{if } p \text{ odd, } m \text{ odd;} \\ \frac{1}{4}[g + h(\frac{1}{2}p - 2, \frac{1}{2}m - 1) + h(\frac{1}{2}p, \frac{1}{2}m)] & \text{if } p \text{ even, } m \text{ even;} \\ \frac{1}{4}[g + 2h(\frac{1}{2}p - 1, \frac{1}{2}m - \frac{1}{2})] & \text{if } p \text{ even, } m \text{ odd.} \end{array} \qquad (2.2)$$

We can also apply his approach to deduce something (manageable) about the case $n = 3$. Let

$$PXA(p, n) = \sum_m PXB(p, m, n)$$

be the total of fixed p-minos of given height n and any width, and define $f_n(p)$ to let those of lesser height wobble up and down inside the b.r.:

$$f_n(p) = \sum_{i=0}^{n-1} (i + 1)\, PXA(p, n - i).$$

Then for given n, f_n satisfies a linear recurrence (as does PXA, but at greater length):

$$\begin{aligned} f_1(p) &= f_1(p-1); \\ f_2(p) &= f_2(p-1) + f_2(p-2) + f_2(p-3); \\ f_3(p) &= 3f_3(p-1) - 4f_3(p-3) + f_3(p-4) \\ &\quad + 2f_3(p-6) - f_3(p-8) - 3f_3(p-9) + f_3(p-10). \end{aligned} \tag{2.3}$$

Besides the lowest p-minos we can deal with the tallest (assuming $b \leqslant a$): the 'stretched' p-minos whose b.r. is maximal, i.e. $(a+1) \times (b+1)$ where $a + b = p - 1$. A simple example of a stretched p-mino is the L-shape with $a + 1$ cells along the bottom and $b + 1$ up the side. (2.4) shows a more complicated example, a 20-mino in a 13×8 b.r.

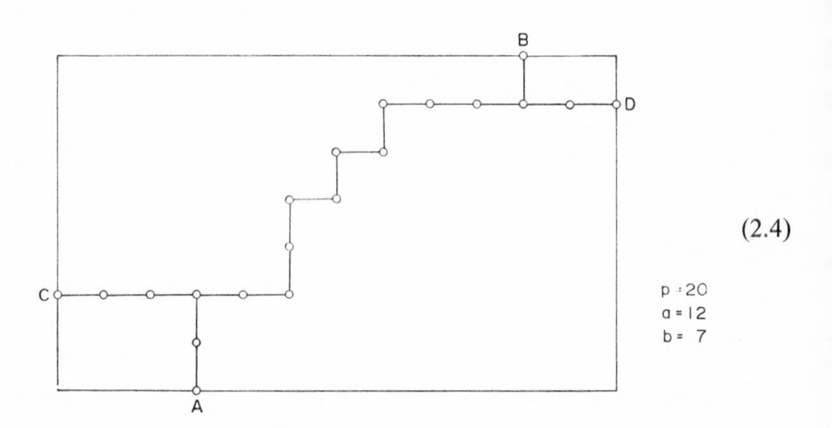

(2.4)

The general stretched p-mino consists of a shortest walk between a pair of cells A and B on the top and bottom of the b.r., together with a pair of straight legs joining the nearest parts of the walk to the sides of the b.r. at C and D. Using the fact that there are just $\binom{a+b}{a}$ shortest walks from $(0, 0)$ to (a, b) we find that if $a, b > 0$,

$$PXB(a+b+1, a+1, b+1) = 8\binom{a+b}{a} - (3ab + a + b + 7). \tag{2.5}$$

Letting g stand for this as above, setting $c = [\frac{1}{2}a]$, $d = [\frac{1}{2}b]$, and discarding a few symmetries we get the free case:

$$PEB(a + b + 1, a + 1, b + 1) =$$

$$\begin{array}{ll} \frac{1}{4}g & \text{if } a \neq b, \text{ both odd;} \\ \frac{1}{4}g + \frac{1}{2}a + \binom{c+d}{c} & \text{if } a \neq b, \text{ only } b \text{ even;} \\ \frac{1}{4}(g-1) + \frac{1}{2}(c+d) + \binom{c+d}{c} & \text{if } a \neq b, \text{ both even;} \\ \frac{1}{8}g + \frac{1}{2}(2^a - c - 1) & \text{if } a = b, \text{ both odd;} \\ \frac{1}{8}(g-1) + \frac{1}{2}\left(2^a + \binom{a}{c}\right) & \text{if } a = b, \text{ both even.} \end{array} \tag{2.6}$$

For a complete, but unilluminating, proof see Lunnon (1969).

3. Computational Strategy

We compute $PEB(p, a+1, b+1)$ rather than $PE(p)$ directly: this has the practical advantage of conveniently decomposing the computation into smaller, independent parts. Sadly, it slows things down as well, but with some effort the time penalty can be held to 50% or so.

The first step is $PXB(p, a + 1, b + 1)$: that is, we wish to count all choices of p cells from an $(a + 1) \times (b + 1)$ lattice rectangle such that:

$$\text{the choice } (p\text{-mino}) \text{ is edge-connected:} \tag{3.1}$$

$$\text{there is at least one cell of it on each edge of the rectangle (b.r.).} \tag{3.2}$$

Our approach, as in other problems of this nature, is fundamentally simple: brute force enumeration.

Enumerating the choices of p things from $s = (a + 1)(b + 1)$ is a well-known combinatorial chore. Let the s things be indexed $1, 2, \ldots, s$, and let $C_1, \ldots, C_p$ be the current choice.

$$\text{Initially set all } C_k := k. \tag{3.3}$$

To get the next choice from the current one, find the maximum i for which $C_i \neq s - p + i$, say $C_i = l - 1$, and for all $j \geqslant i$ reset $C_j := l - i + j$.

When no such i can be found, all choices have been enumerated.

To involve conditions like (3.1), (3.2) in the choosing mechanism (rather than reject fully-made choices which fail to satisfy them) we have to interpret this algorithm more abstractly. Given a current partial choice $C_1, \ldots, C_{q-1}$, where $q \leqslant p$, the values which C_q may assume are just those indices h such that

$$1 \leqslant h \leqslant s \text{ (obviously);} \tag{3.4}$$

there is no level $r < q$ for which, on some previous occasion, $C_1, \ldots, C_r$ had their current values and C_{r+1} was h (we say h is not 'forbidden'); (3.5)

h satisfies any additional local restrictions ('growth criteria'), to be described. (3.6)

Now we index the squares of the b.r., $0 \leqslant x \leqslant a$, $0 \leqslant y \leqslant b$ by assigning to (x, y) the index $(a + 3)(x + 1) + (y + 1)$, e.g. (3.7) where $a = 3$, $b = 2$.

11	12	13
6	7	8

(3.7)

To ensure connectedness (3.1) we choose p-minos from the b.r. by the above algorithm (3.3), (3.4), (3.5) with the additional growth criterion (g.c.):

h must be connected to a cell currently chosen, i.e. there is some $r < q$ for which $h = C_r \pm (a + 3)$ or ± 1. The cell whose neighbourhood to h causes h to be chosen is its 'root'. (3.8)

To touch all the edges (3.1) we add a g.c. which ensures that there are always sufficient cells left to reach them from the current q-mino:

$$\min y + \min x + (a - \max x) + (b - \max y) \leqslant p - q, \tag{3.9}$$

where $\min x$ is the minimum of x over the q-mino, etc.

A new cell h which fails (3.8) may well satisfy it at the next level down, $q + 1$. However, one which fails (3.9) will never satisfy it until some previous level, C_r for $r < q$, is changed. It can therefore be forbidden in the sense of (3.5), since if it were chosen no new p-minos would result.

Having enumerated fixed p-minos, we need a canonical form which distinguishes just one among the up to 8 fixed p-minos which correspond to each free one, (its 'variations'); a new fixed p-mino counts as a new free one if it is in canonical form. Now any fixed p-mino is described by the indices of its cells, arranged for uniqueness in ascending order (3.10), (3.7).

6 7 8 11   6 7 8 13   6 11 12 13   8 11 12 13   (3.10)

Our canonical form is simply the first of these in alphabetical order, e.g. the first one in (3.10). To recognise whether a fixed p-mino is canonical we compare its index description with those of all its variations (all those, that is, which leave the b.r. invariant: e.g. if $b \neq a$ we do not rotate through $\pi/2$). If none are alphabetically earlier it is canonical.

Finally, we turn the tables by throwing away some of the non-canonical fixed p-minos: they can still be counted if we know the symmetry types of the free ones. We use the centre of gravity. Out of all the fixed variations of a given free p-mino, just one, if we're lucky, has its c. of g. in a given quarter of the bounding rectangle (octant of the bounding square if $a = b$): say that nearest the origin (and touching the x-axis).

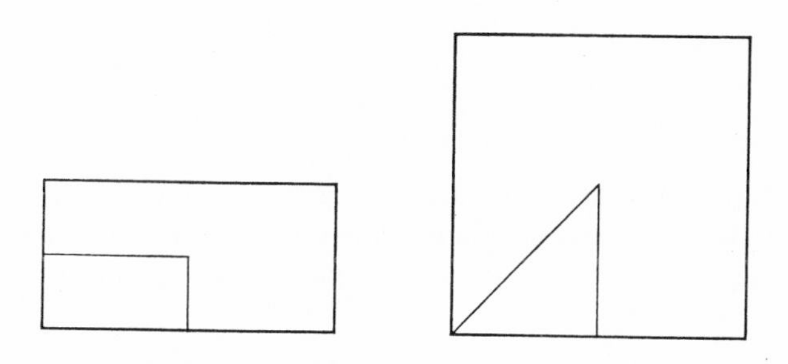

The growth criterion to ensure this is

$$\begin{aligned} &\text{sum}\, x \leqslant \tfrac{1}{2}pa \\ &\text{sum}\, y \leqslant \tfrac{1}{2}pb \\ &(\text{sum}\, x \geqslant \text{sum}\, y \text{ if } a = b); \end{aligned} \tag{3.11}$$

like (3.9) it causes prospective new cells to be forbidden should they fail it. Notice that it is wrong to force the c. of g. of the partial q-mino to lie in the desired region: this would exclude e.g. (3.12), whose c. of g. goes up before it comes down.

(3.12)

The canonical form is now revised. If the c. of g. is properly inside the desired region, the p-mino is now declared canonical. If we are unlucky and the c. of g. is on the edge of the region, it is canonical iff it is alphabetically earliest in the old sense (3.10) among those of its variations which leave the

c. of g. fixed; note that the earliest overall may not have its c. of g. in the region at all. For example, in (3.13) both variations have the c. of g. on the edge of the region (actually, in the centre of the square). The first is canonical.

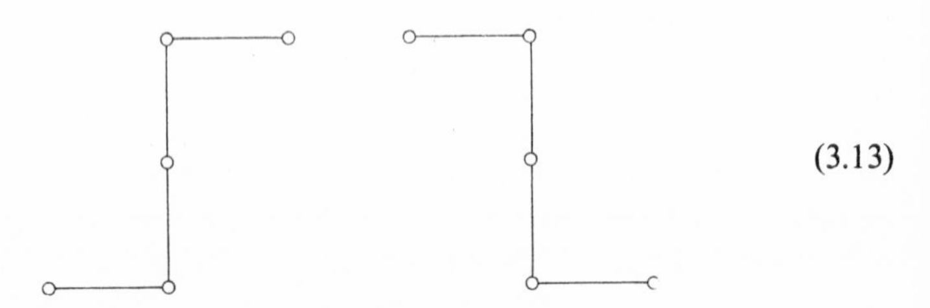

(3.13)

A further small improvement is gained by partially combining (3.9) and (3.11).

4. Computational Tactics

We have adopted a rather complicated data structure for use in this kind of locally-restricted choice-enumeration program. The current partial choice (q-mino, $q \leqslant p$) is represented in 3 independent ways. Squares are referred to by their indices (3.7).

Firstly, there is a 'picture' of it: an **array** *dist* (for 'distance'—a usage which may or may not be explained later), such that

$$\begin{aligned} \mathit{dist}\,[i] &= 0 \text{ if square } i \text{ is a cell of the current } q\text{-mino;} \\ &= X \text{ if } i \text{ is forbidden (say } X = -1); \\ &= 1 \text{ if } i \text{ is in an immediate neighbour of a cell, and not forbidden;} \\ &= 2 \text{ if } i \text{ is anything else.} \end{aligned} \tag{4.1}$$

In particular, we put a ring of squares around the outside of the b.r. edge and permanently forbid them, thus painlessly restraining the contents to the b.r. See (4.2).

i

30	31	32	33	34	35
24	25	26	27	28	29
18	19	20	21	22	23
12	13	14	15	16	17
6	7	8	9	10	11
0	1	2	3	4	5

dist [*i*]

X	X	X	X	X	X
X	2	2	X	2	X
X	2	X	0	2*	X
X	X	X	0	1	X
X	0	0	0	1	X
X	X	X	X	X	X

(4.2)

$q = 5 \qquad a = b = 3$

* temporarily: see text

Secondly we keep in a 'linked stack' the recent history of the choice, necessary to implement (3.5): this is an **array** *bord* (for 'border') such that, if i is a square on the usable border of the q-mino (*dist* $[i] = 1$), then *bord* $[i]$ is the next square along the border. By 'next' we mean—at this stage—next in order of consideration as qth cell.

Lastly there is an **array** *cell*, such that *cell* $[r]$ is the cell currently chosen at level $r \leqslant q$.

(4.3) shows the *bord* and *cell* **arrays** for the q-mino in (4.2). They tell us, for example, that the current 5th cell is 21 and that 16 and 10 are subsequent possible 5th cells on the current partial 4-mino.

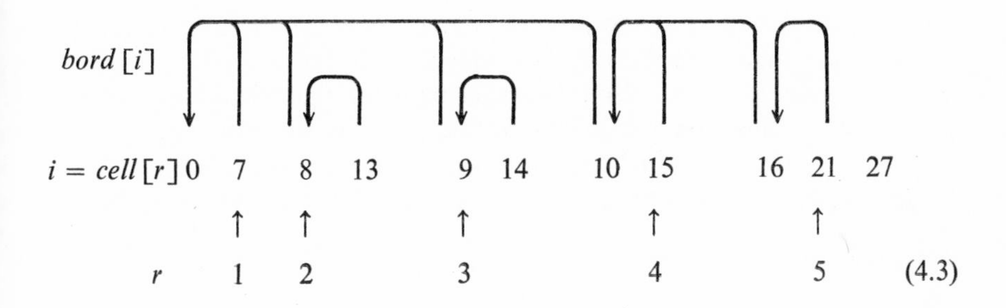

(4.3)

The scheme is manipulated as follows. Suppose $q = 5$, and we have just chosen square 21 as current 5th cell. We look at its neighbours j and select those for which *dist* $[j] = 2$: that is they are neither forbidden (X) nor neighbours of some earlier cell. These are linked on to the end of bord, and their distances *dist* $[j]$ set to 1. (In the case shown only 22 is selected, and *bord* [22] is set to *bord* [21] $= 16$.) Then we set *cell* $[q+1]$ to the last of them (cell [6] = 22) and descend recursively to choose all possible $(q + 1)$-th cells, etc. down to level p. On return we have explored all p-minos grown from the current q-mino; so we set *dist* [*cell*$[q]$] $\equiv$ *dist* [21] to X (forbidden), and then do

$$cell\,[q] := bord\,[cell\,[q]]$$

which fetches the next possible 5th cell, and repeat.

Eventually *cell* $[q] = 0$, and all choices of 5th cell have been exhausted. All cells forbidden at level q (i.e. 21, 16, 10) are released again with their original distances (1), and we ascend a level recursively to choose a new *cell* $[q - 1]$.

Before a cell is chosen it is subjected to the growth criteria (3.9) and (3.11), ((3.8) being implicit in the existing mechanism). If it fails, its distance is set immediately to X and we pass on to the next possibility on the border; it is now forbidden until we return to the level previous, when all such forbidden squares are restored. The max x, sum x etc. required to evaluate the criteria are reset from the previous level, not recomputed from scratch each time.

When a complete p-mino has been constructed, we may have to investigate its canonicity (3.10). Notice that we do not construct the whole of each (c. of g. fixing) variation: for example to check that it is alphabetically no later than its y-axis reflection we compare

$$dist\,(x, y) \text{ with } dist\,(a - x, y)$$

in order of index until one is zero and the other is not, say at (x_1, y_1). If $dist$ $(x_1, y_1) = 0$ then the p-mino is earlier; if $dist$ $(a - x_1, y_1) = 0$ then the reflection is earlier, and the p-mino is not canonical. In Fig. (3.13) this discloses the answer on the very first square.

Occasionally the comparison does pass right through without finding any difference. In this case the p-mino is the same as its variation, i.e. it is symmetric. The tests which pass through decide its symmetry type, and 8/(the number of such tests) equals the number of fixed p-minos for the free p-mino.

Finally, the first cell of a p-mino may always be chosen on the x-axis, since the p-mino must eventually touch that edge anyway. This means that min $x = 0$ in (3.9) etc.

5. Performance; Results

The table in section 9 should be self-explanatory. The first line of each section below the heading shows $PE(p)$ and $PX(p)$ for the specified p, followed by the subtotals for each symmetry type catalogued in (1.4). Below are the analogous totals $PEB(p, m, n)$, $PXB(P, M, n)$ etc. for the various $m \times n$ bounding rectangles. At the end of the section are a few incomplete results for larger p.

Most of the table was produced by a carefully hand-coded version of the algorithm described in Sections 3 and 4, plus a few bells and whistles. It ran for about 175 hours (2.10^{11} instructions) on the Chilton Atlas I in background mode, using 12 blocks (5%) of the available main store (of which about 5 were core-resident). A dump/restart package dumped it onto disc every $2\frac{1}{2}$ minutes for easy restarting. $p = 1(1)18$.

With the most powerful machine currently available, using a combination of Read's method and the blind alley technique (see Sections 6 and 7), we might push the computation as far as $p = 21$.

The only other recent computation known to us is Parkin *et al.* (1967) for $p \leqslant 15$, using a CDC 6600. His method is also choice enumeration, but he uses a more complicated canonical form which probably accounts for his program being somewhat slower than ours. We confirm his results for $p \leqslant 14$; for $p = 15$ they are seen to be in error by comparing the computed values of *PEB* with our formula for stretched p-minos (2.6). He also presents counts of various topological classes of p-minos.

Regarding the reliability of our own results, we must emphasise that it would be of great interest to confirm them independently. Our program is basically simple, and agreement over the range $p = 1(1)12$ convinces us that it works. (But a faster and more complex program of ours, at one stage in its development, worked for $p = 1(1)8$ and failed for $p = 9$.) The machine produced several parity failures (hardware detected) in the course of the run. We would expect frequent undetected errors to produce stupid answers or wild misbehaviour; no such events have occurred.

We have discovered one error so far, using the fact that it is possible to predict $PX(p)$ from previous values correct to a few hundred. (We hope to describe this procedure in a later paper.) It turned out that $PX(18)$ was about 194,000 too small; by repeating the process on $PXA(p, n)$ for various n—where it is unfortunately much less sensitive—the error was traced to $PXB(18, 8, 8)$, which was correctly recomputed. We are now happy that our results are correct to within a few hundred, $p \leqslant 18$.

6. Improvements; Blind Alleys

The growth criteria (3.9) and (3.11) are not ideal: for $p = 12$, 82% of cells actually chosen did not lead to any new p-minos, even fixed ones: that is, a q-mino resulted which could not possibly touch all the edges, and have its c. of g. in the desired region, of the b.r. At any rate (3.9) can be improved.

Its weakness is that it assumes the q-mino can grow in straight lines to the edges, taking no account of forbidden squares which may be blocking the way. They can be taken into account as follows.

Firstly, we extend the meaning of the **array** *dist* (4.2):

$$
\begin{aligned}
dist\,[i] &= \text{the length of the shortest walk from square } i \text{ to the } q\text{-mino, if such a walk exists;} \\
&= U \text{ if no such walk exists (say } U = 1000); \\
&= X \text{ if } i \text{ is forbidden, as before.}
\end{aligned}
\tag{6.1}
$$

Secondly, we insist that when possible next-level cells are selected and linked on the end of *bord* (see Section 4), this is done in anticlockwise order from their root (3.8). This organises the forbidden squares into an essentially continuous rind around the q-mino and makes it possible to find a minimal path to the edges, consisting of (6.2):

X	X	X	X	X	X	X	X		
X	9	8	7	6	5	6	X		
X	8	7	6	5	4	5	X		
X	9	X	5	4	3	4	X	$p \geqslant 16$	
X	X	0	X	X	2	3	X	$q = 6$	(6.2)
X	X	0	0	0	1	2	X	$a = 5$	
X	U	X	X	0	1	2	X	$b = 6$	
X	U	U	X	0	1	2	X		
X	X	X	X	X	X	X	X		

a shortest path from the q-mino to the left edge $x = 0$ of the b.r. (length 8 in the figure) plus

a straight leg from the extended figure to the top of $y = b$ (length 1) plus

a straight leg to the right edge $x = a$ (length 1). (6.3)

The total length of the path (10 in the figure) must be less than $p - q$ for growth to continue; this criterion replaces (3.9). Variations in the procedure occur when the q-mino already touches some edges.

Thirdly, when a new cell is chosen, or forbidden, or released at the end of a level (Section 4), not only its own distance but those of all squares whose shortest paths to the q-mino passed or pass through it must be updated, by a 'treeing-out' process (see Lunnon (1969)).

Another kind of blind alley occurs when there are simply not enough squares left. To eliminate this situation, ensure that

$$(a + 1)(b + 1 \geqslant p + u + x, \tag{6.4}$$

where u squares are inaccessible and x are forbidden.

We have not solved the corresponding problem for the c. of g.

7. Read's Method

Read (1962) describes an entirely different approach, which we summarise here for completeness. It consists of a computational method followed by a twist which gives theoretical results.

Again, we are computing $PXB(p, m, n)$. Suppose n is fixed, say $n = 2$. The b.r. may be broken up into columns of n squares: each possible column (and a couple of impossible ones) appear in (7.1).

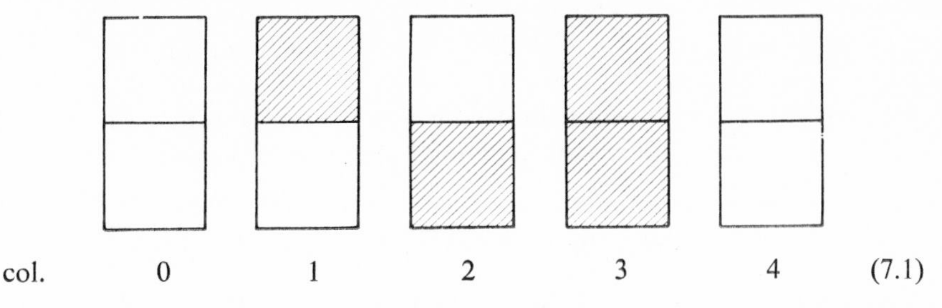

Any fixed p-mino starts with one of columns 1, 2, 3. Its second column depends only on the first: if the first is col. 1, the second may be either col. 1 or col. 3, for example. The third depends only on the second, and so on. A computation can easily be conceived (and programmed) which keeps tallies $S_m(q, c)$ of the totals of fixed q-minos which end in column c after m steps (columns), for each column c and each $q \leqslant p$. On the $(m + 1)$-th step, for all pairs of columns c and d such that d may follow c, and all q, it performs

$$S_{m+1}(q + k_c, d) \text{ adds } S_m(q, c) \tag{7.2}$$

where column c contains k_c cells.

During the course of the computation objects arise which are of actual height less than m, or are disconnected ($m \geqslant 3$). They can be eliminated.

All this can be expressed by polynomial matrices. Let T be such that $T_{ij} = y^k$ if column i may be followed by column j, which has k cells, and let $T_{ij} = 0$ otherwise (7.3). Column 0 in (7.1) is for painlessly starting things off and column 4 for stopping them. Let $U = (1, 0, \ldots, 0, 1)$. It is easily verified that

$$UT^m U'$$

is a polynomial in y in which the coefficient of y^p is essentially $PXB(p, m, n)$; (actually other PXB for lower m are added in, cf. the definition of f_n in Section 2).

$$\begin{matrix} 0 & y & y & y^2 & 1 \\ 0 & y & 0 & y^2 & 1 \\ 0 & 0 & y & y^2 & 1 \\ 0 & y & y & y^2 & 1 \\ 0 & 0 & 0 & 0 & 0 \end{matrix} \tag{7.3}$$

This is the end of the computational part. Now Read introduces another unknown x and considers the formal sum

$$\sum_m x^m UT^m U' = U(I - Tx)^{-1} U'$$

$$= \sum_{m,p} PXB(p, m, n)\, x^m y^p \text{ more or less.} \tag{7.4}$$

Now $(I - Tx)^{-1}$ can be (in principle) exhibited as a matrix of rational functions in x and y; for example, for $n = 2$ after some reduction it becomes

$$\frac{1}{(1 - xy)(1 - xy^2) - (xy)(2xy^2)} \begin{pmatrix} 1 - xy^2 & 2xy^2 \\ xy & 1 - xy \end{pmatrix}$$

Expanding the inverse of the denominator (which is simply $|I - Tx|$) by the binomial theorem we get PXB as a sum of binomial coefficients.

This is how Read proves (2.1). It would be an interesting exercise in algebraic manipulation to compute similar formulae for $PXB(p, m, 3)$ etc.

To prove the recurrences (2.3) we observe that $|I - T|$ is essentially the auxiliary polynomial of the linear recurrence for $f_n(p)$. (Actually it is the wrong way round and contains spurious factors). For $n \leqslant 3$ we evaluated the determinants by hand and removed the extra factors. The dominant roots r' of these polynomials are lower bounds on the ratio r of Section 8, but not very good ones: we show the first few in (7.5).

n	r'
1	1·0000
2	1·8393
3	2·3972
4	2·7667

(7.5)

Returning to the computational side, the snag is that the number of columns grows rapidly with n; we call it $MAT(n)$ and show a few values in (7.6). In practice it can be halved by omitting one of each pair of mirror-image columns.

n	1	2	3	4	5	6	7	8	9	10
$MAT(n)$	3	5	10	22	52	128	324	836	2189	5799

Two tallies (old and new) $S(q, c)$ must be kept for each $q \leqslant p$ and each column c, a total of $p.MAT(n)$; the operation of evaluating T_{ij} and possibly incrementing S (7.2) must be performed $2^n\, MAT(n)\, p^2/4$ times. For $p = 20$ this is 120,000 words of store and 10^9 times round the inner loop, which might well be 100 instructions long, making 10^{11} instructions. A formidable undertaking.

To find symmetrical p-minos and deduce PEB from PXB, the best way seems to be brute force (Sections 3 and 4) modified to construct only symmetrical ones (much faster). A mixture of methods for different rectangles is also a possibility, since Read finds stretched (or almost stretched) p-minos indigestible, whereas they can quickly be enumerated because there aren't many of them.

8. Associated Problems: 'the Ratio', Other Shapes

Some effort has been expended (Eden (1961), Klarner (1965), Klarner (1967) and others) on estimating $PE(p)$; or rather $PX(p)$, since it is known that

$$PE(p) \sim PX(p)/8. \tag{8.1}$$

(See Lunnon (1969) for another unilluminating—but apparently unique—proof.)

It is known that

$$\sqrt[p]{PX(p)} \to r, \text{ the 'ratio'}, \tag{8.2}$$

where it can easily be shown that

$$3{\cdot}20 < r \leqslant 6{\cdot}75. \tag{8.3}$$

With more difficulty, and using a computer, this can be improved to

$$3{\cdot}72 < r < 4{\cdot}5 \tag{8.4}$$

The lower bound is due to Klarner (1967), the upper to Conway and Guy at Cambridge (unpublished), whose proof I have not seen.

The proof that $3{\cdot}20 < r$ involves counting 'board-piles', fixed p-minos cut by every horizontal lattice line in a single connected 'board' of cells—e.g. (8.5).

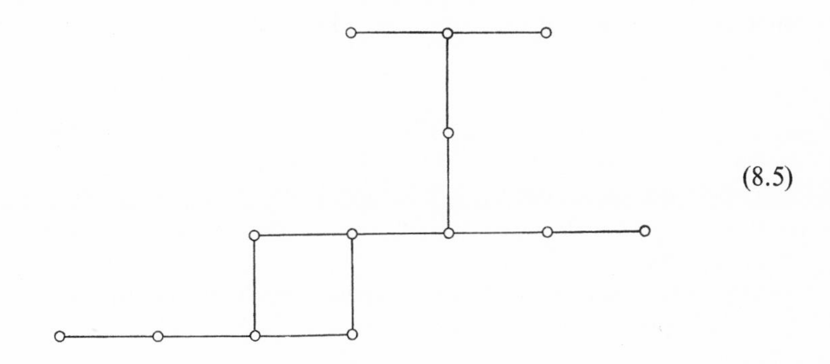

(8.5)

Guy reports that building up p-minos out of boards together with finitely many more complex shapes (e.g. U-shape) does not improve the bound. We have counted 'board-pair-piles', cut by every horizontal line in at most two boards of cells—e.g. (8.6)—for $p \leqslant 20$ and conjecture that these would give a lower bound of about 3·74; however the computations are much larger than those involved in (8.4).

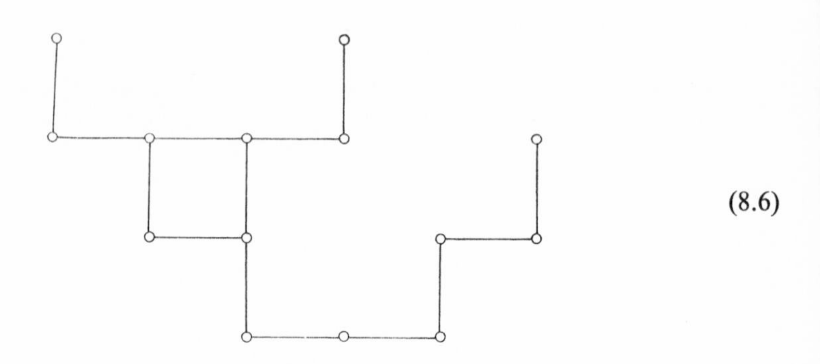

The short table (8.8) lists some of the functions arising in bounding PX. BX is the total of board-piles and CX of board-pair-piles, SX the total of 'staircases' (which can wind around any point indefinitely) used in showing $r \leqslant 6{\cdot}75$, and DX the numbers arising in Klarner's proof of $r > 3{\cdot}72$.

It is conjectured that in fact

$$PX(p)/PX(p-1) \uparrow r. \tag{8.7}$$

If this monotonicity could be proved, we would immediately have $r > 3{\cdot}84$. We formerly thought $r = 4$ probable, but on the evidence of the extrapolation procedures mentioned above (Section 5) we do so no longer. In fact, we consider that

$$PX(p) \sim (4{\cdot}06 \pm 0{\cdot}000)^p \times p^{(-0{\cdot}98 \pm 0{\cdot}02)} \times \text{constant}.$$

The exponent of p could just possibly be -1.

Instead of constructing polyominoes with squares, we could use the hexagonal or triangular lattices, or the hypercubic lattice in d dimensions, (or indeed many more exotic possibilities). We have investigated these and present some new counts (Lunnon, 1971).

Square Polyominoes (8.8)

p	free $PE(p)$	fixed $PX(p)$	board-pile $BX(p)$	board-pair $CX(p)$	Klarner $DX(p)$	staircase $SX(p)$
1	1	1	1	1	1	1
2	1	2	2	2	2	2
3	2	6	6	6	6	6
4	5	19	19	19	19	22
5	12	63	61	63	63	91
6	35	216	196	216	216	408
7	108	760	629	760	756	1938
8	369	2725	2017	2723	2681	9614
9	1285	9910	6466	9880	9600	49335
10	4655	36446	20727	36168	34626	260130
11	17073	135268	66441	133237	125582	1402440
12	63600	505861	212980	492993	457425	7702632
13	238591	1903890	682721	1829670	1671854	42975796
14	901971	7204874	2188509	6804267	6127385	243035536
15	3426576	27394666	7015418	25336611	22507654	1390594458
16	13079255	104592937	22488411	94416842	82830896	8038677054
17	50107911	400795860	72088165	351989967	305299746	46892282815
18	192622052	1540820542	231083620	1312471879	1126742108	275750636070
19			740754589	4894023222	4162932807	1633292229030
20			2374540265	18248301701	15395008353	9737153323590
21					56978046025	

9. Table

P	M	N	FREE	FIXED	G	R	DD	SS	R2	D	S	I
1			1	1	1							
	1	1	1	1	1							
2			1	2				1				
	2	1	1	1				1				
3			2	6				1		1		
	3	1	1	1				1				
	2	2	1	4						1		
4			5	19	1			1	1		1	1
	4	1	1	1				1				
	2	2	1	1	1							
	3	2	3	8					1		1	1
5			12	63	1			1	1	2	2	5
	5	1	1	1				1				
	3	2	2	6							1	1
	4	2	3	12								3
	3	3	6	25	1				1	2	1	1
6			35	216				2	5	2	6	20
	6	1	1	1				1				
	3	2	1	1				1				
	4	2	6	18					1		2	3
	5	2	5	16					1		1	3
	3	3	7	44						2	1	4
	4	3	15	50					3		2	10
7			108	760			1	3	4	7	9	84
	7	1	1	1				1				
	4	2	2	8								2
	5	2	11	38							3	8
	6	2	5	20								5
	3	3	7	32			1	1		1	2	2
	4	3	39	154							1	38
	5	3	25	83				1	4		3	17
	4	4	18	120						6		12
8			369	2725	1	1	1	4	18	5	23	316
	8	1	1	1				1				
	4	2	1	1				1				
	5	2	10	32					2		2	6
	6	2	19	66					1		4	14
	7	2	7	24					1		1	5
	3	3	3	9	1					1	1	
	4	3	59	212				2	3		6	48
	5	3	96	376							4	92
	6	3	35	124					5		3	27
	4	4	77	584		1	1		1	4		70
	5	4	61	230					5		2	54

P	M	N	FREE	FIXED	G	R	DD	SS	R2	D	S	I
9			1285	9910	2			4	19	26	38	1196
	9	1	1	1				1				
	5	2	3	10							1	2
	6	2	22	88								22
	7	2	28	102							5	23
	8	2	7	28								7
	3	3	1	1	1							
	4	3	42	158							5	37
	5	3	210	784				2	8		17	183
	6	3	188	750							1	187
	7	3	49	173				1	6		4	38
	4	4	181	1396						13		168
	5	4	383	1526							3	380
	6	4	97	388								97
	5	5	73	497	1				5	13	2	52
10			4655	36446			1	8	73	22	90	4461
	10	1	1	1				1				
	5	2	1	1				1				
	6	2	15	50					2		3	10
	7	2	52	192					4		4	44
	8	2	40	146					1		6	33
	9	2	9	32					1		1	7
	4	3	21	62				2	2		6	11
	5	3	255	987				1	1		14	239
	6	3	550	2133				3	11		18	518
	7	3	332	1316							6	326
	8	3	63	230					7		4	52
	4	4	266	2038			1		3	12	6	244
	5	4	1304	5154					17		14	1273
	6	4	822	3276					6			816
	7	4	155	602					7		2	146
	5	5	529	4180						10	3	516
	6	5	240	932					11		3	226
11			17073	135268			2	10	73	91	147	16750
	11	1	1	1				1				
	6	2	3	12								3
	7	2	45	170							5	40
	8	2	90	360								90
	9	2	53	198							7	46
	10	2	9	36								9
	4	3	4	12							2	2
	5	3	212	778				4	7		22	179
	6	3	954	3802							7	947
	7	3	1231	4803				3	20		36	1172
	8	3	529	2114							1	528
	9	3	81	295				1	8		5	67
	4	4	251	1952						14		237
	5	4	2847	11328							30	2817
	6	4	3548	14192								3548
	7	4	1551	6194							5	1546
	8	4	220	880								220
	5	5	2413	18944			2		20	48	19	2324
	6	5	2366	9458							3	2363
	7	5	410	1591				1	18		5	386
	6	6	255	1924						29		226

P	M	N	FREE	FIXED	G	R	DD	SS	R2	D	S	I
12			63600	505861	3	3	3	15	278	79	341	62878
	12	1	1	1				1				
	6	2	1	1				1				
	7	2	21	72					3		3	15
	8	2	119	450					4		9	106
	9	2	158	608					6		6	146
	10	2	69	258					1		8	60
	11	2	11	40					1		1	9
	4	3	1	1				1				
	5	3	103	370				2	1		17	83
	6	3	1184	4622				2	17		37	1128
	7	3	2800	11127				1	1		34	2764
	8	3	2406	9490				4	25		36	2341
	9	3	800	3184							8	792
	10	3	99	368					9		5	85
	4	4	168	1232	2	1	1	1	2	11	7	143
	5	4	4441	17598				2	30		50	4359
	6	4	10323	41196					23		25	10275
	7	4	8239	32824					40		26	8173
	8	4	2680	10704					8			2672
	9	4	313	1230					9		2	302
	5	5	7375	58665	1	1	1		2	50	27	7293
	6	5	13161	52488					53		25	13083
	7	5	4738	18936							8	4730
	8	5	646	2538					19		4	623
	6	6	2835	22576		1	1		5	18		2810
	7	6	908	3588					19		3	886
13			238591	1903890	2	2	3	17	283	326	564	237394
	13	1	1	1				1				
	7	2	4	14							1	3
	8	2	73	292								73
	9	2	257	1002							13	244
	10	2	238	952								238
	11	2	86	326							9	77
	12	2	11	44								11
	5	3	33	101				3	2		9	19
	6	3	964	3814							21	943
	7	3	4634	18278				4	37		86	4507
	8	3	6818	27252							10	6808
	9	3	4313	17042				4	38		61	4210
	10	3	1142	4566							1	1141
	11	3	121	449				1	10		6	104
	4	4	66	488						10		56
	5	4	5008	19912							60	4948
	6	4	21995	87980								21995
	7	4	29442	117616							76	29366
	8	4	16821	67284								16821
	9	4	4327	17294							7	4320
	10	4	415	1660								415
	5	5	17041	135325	1	2	3	3	44	104	89	16795
	6	5	51133	204466							33	51100
	7	5	30998	123652					105		65	30828
	8	5	8683	34726							3	8680
	9	5	979	3845				1	28		6	944
	6	6	18533	147656						152		18381
	7	6	11952	47798							5	11947
	8	6	1553	6212								1553
	7	7	950	7265	1				19	60	3	867

P	M	N	FREE	FIXED	G	R	DD	SS	R2	D	S	I
14			901971	7204874			5	30	1076	301	1294	899265
	14	1	1	1				1				
	7	2	1	1				1				
	8	2	28	98					3		4	21
	9	2	237	912					9		9	219
	10	2	505	1970					6		19	480
	11	2	360	1408					8		8	344
	12	2	106	402					1		10	95
	13	2	13	48					1		1	11
	5	3	6	15				1			3	2
	6	3	546	2068				6	12		37	491
	7	3	5497	21825				1	6		74	5416
	8	3	14182	56382				2	57		113	14010
	9	3	14722	58755				1	1		64	14656
	10	3	7171	28459				5	45		60	7061
	11	3	1580	6300							10	1570
	12	3	143	538					11		6	126
	4	4	20	116			2		1	3	4	10
	5	4	4168	16440				6	32		75	4055
	6	4	36035	143866					50		87	35898
	7	4	79155	316126					112		135	78908
	8	4	71742	286748					50		60	71632
	9	4	31576	126078					73		40	31463
	10	4	6634	26516					10			6624
	11	4	551	2178					11		2	538
	5	5	30320	241550			1	2	10	122	116	30069
	6	5	153122	611859				3	159		151	152809
	7	5	143230	572705				1	8		98	143123
	8	5	65236	260616					124		40	65072
	9	5	14894	59556							10	14884
	10	5	1415	5592					29		5	1381
	6	6	86974	695088			2		33	140		86799
	7	6	89212	356514					126		41	89045
	8	6	23215	92820					20			23195
	9	6	2555	10156					29		3	2523
	7	7	13402	107052						36	5	13361
	8	7	3417	13582					39		4	3374
15			3426576	27394666			6	35	1090	1186	2148	3422111
	15	1	1	1				1				
	8	2	4	16								4
	9	2	119	462							7	112
	10	2	591	2364								591
	11	2	895	3530							25	870
	12	2	498	1992								498
	13	2	127	486							11	116
	14	2	13	52								13
	5	3	1	1				1				
	6	3	187	708							20	167
	7	3	4745	18671				7	37		107	4594
	8	3	22011	87964							40	21971
	9	3	36920	147021				5	103		219	36593
	10	3	28762	115022							13	28749
	11	3	11304	44893				5	62		92	11145
	12	3	2107	8426							1	2106
	13	3	169	635				1	12		7	149
	4	4	3	16						2		1
	5	4	2439	9658							49	2390
	6	4	45748	182992								45748
	7	4	167354	668840							288	167066

P	M	N	FREE	FIXED	G	R	DD	SS	R2	D	S	I
	8	4	230998	923992								230998
	9	4	156007	623728							150	155857
	10	4	55084	220336								55084
	11	4	9751	38986							9	9742
	12	4	698	2792								698
	5	5	42670	339488			4	6	66	166	221	42207
	6	5	366832	1466938							195	366637
	7	5	517302	2067669				7	354		405	516536
	8	5	348090	1392272							44	348046
	9	5	126496	505362					213		98	126185
	10	5	24214	96850							3	24211
	11	5	1991	7867				1	40		7	1943
	6	6	323331	2584696						488		322843
	7	6	483329	1933180							68	483261
	8	6	193911	775644								193911
	9	6	42154	168602							7	42147
	10	6	3965	15860								3965
	7	7	111296	887992			2		135	406	50	110703
	8	7	54983	219922							5	54978
	9	7	6003	23859				1	68		7	5927
	8	8	3473	27288						124		3349
16			13079255	104592937	5	12	14	60	4125	1117	4896	13069026
	16	1	1	1				1				
	8	2	1	1				1				
	9	2	36	128					4		4	28
	10	2	429	1666					9		16	404
	11	2	1353	5336					19		19	1315
	12	2	1493	5890					8		33	1452
	13	2	690	2720					10		10	670
	14	2	151	578					1		12	138
	15	2	15	56					1		1	13
	6	3	47	149				3	3		12	29
	7	3	2833	11125				5	6		90	2732
	8	3	26097	103856				8	79		175	25835
	9	3	70760	282621				1	10		198	70551
	10	3	84925	338896				2	140		259	84524
	11	3	52245	208767				1	1		104	52139
	12	3	16997	67648				6	71		90	16830
	13	3	2752	10984							12	2740
	14	3	195	740					13		7	175
	4	4	1	1	1							
	5	4	1008	3892				4	18		46	940
	6	4	45289	180736				8	67		131	45083
	7	4	285375	1140300				2	221		376	284776
	8	4	594488	2377060					154		292	594042
	9	4	585305	2340070					281		294	584730
	10	4	310890	1243156					87		115	310688
	11	4	91127	364164					116		56	90955
	12	4	13852	55384					12			13840
	13	4	885	3510					13		2	870
	5	5	47344	376953	3	2	3	4	15	173	243	46901
	6	5	720244	2879126				6	343		573	719322
	7	5	1524442	6096423				3	58		610	1523771
	8	5	1459163	5835024				4	483		325	1458351
	9	5	763678	3054369				1	10		160	763507
	10	5	229573	917690					242		59	229272
	11	5	37708	150808							12	37696
	12	5	2715	10766					41		6	2668
	6	6	991660	7930294		8	9		119	540	62	990922

P	M	N	FREE	FIXED	G	R	DD	SS	R2	D	S	I
	7	6	2097534	8388536					526		274	2096734
	8	6	1182179	4728414					151			1182028
	9	6	390229	1560298					250		59	389920
	10	6	72565	290200					30			72535
	11	6	5985	23852					41		3	5941
	7	7	675981	5406161	1	1	1		6	336	75	675561
	8	7	500827	2002570					309		60	500458
	9	7	105600	422376							12	105588
	10	7	10001	39856					69		5	9927
	8	8	59728	477464		1	1		19	68		59639
	9	8	12859	51290					69		4	12786
17			50107911	400795860	4	7	9	64	4183	4352	8195	50091097
	17	1	1	1				1				
	9	2	5	18							1	4
	10	2	172	688								172
	11	2	1248	4942							25	1223
	12	2	2709	10836								2709
	13	2	2343	9290							41	2302
	14	2	902	3608								902
	15	2	176	678							13	163
	16	2	15	60								15
	6	3	6	18							3	3
	7	3	1173	4486				8	17		74	1074
	8	3	22883	91342							95	22788
	9	3	106490	424765				9	174		410	105897
	10	3	193669	774538							69	193600
	11	3	177886	710186				6	224		446	177210
	12	3	89146	356552							16	89130
	13	3	24660	98180				6	92		129	24433
	14	3	3504	14014							1	3503
	15	3	225	853				1	14		8	202
	5	4	271	1050							17	254
	6	4	34112	136448								34112
	7	4	395338	1580262							545	394793
	8	4	1256623	5026492								1256623
	9	4	1764700	7057196							802	1763898
	10	4	1331013	5324052								1331013
	11	4	577936	2311232							256	577680
	12	4	143749	574996								143749
	13	4	19119	76454							11	19108
	14	4	1085	4340								1085
	5	5	41330	328380	2	3	3	9	64	184	291	40774
	6	5	1168734	4673698							619	1168115
	7	5	3766042	15058973				11	868		1713	3763450
	8	5	5039358	20156838							297	5039061
	9	5	3630653	14519109				9	931		807	3628906
	10	5	1547398	6189482							55	1547343
	11	5	395198	1579762					378		137	394683
	12	5	56623	226486							3	56620
	13	5	3630	14393				1	54		8	3567
	6	6	2567828	20537800						1206		2566622
	7	6	7630115	30519318							571	7629544
	8	6	5799584	23198336								5799584
	9	6	2633896	10535342							121	2633775
	10	6	737198	2948792								737198
	11	6	119428	477694							9	119419
	12	6	8689	34756								8689
	7	7	3334120	26662289	1	4	6	2	599	1695	354	3331459
	8	7	3356103	13424246							83	3356020

P	M	N	FREE	FIXED	G	R	DD	SS	R2	D	S	I
	9	7	1047667	4189208				589			141	1046937
	10	7	191974	767886							5	191969
	11	7	16025	63861				1	110		8	15906
	8	8	587349	4694728						1016		586333
	9	8	241977	967894							7	241970
	10	8	22827	91308								22827
	9	9	13006	102745	1				69	251	4	12681
18			192622052	1540820542			20	117	15939	4212	18612	192583152
	18	1	1	1				1				
	9	2	1	1				1				
	10	2	45	162					4		5	36
	11	2	720	2816					16		16	688
	12	2	3192	12642					19		44	3129
	13	2	5097	20256					33		33	5031
	14	2	3531	14002					10		51	3470
	15	2	1180	4672					12		12	1156
	16	2	204	786					1		14	189
	17	2	17	64					1		1	15
	6	3	1	1				1				
	7	3	324	1199				3	2		42	277
	8	3	14761	58490				10	66		196	14489
	9	3	124513	497268				4	27		359	124123
	10	3	353037	1410450				12	273		558	352194
	11	3	471268	1884183				1	14		429	470824
	12	3	345168	1379100				2	284		499	344383
	13	3	144905	579307				1	1		154	144749
	14	3	34644	138097				7	103		126	34408
	15	3	4396	17556							14	4382
	16	3	255	974					15		8	232
	5	4	55	186				2	4		10	39
	6	4	19282	76754					64		123	19095
	7	4	444276	1775266				10	300		604	443362
	8	4	2217704	8868708					342		712	2216650
	9	4	4421955	17684044					758		1130	4420067
	10	4	4597056	18385978					360		763	4595933
	11	4	2784608	11136190					581		540	2783487
	12	4	1015049	4059540					134		194	1014721
	13	4	218608	873946					169		74	218365
	14	4	25758	103004					14			25744
	15	4	1331	5290					15		2	1314
	5	5	27764	220332			4	6	16	160	254	27324
	6	5	1574307	6293542				26	524		1280	1572477
	7	5	7903297	31607964				10	201		2396	7900690
	8	5	14767740	59064860				6	1421		1620	14764693
	9	5	14284582	57135593				3	99		1264	14283216
	10	5	8214249	32853411				5	1174		611	8212459
	11	5	2941738	11766441				1	12		242	2941483
	12	5	650544	2601174					419		82	650043
	13	5	82476	329876							14	82462
	14	5	4746	18860					55		7	4684
	6	6	5682531	45451572			12		337	1439	375	5680368
	7	6	23875284	95494783				3	1692		1480	23872109
	8	6	24038883	96153722					711		194	24037978
	9	6	14344365	57373826					1288		529	14342548
	10	6	5457210	21828266					287			5456923
	11	6	1323538	5293116					439		79	1323020
	12	6	189324	757212					42			189282
	13	6	12354	49300					55		3	12296
	7	7	13980630	111835470			2	1	69	1726	593	13978239

P	M	N	FREE	FIXED	G	R	DD	SS	R2	D	S	I
	8	7	18321923	73283704					1539		455	18319929
	9	7	7644454	30577329				1	22		220	7644211
	10	7	2062794	8249710					650		83	2062061
	11	7	333324	1333268							14	333310
	12	7	24755	98786					111		6	24638
	8	8	4259395	34071432			2		176	753		4258464
	9	8	2557203	10227332					658		82	2556463
	10	8	463690	1854620					70			463620
	11	8	38895	155350					111		4	38780
	9	9	258483	2067300						134	7	258342
	10	9	48632	194240					139		5	48488
19	19	1	1	1				1				
	10	2	5	20								5
	11	2	249	978							9	240
	12	2	2356	9424								2356
	13	2	7235	28814							63	7172
	14	2	8859	35436								8859
	15	2	5113	20330							61	5052
	16	2	1482	5928								1482
	17	2	233	902							15	218
	18	2	17	68								17
	7	3	64	206				4	3		16	41
	8	3	6730	26674							123	6607
	9	3	112111	447024				14	181		508	111408
	10	3	514669	2058246							215	514454
	11	3	1007794	4027835				13	517		1134	1006130
	12	3	1043651	4174392							106	1043545
	13	3	629639	2516117				7	418		791	628423
	14	3	225722	902850							19	225703
	15	3	47445	189159				7	128		172	47138
	16	3	5413	21650							1	5412
	17	3	289	1103				1	16		9	263
	5	4	6	20							2	4
	6	4	7852	31408								7852
	7	4	399618	1597288							592	399026
	8	4	3280863	13123452								3280863
	9	4	9384420	37532960							2360	9382060
	10	4	13265780	53063120								13265780
	11	4	10831259	43321472							1782	10829477
	12	4	5439572	21758288								5439572
	13	4	1700319	6800480							398	1699921
	14	4	321540	1286160								321540
	15	4	33983	135906							13	33970
	16	4	1592	6368								1592
	5	5	14083	111100			4	8	39	133	201	13698
	6	5	1752470	7007536							1172	1751298
	7	5	14197830	56778486				38	1599		4761	14191432
	8	5	37349883	149396746							1393	37348490
	9	5	53448935	213781108				10	3033		4268	53441624
	10	5	36158187	144631830							459	36157728
	11	5	17238758	68948067				11	2042		1424	17235281
	12	5	5304151	21216472							66	5304085
	13	5	1031536	4124556					612		182	1030742
	14	5	116998	467986							3	116995
	15	5	6112	24287				1	70		9	6032
	6	6	7863367	62899676						1815		7861552
	7	6	65184907	260733640							2994	65181913
	8	6	86525013	346100052								86525013
	9	6	66223132	264889964							1282	66221850

P	M	N	FREE	FIXED	G	R	DD	SS	R2	D	S	I
	10	6	32587346	130349384								32587346
	11	6	10657040	42627772							194	10656846
	12	6	2274122	9096488								2274122
	13	6	290518	1162050							11	290507
	14	6	17081	68324								17081
	12	7	556431	2225714							5	556426
	13	7	37156	148271				1	166		9	36980
	11	8	847157	3388610							9	847148
	12	8	63584	254336								63584
	10	9	1041580	4166306							7	1041573
	11	9	87580	349799				1	250		9	87320
	10	10	48840	388692						507		48333
20	20	1	1	1				1				
	10	2	1	1				1				
	11	2	55	200					5		5	45
	12	2	1141	4482					16		25	1100
	13	2	6796	27008					44		44	6708
	14	2	15041	59906					33		96	14912
	15	2	14733	58728					51		51	14631
	16	2	7195	28610					12		73	7110
	17	2	1862	7392					14		14	1834
	18	2	265	1026					1		16	248
	19	2	19	72					1		1	17
	7	3	8	21				1			4	3
	8	3	2207	8518				10	25		115	2057
	9	3	75982	302954				8	32		443	75499
	10	3	598863	2392971				11	378		846	597628
	11	3	1756745	7024791				5	57		1030	1755653
	12	3	2563739	10250641				17	723		1409	2561590
	13	3	2143813	8573603				1	18		805	2142989
	14	3	1090162	4357914				2	507		857	1088796
	15	3	339710	1358407				1	1		214	339494
	16	3	63480	253278				8	141		168	63163
	17	3	6592	26336							16	6576
	18	3	323	1240					17		9	297
	16	4	44032	176096					16			44016
	17	4	1905	7582					17		2	1886
	15	5	162238	648920							16	162222
	16	5	7740	30802					71		8	7661
	15	6	23234	92788					71		3	23160
	14	7	54286	216796					167		7	54112

References

Eden, M. (1961). "A Two-dimensional Growth Process". Proc. 4th Berkeley Symp. on Math. Stats. and Prob., 223–239, Univ. Calif. Press.

Klarner, D. A. (1965). Some Results Concerning Polyominoes. *Fib. Quart*. **3,** 9–20.

Klarner, D. A. (1967). Cell Growth Problems. *Canad. J. Math*. **19,** 851–863.

Lunnon, W. F. (1969). Ph.D. Thesis, University of Manchester.

Lunnon, W. F. (1971). *In* "Computing and Graph Theory" (Ed. Read, E. C.). Academic Press, London and New York.

Parkin, T. R., Lander, L. J., Parkin, D. R. (1967). "Polyomino Enumeration Results". Presented at SIAM Fall Meeting, Santa Barbara.

Read, R. C. (1962). Contributions to the Cell Growth Problem. *Canad. J. Math.* **14,** 1–20.

Combinatorial Analysis with Values in a Semigroup

A. M. MACBEATH

University of Birmingham, England

1. Background

The analogy between the formula for the partition function

$$\sum_{n=0}^{\infty} p(n)\, x^n = \prod_{n=1}^{\infty} (1 - x^n)^{-1} \tag{1}$$

and Euler's product formula

$$\sum_{n=1}^{\infty} n^{-s} = \prod_{p} (1 - p^{-s})^{-1} \tag{2}$$

expressing the fundamental theorem of arithmetic, suggests that both might be dealt with as part of a common theory. In fact, if E is a subset of the natural numbers we can consider the E-partitions to be finite subsets of E together with assignments of a multiplicity to each element. If the weight of a partition is the *sum* of its elements, repeated according to multiplicity, the generating function is a power series and we have a formula such as

$$\sum_{n=0}^{\infty} p_E^{+}(n)\, x^n = \prod_{n \in E} (1 - x^n)^{-1}.$$

If the weight of a partition is the *product* of its elements, repeated according to multiplicity, then the appropriate generating function is a Dirichlet series and we have the formula

$$\sum p_E^{\times}(n)\, n^{-s} = \prod_{n \in E} (1 - n^{-s})^{-1}.$$

As recognised by Lehmer (1931), the appropriate algebraic background is the theory of semigroups. Power series correspond to the additive semigroups of natural numbers, Dirichlet series are used when the multiplicative semigroups of natural numbers occur. In general, one may consider a commutative semigroup Γ such that each element of Γ can be expressed as a product of two elements in only a finite number of ways. If we are given a set E and a map $\nu : E \to \Gamma$ such that $\nu^{-1}(\gamma)$ is a finite set for each $\gamma \in \Gamma$ (ν is a "proper map"), then the generating function for (E, ν, Γ) is the formal sum

$$\sum_{e \in E} \nu(e) = \sum_{\gamma \in \Gamma} |\nu^{-1}(\gamma)| \gamma,$$

where the absolute value sign denotes the number of elements in the set. A good deal of the formalism of ordinary power series manipulations can be generalised to this situation. Lehmer has applied it in particular to the semigroup of natural numbers with l.c.m. as operation. The beautiful theorem of Pólya (1937) can also be expressed in this setting (see Riordan (1958)). The Dirichlet series form of Pólya's theorem is due to Lloyd (1968).

2. Application to the Theory of Trees

We give an instance where the set E to be enumerated is the set of root-trees (see Riordan, 1958), and the semigroup Γ is such that the number of root-trees with a given number of vertices and a given automorphism group can be read off from the generating function.

By a *permutation group* we mean a pair (G, A) consisting of a finite set A and a finite subgroup G of the symmetric group of permutations of A. If $\sigma = (G, A)$ and $\tau = (H, B)$ we define $\sigma\tau = (G \times H, A \cup B)$, where the permuted set is the disjoint union of A and B and G acts on A, fixing all of B, while H acts on B fixing all of A. We also have the *wreath product* $\sigma \wr \tau$, defined in Hall (1959).

A root-tree is a graph without circuits and with one distinguished vertex called the *root*. An *automorphism* of a root-tree is a permutation f of its vertices such that both f and f^{-1} map pairs of vertices joined by an edge into pairs joined by an edge, and map the root on itself. If we remove from a root-tree T the root and all the edges incident with the root, then the (disconnected) graph that remains is a root-tree-partition, say

$$T_1^{n_1} T_2^{n_2} \dots T_k^{n_k},$$

the roots being the vertices which were, in T, joined by an edge to the root of T. Any automorphism of T must permute among themselves the n_i trees T_i, for each i. Further, the ways of mapping one of the trees T_i into another isomorphic one T_i' are all given from a standard one by applying an automorphism of T_i'. Thus, if $A(T)$ denotes the automorphism group of T, we have

$$A(T) = (A(T_1) \wr \sigma_{n_1})(A(T_2) \wr \sigma_{n_2}) \ldots (A(T_k) \wr \sigma_{n_k}), \tag{3}$$

where σ_n denotes the symmetric group on n elements acting on these elements in the normal way. (Here the group $A(T)$ is represented as a permutation group acting on the *free* vertices of T, i.e. those vertices, other than the root, which are incident with only one edge; the root counts as a free vertex of the tree with one vertex and no edges, but not of any other tree).

We cannot enumerate the trees by their automorphism group alone, for it is easy to deduce from what follows that there are infinitely many trees with each given group. Hence we enumerate them simultaneously by automorphism group and number of vertices. If $\mathcal{T}$ denotes the set of all root-trees (strictly the set of all isomorphism classes of root-trees), and, for each $T \in \mathcal{T}$, $A(T)$ is the automorphism group, $n(T)$ the number of vertices, then our formula is

$$\sum_{T \in \mathcal{T}} A(T)\, x^{n(T)} = x \prod_{T \in \mathcal{T}} \left(1 + \sum_{\nu=1}^{\infty} (A(T) \wr \sigma_\nu)\, x^{\nu n(T)}\right) \tag{4}$$

This follows at once from (3), the factor x at the beginning, arising because T has one more vertex (the root) than the total number of vertices of the tree-partition

$$T_1^{n_1} \ldots T_k^{n_k}$$

A more conventional numerical formula, entirely equivalent to (4), is

$$\sum t(g, n)\,(g, x^n) = x \prod_{g,n} \left(1 + \sum_{\nu=1}^{\infty} (g \wr \sigma_\nu, x^{\nu n})\right)^{t(g,n)}, \tag{5}$$

where $t(g, n)$ is the number of root-trees with n vertices the automorphism group g, taken as permutation group on the vertices. The semigroup here used is the product $\Gamma \times N^+$, consisting of pairs (g, n) where g is a permutation group and n is a natural number. The permutation groups combine by "direct sum" as indicated above while N^+ is the additive semigroup. However, the possibility of permuting trees in the partition leads to the intrusion of the wreath product in (5), so that the analogy with (1) is not complete.

Since the terms on the right of (5) contributing to x^n involve only $t(g', n')$ for $n' \leqslant n - 1$, the formula enables us to derive $t(g, n)$ recursively from those with lower n, starting from $t(\{1\}, 1) = 1$, $t(g, 1) = 0$ if $g \neq \{1\}$. Here $\{1\}$ denotes the trivial group consisting of the unit element only.

If we ignore the automorphism group by setting $g = 1$ everywhere we obtain Cayley's formula (1889–1897) for the number $t(n)$ of root-trees with n vertices.

$$\sum t(n)\, x^n = x \prod_{n=1}^{\infty} (1 - x^n)^{-t(n)}. \tag{6}$$

Again, if we select from (5) only the trees with trivial automorphism group ($g = \{1\}$), we derive, for the number $t'(n)$ of such trees, the formula

$$\sum t'(n)\, x^n = x \prod_{n=1}^{\infty} (1 + x^n)^{t'(n)}. \tag{7}$$

This formula, easily derived directly, also appears to be new.

References

Cayley, A. (1889–97). *Collected Math. Papers, Cambridge.* **3,** 242; **9,** 202, 427; **11,** 365; **13,** 26.

Hall, M. (1959). "The Theory of Groups". Macmillan, New York.

Lehmer, D. H. (1931). Arithmetic of Double Series. *Trans. Amer. Math. Soc.* **33,** 954–957.

Lloyd, E. K. (1968). Pólya's Theorem in Combinatorial Analysis applied to enumerate multiplicative partitions. *J. London Math. Soc.* **43,** 224–230.

Pólya, G. (1937). Kombinatorische Anzahlbestimmungen für Gruppen, Graphen und chemische Verbindungen. *Acta Math.* **68,** 145–252.

Riordan, J. (1958). "An Introduction to Combinatorial Analysis". Wiley, New York.

The Use of Computers in Search of Identities of the Rogers-Ramanujan Type

GEORGE E. ANDREWS*

Pennsylvania State University, University Park, Pennsylvania, U.S.A.

1. Introduction

The study of identities of the Rogers–Ramanujan type has until recently been one related to a few striking but isolated results. There are now known to exist many general theorems in this area some of which are described in Sections 3 and 4. Our object here is to suggest the role a computer might play in the search for a complete characterization of partition identities of the Rogers–Ramanujan type. In Section 5, three problems are posed whose solution might possibly be indicated with the aid of a computer. Also in Section 5, a conjecture concerning asymptotic properties of the partition functions under consideration is given. If this conjecture is correct, its proof should yield information which would make the computer's task much simpler. In Section 6, we analyze how the computer may be used to approach our problems. In Section 7 we prove a partition theorem whose truth was suggested in Section 6.

2. History up to 1960

The study of partition theorems of the Rogers–Ramanujan type dates back to Euler (1748) who proved the following partition theorem.

THEOREM 1. *The number of partitions of a natural number, n, into distinct parts equals the number of partitions of n into odd parts.*

For example, 5 may be partitioned into distinct parts in three ways, 5, $4+1$, $3+2$ and into odd parts in three ways, 5, $3+1+1$, $1+1+1+1+1$.

* Partially supported by National Science Foundation grant GP–8075 and by a grant from the Atlas Symposium No. 2.

As is well-known (cf. Hardy and Wright, 1960), Theorem 1 may be deduced from the simple infinite product identity

$$\prod_{n=1}^{\infty}\left(1+q^{n}\right)=\prod_{n=1}^{\infty}\left(1-q^{2n-1}\right)^{-1}.$$

Up until the early 1900's very little else was discovered in the way of partition theorems of this type.

In 1916, two new theorems of this type were described by MacMahon (1916).

THEOREM 2. *The number of partitions of n with minimal difference 2 between summands is equal to the number of partitions of n into parts of the forms $5m+4$ or $5m+1$.*

THEOREM 3. *The number of partitions of n with minimal difference 2 between summands and each part $\geqslant 2$ is equal to the number of partitions of n into parts of the forms $5m+2$ and $5m+3$.*

MacMahon remarks that these theorems are deduced from the two identities for the generating functions due to Ramanujan.

$$1+\sum_{n=1}^{\infty}\frac{q^{n^2}}{(1-q)\dots(1-q^{n})}=\prod_{n=1}^{\infty}(1-q^{5n-4})^{-1}(1-q^{5n-1})^{-1}; \quad (2.1)$$

$$1+\sum_{n=1}^{\infty}\frac{q^{n^2+n}}{(1-q)\dots(1-q^{n})}=\prod_{n=1}^{\infty}(1-q^{5n-3})^{-1}(1-q^{5n-2})^{-1}. \quad (2.2)$$

MacMahon (1916) does not have a proof for these results; however, he states on page 33:

"This most remarkable theorem has been verified as far as the coefficient of q^{89} by actual expansion so that there is practically no reason to doubt its truth; but it has not yet been established."

This indicates a "rule-of-thumb" procedure for verifying partition identities of the type described in Theorems 2 and 3. A hand-computation in simple cases, or a machine computation can be used to check results like Theorems 2 and 3 for small values of n. While nothing is proved, nonetheless, the results are sufficiently likely that one may undertake to find a proof.

Actually (2.1) and (2.2) had been proved much earlier by Rogers (1894); also, Theorems 2 and 3 were discovered and proved independently by Schur (1917).

The only other major theorem of this type discovered prior to 1960 was given by Schur (1926) (cf. Andrews, 1967d, 1968a).

THEOREM 4. *The number of partitions of n with minimal difference* 3 *between summands and no consecutive multiples of* 3 *as summands is equal to the number of partitions of n into parts of the forms* $6m + 1$ *and* $6m + 5$.

Apart from these results, a rather rich theory of basic hypergeometric series was developed by W. N. Bailey and his students; numerous analytic results similar to (2.1) and (2.2) were found (cf. Slater, 1952). On the combinatorial side of things Lehmer (1946) and Alder (1948) proved that no other theorems of as simple a character as Theorems 1–4 could exist.

3. Recent Results Related to Well-poised, Hypergeometric Series

Gordon (1961) was the first to score a major breakthrough in this study. He proved the following result.

THEOREM 5. *Let* a *and* k *be integers* $0 < a \leqslant k$. *Let* $A_{k,a}(n)$ *denote the number of partitions of n into parts not of the forms* $(2k+1)m$, $(2k+1)m \pm a$. *Let* $B_{k,a}(n)$ *denote the number of partitions of n of the form*

$$n = b_1 + \ldots + b_s,$$

where $b_i \geqslant b_{i+1}$, $b_i - b_{i+k-1} \geqslant 2$, *and* 1 *appears as a summand at most* $a - 1$ *times. Then* $B_{k,a}(n) = A_{k,a}(n)$.

For $k = a = 2$ we obtain Theorem 2, and $k = 2$, $a = 1$ yields Theorem 3.

Subsequent to this Andrews in a series of papers (1966), (1967a), (1967b), (1967c), (1969b) extended Gordon's results by means of generating functions of the basic hypergeometric type. Andrews (1969b) gives the following result which contains much of the previous work.

THEOREM 6. *Let* $0 \leqslant d - 1 \leqslant a \leqslant k$ *all be integers, and* $2d - 3 \leqslant k$ *for* $d > 3$. *Let* $B_{d,k,a}(n)$ *denote the number of partitions of n of the form* $n = b_1 + \ldots b_s$, *where* $b_i \geqslant b_{i+1}$, *only parts divisible by d may be repeated,* $b_i - b_{i+k-1} \leqslant d$ *with strict inequality if* $d|b_i$ *and at most* $a - 1$ *parts are* $\leqslant d$.

If d is odd, define $A_{d,k,a}(n)$ *to be the number of partitions of n into parts such that only parts divisible by d may be repeated and no part is* $\equiv 0$, $\pm\left(a - \tfrac{1}{2}(d-1)\right) d \pmod{(2k - d + 2)d}$.

If d is even, define $A_{d,k,a}(n)$ *to be the number of partitions of n into parts where only parts divisible by* $d/2$ *may be repeated, no part is* $\equiv d \pmod{2d}$, *and no part is* $\equiv 0$, $\pm(2a - d + 1)\tfrac{1}{2}d \pmod{(2k - d + 2)d}$. *Then*

$$A_{d,k,a}(n) = B_{d,k,a}(n).$$

The generating function related to $B_{d,k,1}(n)$ is (after suitable specialization of parameters)

$$\frac{\prod_{j=1}^{d-1}\prod_{m=1}^{\infty}(1+xqa_j^{-1})}{\prod_{m=1}^{\infty}(1-xq^m)}$$

$$\times\ {}_{d+2}\Phi_{d+1}\left[\begin{matrix} x, q\sqrt{x}, -q\sqrt{x}, a_1, \dots, a_{d-1}; q\dfrac{(-1)^d x^k q^{\frac{1}{2}(2k-d+2)n-1+\frac{1}{2}d}}{a_1 \dots a_{d-1}}, \\ \sqrt{x}, -\sqrt{x}, xq/a_1, \dots, xq/a_{d-1} \end{matrix}\right]$$

in the standard notation (Slater, 1966).

The above is a very general well-poised basic hypergeometric series whose function-theoretic properties are studied in great detail for $d = 4$, 5, and 6 (see Slater, 1966). Thus the partition identities of this section are related to a well explored area of basic hypergeometric series.

This particular aspect of partition theory is still far from complete. In particular, numerical evidence indicates that the condition $2d - 3 \leqslant k$ in Theorem 6 is unnecessary.

In the next section we shall study results where the generating functions are not nearly so well-known.

4. Partitions with Difference Conditions

This section is devoted to results on partitions in which restrictions relate to the difference between *adjacent* summands. Thus Theorems 2, 3, and 4 are of this type as are the negative results of Lehmer (1946) and Alder (1948).

Göllnitz (1967) and Gordon (1965) were the first to give results of this type beyond Theorems 2, 3, and 4.

DEFINITION 1. *Let* $A(a_1, a_2, \dots, a_s; M, n)$ *denote the number of partitions of* n *into distinct parts congruent to some* $a_i \pmod{M}$.

DEFINITION 2. *If* C *denotes a set of positive integers and if* $f = 0, 1$, *let* $B_f(b_1, \dots, b_M; C; n)$ *denote the number of partitions of* n *of the form* $n = e_1 + \dots + e_t$ *where* $e_i - e_{i+1} \geqslant b_j$ *if* $e_{i+f} \equiv j \pmod{M}$ *and no* e_i *is in* C.

DEFINITION 3. *Let* $C(a_1, \dots, a_s; M; n)$ *denote the number of partitions of* n *into parts congruent to some* $a_i \pmod{M}$.

THEOREM 7. (Gordon, 1965; Göllnitz, 1967)

$$B_0(2,3;2;\varnothing;n) = C(1,4,7;8;n);$$
$$B_0(2,3;2;\{1,2\};n) = C(3,4,5;8;n).$$

THEOREM 8. (Göllnitz, 1967)

$$B_0(3,2;2;\varnothing;n) = A(1,2,4;4;n) = C(1,5,6;8;n);$$
$$B_0(3,2;2;\{1\};n) = A(2,3,4;4;n) = C(2,3,7;8;n).$$

THEOREM 9. (Göllnitz, 1967)

$$B_0(7, 6, 7, 6, 6, 7; 6; \{1; 3\}; n) = A(2, 4, 5; 6; n) = C(2, 5, 11; 12; n).$$

More recently Andrews (1968b, 1969a) has shown that very general results of this nature exist.

THEOREM 10. (Andrews ,1968b). If $M = 2^n - 1$,

$$B_0(\gamma_1, \ldots, \gamma_M; M; \mathscr{C}; N)$$
$$= A(M - 2^{n-1}, M - 2^{n-2}, \ldots, M - 2, M - 1; M; N),$$

where $\mathscr{C}$ is the set of all N for which the right hand side of the above equation$=0$, and $\gamma_j = (2^n - 1)\omega(\beta(-j)) + v(\beta(-j)) - \beta(-j)$, with $\beta(m)$ the least positive residue of m (mod M), $\omega(m)$ the number of powers of 2 appearing in the binary number representation of m, and $v(m)$ the least power of 2 appearing.

THEOREM 11. (Andrews, 1969a) *In the notation of Theorem* 10,

$$B_1(\delta_1, \ldots, \delta_M; M; \phi; N) = A(1, 2, \ldots, 2^{n-1}; M; N),$$

where $\delta_j = (2^n - 1)\omega(j) + v(j) - j$.

It should be remarked that these theorems can be refined to give stronger theorems which take into account the number of parts in the partitions. As an example we give a generalization of Theorem 4 due to Gleissberg (1928).

THEOREM 12. *Let $A_s(n)$ denote the number of partitions of n into s distinct parts not divisible by* 3. *Let $B_s(n)$ denote the number of partitions of n into s parts (parts divisible by* 3 *are counted twice) of the form $n = e_1 + \ldots + e_t$, $e_i - e_{i+1} \geqslant 3$ with strict inequality if $3|e_i$. Then $A_s(n) = B_s(n)$.*

Theorems 10 and 11 are susceptible to similar generalizations.

DEFINITION 3. *Let* $A(v_1, v_2, \ldots, v_s; a_1, \ldots, a_s; M; n)$ *denote the number of partitions of n into* $\sum_{j=1}^{s} v_j$ *distinct parts where* v_i *parts are* $\equiv a_i \pmod{M}$.

THEOREM 13. *If* $M = 2^n - 1$ *and* $\delta_1, \ldots, \delta_M$ *are as defined in Theorem* 11, *let* $B_1{}^{\#}(v_1, v_2, \ldots, v_n; \delta_1, \ldots, \delta_M; M; N)$ *denote the number of partitions of the form* $n = e_1 + \ldots + e_t$, *where* $e_i - e_{i+1} \geqslant \delta_j$ *if* $e_{i+1} \equiv j \pmod{M}$, *and if we write* $e_i = \varepsilon_i M + \rho_i$ *where* ρ_i *is the least positive residue of* $e_i \pmod{M}$ *and if* $\rho_i = 2^{\xi_1(i)} + \ldots + 2^{\xi_j(i)}$, *where* $\xi_1(i) > \ldots > \xi_j(i)$, *then exactly* v_1 *of all the* $\xi_k(h)$ *should be zero,* v_2 *of all the* $\xi_k(h)$ *should be* 1, ..., *and* v_n *of all the* $\xi_k(h)$ *should be* $n - 1$. *Then*

$$A(v_1, v_2, \ldots, v_s; 1, 2, \ldots, 2^{n-1}; M; N) = B_1{}^{\#}(v_1, \ldots, v_n; \xi_1, \ldots, \xi_M; M; N).$$

This theorem generalizes Theorem 11. Its proof does not appear in the literature; however, it follows from introducing n parameters, $\tau_1, \tau_2, \ldots, \tau_n$ into the functions discussed by Andrews (1969a) (see equations (3.11) and (3.15) in Andrews (1969a) where now $\alpha(i) = 2^{i-1}$):

$$\prod_{t=0}^{\infty} (1 - xq^{Mt}) \sum_{m=0}^{\infty} x^m \prod_{j=0}^{n-1} \frac{(1 + \tau_1 q^{Mj+1})(1 + \tau_2 q^{Mj+2}) \ldots (1 + \tau_n q^{Mj+2^{n-1}})}{(1 - q^{Mj+M})},$$

with $M = 2^n - 1$.

A similar generalization holds for Theorem 10.

5. Problems Concerning Partition-theoretic Identities

After the discussion of the previous section concerning the various known partition identities, the following questions are quite natural. Furthermore these seem to be the next steps in the study of such identities.

PROBLEM 1. *Let M be a positive integer. What are the sets of positive integers* $\{a_1, \ldots, a_s\}$, $\{b_1, \ldots, b_M\}$, *and* $\{c_1, \ldots, c_t\} = \mathscr{C}$ *such that*

$$A(a_1, \ldots, a_s; M; n) = B_0(b_1, \ldots, b_M; M; \mathscr{C}; n)?$$

First we remark that if we know an answer to Problem 1 for a particular M, then the sets $\{a_1, \ldots, a_s, a_1 + M, \ldots, a_s + M, \ldots, a_1 + (k-1)M, \ldots, a_s + (k-1)M\}$, $\{b_1', \ldots, b_{kM}'\}$ where $b_j' = b_i$ for

$j \equiv i \pmod{M}$, and $\mathscr{C}$ work for kM. Such a solution is clearly trivial for kM. From the previous section we note that non-trivial solutions are known for

$$M = 1, 3, 4, 6, 7, 15, 31, \ldots, 2^n - 1, \ldots.$$

PROBLEM 2. *Same as Problem* 1 *except that* $B_0(b_1, \ldots, b_M; M; \mathscr{C}; n)$ *is replaced by* $B_1(b_1, \ldots, b_M; M; \mathscr{C}; n)$.

As with problem 1, we know non-trivial results for $m = 1, 3, 4, 6, 7, 15, 31, \ldots, 2^n - 1, \ldots$.

PROBLEM 3. *Same as Problem* 1 *except that we now specify* $a_j = M - 2^{j-1}$,

$$\mathscr{C} = \{m | A(a_1, \ldots, a_s; M; m) = 0\}.$$

All the cases known in answer to Problem 1 are of this form.

It would greatly aid the search for partition-theoretic identities if something were known concerning the asymptotic properties of the partition functions described in Problems 1, 2, and 3.

By Ingham (1941), it appears likely that

$$A(a_1, \ldots, a_s; M; n) \sim Cn^{-3/4} e^{\pi/4\sqrt{sn/3M}},$$

where C may be explicitly determined; indeed one need only prove that $A(n)$ is monotone to fulfill Ingham's conditions.

This leads us to the following

CONJECTURE. *For the* $B_f(b_1, \ldots, b_M; M; \mathscr{C}; n)$ *defined in any of Problems* 1, 2, *or* 3, *there exist* α *and* β *such that*

$$B_f(b_1, \ldots, b_M; M; \mathscr{C}; n) \sim \alpha n^{-3/4} e^{\beta\sqrt{n}}.$$

If this conjecture is correct, and if one can compute α and β explicitly, then for $A(a_1, \ldots, a_s; M; n)$ to equal $B(b_1, \ldots, b_M; M; \mathscr{C}; n)$ we must have $\alpha = C$ and $\beta = \frac{\pi}{4}\sqrt{s/3M}$. Such knowledge would allow one to discard many incorrect alleged identities *a priori*.

6. Possible Techniques for Solving Problems 1–3

Let us consider Theorem 12 of the previous section. We suppose that we are considering $A_s(n)$ and wish to *define* difference conditions for a partition function $B_s^*(n)$ so that $A_s(n) = B_s^*(n)$.

It is very simple to construct a short table for $A_s(n)$:

$n \backslash s$	1	2	3	4	5
1	1	0	0	0	0
2	1	0	0	0	0
3	0	1	0	0	0
4	1	0	0	0	0
5	1	1	0	0	0
6	0	2	0	0	0
7	1	1	1	0	0
8	1	1	1	0	0
9	0	3	0	0	0
10	1	1	2	0	0
11	1	2	2	0	0
12	0	4	1	1	0
13	1	2	4	0	0
14	1	2	4	1	0
15	0	5	2	2	0

Now we look for three natural numbers γ_1, γ_2, and γ_3, such that if $B_s{}^*(n)$ denotes the number of partitions of n into s parts (parts $\equiv 0 \pmod 3$ are counted twice) of the form $n = b_1 + b_2 + \ldots + b_t$, $b_i - b_{i+1} \geqslant \gamma_j$ if $b_i \equiv j \pmod 3$, then $A_s(n) = B_s{}^*(n)$ in the given range.

Since all singleton partitions are admissible, the values of $B_s{}^*(n)$ for $n \leqslant 4$ imply $\gamma_3 > 2$, $\gamma_2 > 1$, $\gamma_1 > 0$.

For $n = 5$, we see that we must have as admissible partitions 5 and $4 + 1$ ($3 + 2$ is inadmissible since $A_3(5) = 0$), and all other possible partitions of 5 are inadmissible due to the previous conditions on the γ's. Thus $\gamma_1 \leqslant 3$.

For $n = 6$, we see that 6, $5 + 1$ and $4 + 2$ are all possibly admissible (one of these must eventually be excluded however).

For $n = 7$, we see that 7, $6 + 1$, $5 + 2$, are the only possibly admissible partitions. Thus $\gamma_2 \leqslant 3$, $\gamma_3 \leqslant 5$.

We now have the possibilities $\gamma_1 = 1, 2, 3$, $\gamma_2 = 2, 3$, $\gamma_3 = 3, 4, 5$. Thus there are now 18 possible choices for the triple $(\gamma_1, \gamma_2, \gamma_3)$. It is now a simple matter to program a computer to evaluate $B_s{}^*(n)$ for $1 \leqslant s \leqslant 5$, $1 \leqslant n \leqslant 15$ for each possible choice of $(\gamma_1, \gamma_2, \gamma_3)$. Indeed in this case a hand computation will suffice. The results of this computation yield (3, 3, 4) and (3, 2, 5) as the only triples for which $B_s{}^*(n) = A_s(n)$ in the given range. The first triple produces Theorem 12. The second triple has not been discussed previously; we shall show in the next section that it also produces a valid partition theorem.

After our search on the computer has been completed we are still not certain that the proposed definition of $B_s{}^*(n)$ is always identical with $A_s(n)$.

For example, the above approach was applied to $A_s^0(n)$ the number of partitions of n into s distinct parts $\equiv 16, 0, 12$, or $13 \pmod{14}$. By our search technique one finds that $B_s^0(n)$ seems to be the number of partitions of n into s parts (where each part $\equiv 6, 10, 12, 13 \pmod{14}$ is counted once, each part $\equiv 2, 4, 5, 8, 9$, or $11 \pmod{14}$ is counted twice, and each part $\equiv 0, 1, 3, 7 \pmod{14}$ is counted three times) of the form $n = b_1 + \ldots + b_t$, $b_i - b_{i+1} \geqslant \gamma_j$ if $b_i \equiv j \pmod{14}$, and $\gamma_1 = \gamma_3 = \gamma_7 = \gamma_{14} = 29$, $\gamma_2 = \gamma_4 = \gamma_5 = 20$, $\gamma_6 = \gamma_{10} = \gamma_{12} = \gamma_{13} = 14$, $\gamma_9 = 23$, $\gamma_8 = \gamma_{11} = 24$.

Unfortunately all goes well for $n \leqslant 92$; but $A_5(93) \neq B_5(93)$.

Thus the search may indeed not produce a generally valid partition-theoretic identity; however, since such searches produced all the general results in Section 4, it may be hoped that it will be possible to discover other new results. As we shall see in the next section, one such new result has been discovered by the simple search we have outlined here.

7. Proof of Theorem 14

The following result is the second possible result related to the search described in Section 6 (the first possible result was Theorem 12).

THEOREM 14. *Let $A(n)$ denote the number of partitions of n into distinct parts not divisible by* 3. *Let $B(n)$ denote the number of partitions of n of the form $n = e_1 + \ldots + e_t$, where $e_i - e_{i+1} \geqslant 3$, 2 or 5 when $e_i \equiv 1, 2$, or $3 \pmod 3$ respectively. Then $A(n) = B(n)$.*

Proof. Following the approach of Andrews (1967d) to Schur's theorem, we let $\pi(m, n)$ denote the number of partitions of n of the type enumerated by $B(n)$ with the added restriction that all parts are $\leqslant m$. Define the polynomial

$$d_m(q) = d_m = 1 + \sum_{n=1}^{\infty} \pi(m, n) q^n. \tag{7.1}$$

Then as in Andrews (1967d),

$$d_{3m+2} = d_{3m+1} + q^{3m+2} d_{3m}; \tag{7.2}$$

$$d_{3m+1} = d_{3m} + q^{3m+1} d_{3m-2}; \tag{7.3}$$

$$d_{3m} = d_{3m-1} + q^{3m} d_{3m-5}. \tag{7.4}$$

By (7.3),

$$d_{3m} = d_{3m+1} - q^{3m+1} d_{3m-2}, \tag{7.5}$$

and by substituting (7.5) into (7.2), we obtain

$$d_{3m+2} = (1 + q^{3m+2}) d_{3m+1} - q^{6m+3} d_{3m-2}. \tag{7.6}$$

Now substituting (7.5) and (7.6) into (7.4), we have

$$d_{3m+1} = (1 + q^{3m-1} + q^{3m+1})d_{3m-2} + q^{3m}(1 - q^{3m-3})d_{3m-5}. \qquad (7.7)$$

This equation is valid for $m > 0$. Now define

$$s_m = d_{3m+1}\Big/\prod_{j=1}^{m} (1 - q^{3j}), \qquad s_0 = 1 + q. \qquad (7.8)$$

Thus if we multiply (7.7) by $\prod_{j=1}^{m-1} (1 - q^{3j})^{-1}$, we obtain

$$(1 - q^{3m})s_m = (1 + q^{3m-1} + q^{3m+1})s_{m-1} + q^{3m}\, s_{m-2}. \qquad (7.9)$$

If we let $F(x) = \sum_{m=0}^{\infty} s_m x^m$, then from (7.7) we have

$$F(x) - F(xq^3) = xF(x) + xq^2F(xq^3) + xq^4F(xq^3) + x^2q^6F(xq^3). \qquad (7.10)$$

Hence

$$F(x) = \frac{(1 + xq^2)\,(1 + xq^4)}{(1 - x)}\, F(xq^3). \qquad (7.11)$$

Repeated iteration yields

$$F(x) = F(0) \prod_{n=0}^{\infty} \frac{(1 + xq^{3n+2})\,(1 + xq^{3n+4})}{(1 - xq^{3n})} \qquad (7.12)$$

$$= (1 + q) \prod_{n=0}^{\infty} \frac{(1 + xq^{3n+2})\,(1 + xq^{3n+4})}{(1 - xq^{3n})}.$$

Now by Appell's comparison theorem (Dienes (1957))

$$1 + \sum_{n=1}^{\infty} B(n)q^n = \lim_{m=\infty} d_m$$

$$= \prod_{n=1}^{\infty} (1 - q^{3n}) \lim_{m\to\infty} s_m$$

$$= \prod_{n=1}^{\infty} (1 - q^{3n}) \lim_{x\to 1-} (1 - x)F(x)$$

$$= \prod_{n=0}^{\infty} (1 + q^{3n+1})\,(1 + q^{3n+2})$$

$$= 1 + \sum_{n=1}^{\infty} A(n)\, q^n. \qquad (7.13)$$

Comparing coefficients on both sides of this equation, we obtain the desired result.

References

Alder, H. L. (1948). The nonexistence of certain identities in the theory of partitions and compositions. *Bull. Amer. Math. Soc.* **54,** 712–22.

Andrews, G. E. (1966). An analytic proof of the Rogers–Ramanujan–Gordon identities. *Amer. J. Math.* **88,** 844–846.

Andrews, G. E. (1967a). Partition theorems related to the Rogers–Ramanujan identities. *J. Comb. Th.* **2,** 422–430.

Andrews, G. E. (1967b). Some new partition theorems. *J. Comb. Th.* **2,** 431–436.

Andrews, G. E. (1967c). A generalization of the Göllnitz–Gordon partition theorems. *Proc. Amer. Math. Soc.* **18,** 945–952.

Andrews, G. E. (1967d). On Schur's second partition theorem. *Glasgow Math. J.* **8,** 127–132.

Andrews, G. E. (1968a). On partition functions related to Schur's second partition theorem. *Proc. Amer. Math. Soc.* **19,** 441–444.

Andrews, G. E. (1968b). A new generalization of Schur's second partition theorem. *Acta Arithmetica.* **14,** 429–434.

Andrews, G. E. (1969a). A general theorem on partitions with difference conditions. *Amer. J. Math.* **91,** 18–24.

Andrews, G. E. (1969b). A generalization of the classical partition theorems. *Trans. Amer. Math. Soc.* **145,** 205–221.

Dienes, P. (1957). "The Taylor Series". Dover, New York.

Euler, L. (1748). "Introductio in Analysis Infinitorum", Vol. I. Lausanne.

Gleissberg, W. (1928). Über einen Satz von Herrn I. Schur. *Math. Z.* **28,** 372–382.

Göllnitz, H. (1967). Partitionen mit Differenzbedingungen. *J. f. reine u. angew. Math.* **225,** 154–190.

Gordon, B. (1961). A combinatorial generalization of the Rogers–Ramanujan identities. *Amer. J. Math.* **83,** 393–399.

Gordon, B. (1965). Some continued fractions of the Rogers–Ramanujan type. *Duke Math. J.* **31,** 741–748.

Hardy, G. H. and Wright, E. M. (1960). "An Introduction to the Theory of Numbers", 4th ed., Oxford University Press, Oxford.

Ingham, A. E. (1941). A Tauberian theorem for partitions. *Annals of Math.* **42,** 1075–1090.

Lehmer, D. H. (1946). Two nonexistence theorems on partitions. *Bull. Amer. Math. Soc.* **52,** 538–544.

MacMahon, P. A. (1916). "Combinatory Analysis". Vol. II. Cambridge University Press.

Rogers, L. J. (1894). Second memoir on the expansion of certain infinite products. *Proc. London Math. Soc.* (1), **25,** 318–343.

Schur, I. J. (1917). Ein Beitrag zur additiven Zahlentheorie und zur Theorie der Kettenbrüche. *Berliner Sitzungsberichte,* No. 23, 301–321.

Schur, I. J. (1926). Zur additiven Zahlentheorie. *Berliner Sitzungsberichte,* 488–495.

Slater, L. J. (1952). Further identities of the Rogers–Ramanujan type. *Proc. London Math, Soc.* (2), **54,** 147–167.

Slater, L. J. (1966). "Generalized Hypergeometric Functions". Cambridge University Press.

Computers in the Theory of Partitions

M. S. CHEEMA*

The University of Arizona, Tucson, Arizona, U.S.A.

1. Introduction

Euler discovered the identity

$$\prod_{i=1}^{\infty} (1 - x^i) = 1 + \sum_{k=1}^{\infty} (-1)^k x^{k(3k \pm 1)/2} \tag{1.1}$$

by computing the first fifty coefficients. The identity (1.1) combined with

$$\prod_{i=1}^{\infty} (1 - x^i)^{-1} = \sum_{n=0}^{\infty} p(n)x^n$$

yield

$$p(n) - p(n-1) - p(n-2) + p(n-5) + \dots$$
$$+ (-1)^k p(n - k(3k \pm 1)/2) + \dots = 0.$$

This gives a very effective method for computing $p(n)$, the number of partitions of n.

Identities, congruences and asymptotic properties of various partition functions have been discovered in many cases from the computed values of these partitions number. More recently, computers have proved useful in discovering some of the new results. The object of this paper is to indicate the possible applications of computers.

2. Multipartition Numbers

The generating function for $q_r(n_1, n_2, \dots, n_s)$ the number of partitions of $(n_1, n_2, \dots, n_s)$ into at most r parts is given by

$$\prod(1 - x_1^{k_1} x_2^{k_2} \dots x_s^{k_s} Z)^{-1} = \sum_{r=0}^{\infty} \phi_r(x_1, x_2, \dots, x_s) Z^r \tag{2.2}$$

* Supported in part by NSF GP–12716.

where the product on the left extends over all $(k_1, \ldots, k_s)$, $k_i \geqslant 0$, $\phi_0(x_1, x_2, \ldots, x_s) = 1$ and

$$\phi_r(x_1, x_2, \ldots, x_s) = \sum q_r(n_1, n_2, \ldots, n_s) x_1^{n_1} x_2^{n_2} \ldots x_s^{n_s}. \tag{2.3}$$

$\phi_r(x_1, x_2, \ldots, x_s)$ satisfy

$$r\phi_r(x_1, x_2, \ldots, x_s) = \sum_{m=1}^{r} \phi_{r-m}(x_1, x_2, \ldots, x_s) \Big/ \prod_{i=1}^{s} (1 - x_i^m). \tag{2.4}$$

Using these recurrences, we can express $\phi_r(x_1, x_2, \ldots, x_s)$ as

$$\phi_r(x_1, x_2, \ldots, x_s) = \frac{\lambda_r(x_1, x_2, \ldots, x_s)}{\sum_{m=1}^{r} \sum_{i=1}^{s} (1 - x_i^m)} \tag{2.5}$$

where $\lambda_r(x_1, x_2, \ldots, x_s) = \sum \lambda_r(m_1, m_2, \ldots, m_s) x_1^{m_1}, x_2^{m_2}, \ldots, x_s^{m_s}$ is a polynomial of degree $\binom{r}{2}$ in each of x_i.

The quantities $\lambda_r(m_1, m_2, \ldots, m_i)$ can be evaluated with the aid of certain s-tuples of permutations. Let

$$\Pi = \begin{pmatrix} 1\,2\,3 & \ldots & r \\ t_1 t_2 t_3 & \ldots & t_r \end{pmatrix}$$

be a permutation on r marks and let $m_1(\Pi)$, $m_2(\Pi)$ be the characteristic numbers of Π defined by

$$m_1 = \sum_{t_i = t_j + 1, i < j} (r - t_j) \tag{2.6}$$

$$m_2 = \sum_{t_{j+1} < t_j} (r - j). \tag{2.7}$$

It has been shown by Cheema (1961) that for all $r > 0$, $\lambda_r(m_1, m_2)$ is the number of permutations on r marks having characteristic numbers m_1, m_2. Thus $\lambda_r(m_1, m_2) \geqslant 0$. Numerical results indicate that $\lambda_r(m_1, m_2, \ldots, m_s) \geqslant 0$. Gordon (1963) has shown that this is the case by showing the $m_1, m_2, \ldots, m_s$ are the characteristic numbers of certain s-tuples of permutations. A scheme for computing $\lambda_r(m_1, m_2)$ on a computer will proceed in the following way

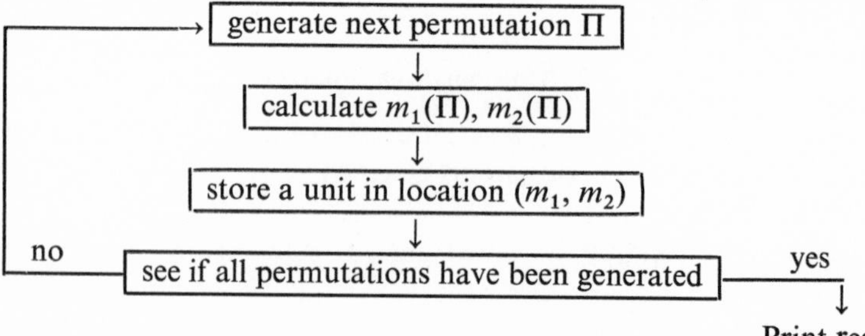

$q_r(n_1, n_2)$ are easily computable using (2.5), and noticing that

$$\sum_{n_1,n_2} q_r(n_1, n_2)x_1^{n_1} x_2^{n_2}$$

$$= \left(\sum_{m_1,m_2} \lambda_r(m_1, m_2)x_1^{m_1} x_2^{m_2}\right)\left(\sum_{n_1,n_2} q_r(n_1)q_2(n_2)x_1^{n_1} x_2^{n_2}\right)$$

which implies

$$q_r(n_1, n_2) = \sum_{\binom{r}{2}\geqslant m_1,m_2\geqslant 0} q_r(n_1 - m_1)q_r(n_2 - m_2)\lambda_r(m_1, m_2), \tag{2.8}$$

where $q_r(n)$ denotes the number of partitions of n into almost r parts. To compute $\lambda_r(m_1, m_2)$, $q_r(n_1, n_2)$ by another method we can use (2.4). This method is quite general and suitable for high speed computing for $s \leqslant 3$. It can be extended to $s > 3$ if the computing machine and coding used allow for arithmetic operations on varibles having four or more subscripts. We use $\{(1 - x_1^m)(1 - x_2^m)\}^{-1}$ as an operator.

Consider a formal power series in two variables x_1, x_2 denoted by $\sum_{i,j=0} a_{ij}x_1^i x_2^j$ whose coefficients may be written as an infinite matrix (a_{ij}). Let

$$\{(1 - x_1^m)(1 - x_2^m)\}^{-1}\left(\sum_{i,j=0}^{\infty} a_{ij}x_1^i x_2^j\right) = \sum_{i,j=0}^{\infty} b_{ij}x_1^i x_2^j. \tag{2.9}$$

The matrix (b_{ij}) is obtained from (a_{ij}) by applying the following row and column operations

$$\begin{pmatrix} b'_{n0} \\ b'_{n1} \\ b'_{n2} \\ \vdots \end{pmatrix} = \begin{pmatrix} a_{n0} \\ a_{n1} \\ a_{n2} \\ \vdots \end{pmatrix} + \begin{pmatrix} a_{n-m,0} \\ a_{n-m,1} \\ a_{n-m,2} \\ \vdots \end{pmatrix} + \dots \begin{pmatrix} a_{n-jm,0} \\ a_{n-jm,1} \\ a_{n-jm,2} \\ \vdots \end{pmatrix}, j = \left[\frac{n}{m}\right], n = 1, 2, \dots \tag{2.0}$$

$$[b_{0,n}, b_{1,n}, \dots] = [b'_{0,n}, b'_{1,n}, \dots] + [b'_{0,n-m}, b'_{1,n-m} + \dots] + \dots$$
$$+ [b'_{0,n-jm}, b'_{1,n-jm}, \dots] \tag{2.11}$$

call $(b_{ij}) = (a_{ij})_{m,m}$.

Thus (2.4) for $s = 2$ yields

$$r(q_r(i, j)) = (q_{r-1}(i, j))_{1,1} + (q_{r-2}(i, j))_{2,2} + \dots + (q_0(i, j))_{r,r} \tag{2,12}$$

where

$$q_0(i,j) = \begin{pmatrix} 100\ldots \\ 000\ldots \\ 000\ldots \\ \ldots\ldots \end{pmatrix}, \qquad q_1(i,j) = \begin{pmatrix} 111\ldots \\ 111\ldots \\ 111\ldots \\ \ldots\ldots \end{pmatrix}.$$

By this technique $q_r(n_1, n_2)$ are easily computible. For $s = 3$, we perform additions on three dimensional arrays. From $q_2(n_1, n_2)$, we obtain $\lambda_r(m_1, m_2)$ by multiplying the formal power series

$$\Sigma q_r(n_1, n_2) x_1^{n_1}, x_2^{n_2}$$

by $(1 - x_1)(1 - x_1^2) \ldots (1 - x_1^r)(1 - x_2)(1 - x_2^2) \ldots (1 - x_2^2)$. This involves subtraction operation on rows and columns of $q_r(n_1, n_2)$. This was used by Cheema to compute $q_r(n_1, n_2)$ for $0 \leqslant n_1, n_2 \leqslant 48$, $2 \leqslant r \leqslant 98$.

The same type of technique can be used for the evaluation of restricted and unrestricted partitions numbers, plane partition numbers and other multipartition numbers. For details see Cheema (1956, 1967), Gupta (1958), MacMahon (1916) and Robertson (1962).

3. Congruence Properties

Ramanujan stated the conjecture that if $p = 5$, 7 or 11 and $24n - 1 \equiv 0 \pmod{p^a}$, $a \geqslant 1$, then $p(n) \equiv 0 \pmod{p^a}$, he had proved the conjecture for certain values of a. Considerable interest was shown for this conjecture as it turned out to be false for $n = 7^3$. Computations by Lehmer (1936) confirmed the validity of the conjecture for certain values of n. Watson (1938) proved the conjecture for all powers of 5 and modified the conjecture for powers of 7. Finally Atkin (1968) settled the conjecture by proving the following.

THEOREM. *If* $24n - 1 \equiv 0 \pmod{5^a 7^b 11^c}$ *where* a, b, c *are nonnegative integers then*

$$p(n) \equiv 0 \pmod{5^a 7^d 11^c} \text{ with } d = \left[\frac{b+2}{2}\right].$$

Recently Atkin and O'Brien (1967) found some properties of $p(n)$ modulo powers of 13. Computers played significant role in these results as modulo computations are extremely fast on computers. Lehmer's computations were based on asymptotic series for $p(n)$.

4. Generating Functions

MacMahon proved that the generating function for plane partitions is given by

$$\sum_{n=0}^{\infty} p_3(n)x^n = \prod_{r=1}^{\infty} (1 - x^r)^{-r} \tag{4.1}$$

and conjectured that the generating function for solid partitions is given by

$$\sum_{n=0}^{\infty} \Pi_4(n)x^n = \prod_{r=1}^{\infty} (1 - x^r)^{-\binom{r+1}{2}}. \tag{4.2}$$

Recently this conjecture has been proved to be false by Atkin, Bratley, Macdonald and Mckay (1967) by generating the solid partitions $p_4(n)$ of n on a computer for $n \leqslant 21$ and noticing that $p_4(n) \neq \Pi_4(n)$ for $n \geqslant 6$. The exact form of the generating function for solid partitions is still not known.

5. Asymptotic Results

Lehmer studied the convergence properties of the Hardy–Ramanujan series for $p(n)$. Using Farey dissections, Hardy and Ramanujan (1918) obtained an infinite series for $p(n)$, the first few terms of which gave the value of $p(n)$ exactly if we neglect the decimal part in the answer. Lehmer showed that the Hardy–Ramanujan series was divergent and Hardy and Ramanujan had been fortunate in breaking at a point where the series gave a correct answer.

The Hardy–Ramanujan series gives

$$p(n) = \frac{(12)^{\frac{1}{2}}}{(24n-1)u} \sum_{k=1}^{[\sqrt{n}]} A_k{}^*(n)(u-k)\exp\left(\frac{u}{k}\right) + O(\log(n)/\sqrt{n}) \tag{5.1}$$

where $u = \Pi(24n-1)^{\frac{1}{2}}/6$.

A change in path of integration enabled Rademacher to replace $(u-k)$ $\exp(u/k)$ in the Hardy–Ramanujan series by $(u-k)\exp(u/k) + (u+k)$ $\exp(-u/k)$ to obtain a convergent series for $P(n)$.

Lehmer also showed that if only $2\frac{n^{\frac{1}{2}}}{3}$ terms of the Hardy–Ramanujan series taken, the resulting series will differ from $p(n)$ by less than $\frac{1}{2}$, provided $n > 600$. Recently asymptotic series for different partition numbers have been obtained by various authors. (See Cheema, 1967; Gordon 1968; Hajis, 1963, 1964; Rademacher, 1937; Robertson, 1960; Todd, 1943; Wright, 1957). Among these are the asymptotic results for the restricted and unrestricted ordinary partition numbers, rowed, plane and multipartition numbers. Again computers should prove useful in the estimates for the error and for proving the convergence properties of some of these series by comparing the exact-values of these numbers with those obtained by using the first term or in some cases the first N terms of these series. The reader should also note the papers of Atkin, Andrews, and Churchouse, presented at this symposium.

References

Atkin, A. O. L., Bratley, P., MacDonald, I. G. and McKay, J. K. S. (1967). Some computations for m-dimensional partitions. *Proc. Cambridge Phil. Soc.* **63,** 1097–1100.

Atkin, A. O. L. (1968). Ramanujan congruences for $P_{-k}(n)$. *Can. J. Math.* **20,** 67–68.

Atkin, A. O. L. (1968). "Congruences for modular forms". Symposium on Computers in Mathematical Research. North Holland Publ. Co.

Atkin, A. O. L. and O'Brien, J. N. (1967). Some properties of $p(n)$ and modulo powers of 13. *Trans. Amer. Math. Soc.* **126,** 442–459.

Cheema, M. S. (1961). Vector partitions and permutation vectors. Dissertation, University of California.

Cheema, M. S. and Haskell, C. T. (1967). Multirestricted and rowed partitions. *Duke Math. J.* **34,** 443–452.

Cheema, M. S. and Motzkin, T. S. Multipartitions and multipermutations. *Proc. Symp. Combinatorics A.M.S.* To be published.

Cheema, M. S. (1956). "Tables of partitions of Gaussian integers", p. 1–67. National Institute of Sciences of India.

Gordon, B. (1963). Two theorems on multipartite partitions. *J. London Math. Soc.,* **38.**

Gordon, B. and Houten, L. (1968). Notes on plane partitions. I. *J. Comb. Theory.* **4,** 72–80.

Gordon, B. and Houten, L. (1968). Notes on plane partitions. II. *J. Comb. Theory.* **4,** 81–99.

Gordon, B. and Houten, L. Asymptotic formulas for restricted plane partitions. *Duke Math. J.* To be published.

Gupta, H., Gwyther, A. E. and Miller, J. C. P. (1958). "Tables of Partitions". Royal Soc. Math. Tables, Vol. 4.

Hajis, P. J. (1963). Partitions into odd summands. *Amer. J. Math.* **85.** 213–222.

Hajis, P. J. (1964). Partitions into odd and unequal parts. *Amer. J. Math.* **86,** 317–324.

Hardy, G. H. and Ramanujan, S. (1918). Asymptotic formulae in combinatory analysis. *Proc. London Math. Soc.* (2), **17,** 75–115.

Lehmer, D. H. (1936). On a conjecture of Ramanujan. *J. London Math. Soc.* **11,** 114–118.

Lehmer, D. H. (1937). On the Hardy–Ramanujan Series for the partition Function. *J. London Math. Soc.* **12,** 171–176.

Lehmer, D. H. (1939). On the remainders and convergence of the series for the partition function. *Trans. Amer. Math. Soc.* **46,** 362–373.

Lehmer, D. H. (1938). On the series for the partition function. *Trans. Amer. Math. Soc.,* **43,** 271–295.

McMahon, P. A. (1916). "Combinatory Analysis", Vols. I and II. Cambridge Univ. Press. Cambridge.

Newman, M. (1956). A table of the coefficients of the powers of $\eta(\tau)$. *Nederl. Akad. Wetensch. Proc. Ser. A.,* **59,** *Indag. Math.* **18,** 204–216.

Rademacher, H. (1937). A convergent series for the partition function. *Proc. Nat. Acad. Sci.* (*U.S.A.*). **23,** 78–84.

Robertson, M. M. (1960). Asymptotic formulae for the number of partitions of a multipartite number. *Proc. Edinburgh Math. Soc.* **12,** 31–40.

Robertson, M. M. (1962). Partitions of large multipartites. *Amer. J. Math.* Vol. 84, No. 1, 16–34.

Todd, J. A. (1943). A table of partitions. *Proc. London Math. Soc.* (2), **48,** 229–240.

Watson, G. N. (1938). Ramanujans Vermutung über Zerfallungszahlen. *J. reine angew. Math.* **179,** 97–128.

Wright, E. M. (1931). Asymptotic partition formulae. I. Plane partitions. *Quart. J. Math.* **2,** 177–189.

Wright, E. M. (1956). Partitions of multipartite numbers. *Proc. Amer. Math. Soc.* **7,** 880–890.

Wright, E. M., (1957). The number of partitions of a large bipartite number. *Proc. London Math. Soc.* (3), **7,** 150–160.

Wright, E. M. (1958). Partitions of large bipartites. *Amer. J. Math.* **80,** 643–658.

Wright, E. M. (1961). Partitions of multipartite numbers into a fixed number of parts. *Proc. London Math. Soc.* (3), **11,** 499–510.

Wright, E. M. (1966). The generating function of solid partitions. *Proc. Roy. Soc. Edinburgh, A,* **67,** 185–195.

Binary Partitions

R. F. CHURCHHOUSE

*Atlas Computer Laboratory, Chilton, Didcot, England**

The classical partition function $p(n)$ has been studied by mathematicians for more than two centuries. Euler, one of the earliest to study it, discovered the generating function

$$\sum_{n=0}^{\infty} p(n)x^n = \prod_{n=1}^{\infty} (1 - x^n)^{-1}. \tag{1}$$

Attempts were made to find estimates of the size of $p(n)$ but it was not until 1917 that Hardy and Ramanujan found, and proved, the asymptotic formula whose first term is given by

$$p(n) \sim \frac{1}{4n\sqrt{3}} \exp\left(\pi\sqrt{\frac{2n}{3}}\right)$$

It is perhaps even more remarkable that nobody before Ramanujan had noticed the congruences satisfied by $p(n)$ whenever n belongs to certain arithmetic progressions, viz.

$$p(5n + 4) \equiv 0 \pmod 5,$$
$$p(7n + 5) \equiv 0 \pmod 7,$$
$$p(11n + 6) \equiv 0 \pmod{11}.$$

A few years ago I had the idea of using a computer to search for congruence properties of these types. It is relatively simple to write a program to do this. I had wondered if there might exist further congruences of the form

$$p(an + b) \equiv 0 \pmod a \tag{2}$$

where $a > 11$. However, at that time I met Dr. Atkin who told me that the prospects were very poor.

A short time later it occurred to me to apply the computer to the calculation

* Present address *Computer Centre, University College, Cardiff, Wales.*

of the number of partitions of n as the sum of powers of 2. Denote this number, the number of binary partitions of n, by $b(n)$. Thus, since

$$5 = 4 + 1 = 2 + 2 + 1 = 2 + 1 + 1 + 1 = 1 + 1 + 1 + 1 + 1,$$

we see that $b(5) = 4$.

By analogy with the Euler product (1) we see that

$$F(x) = \sum_{n=0}^{\infty} b(n)\, x^n = \prod_{n=0}^{\infty} (1 - x^{2^n})^{-1}. \tag{3}$$

Multiplying by $(1 - x)$ we obtain

$$(1 - x)\, F(x) = \prod_{n=1}^{\infty} (1 - x^{2^n})^{-1} = F(x^2), \tag{4}$$

and we deduce at once that

$$b(2n + 1) = b(2n), \tag{5}$$

$$b(2n) = b(2n - 2) + b(n). \tag{6}$$

By means of (5) and (6) we can rapidly compute $b(n)$ on a computer for a wide range of values of n. At the end of the paper I give a table of $b(n)$ for even values of n up to 200. The values for odd n are redundant, from (5).

A study of this table revealed that $b(n)$ possesses some simple congruence properties mod 2, mod 4, and mod 8. Having found these I was able to prove them quite easily. The facts are given by

THEOREM 1. (i) $b(n) \equiv 0 \pmod 2$ *for all* $n \geqslant 2$,

(ii) $b(n) \equiv 0 \pmod 4$ *if and only if* n *or*

$$n - 1 = 4^m(2k + 1)\ (m \geqslant 1),$$

(iii) $b(n) \equiv 0 \pmod 8$ *for no value of* n.

(For a proof see Churchhouse, 1969.) It is astonishing that even these simple congruences had not been noticed before although $b(n)$ had been studied by Euler and others.

No congruences to moduli other than 2, 4, and 8 were found but, in a most fortuitous way, I found a class of congruences of a slightly different type. From (5) and (6)

$$b(2n) = b(n) + b(n - 1) + b(n - 2) + \dots + b(0),$$

$$b(4n) = b(n) + 3b(n - 1) + 5b(n - 2) + \dots + (2n + 1)\, b(0). \tag{7}$$

and so on. It is, in this way, possible to express $b(2^k n)$ as a sum involving $b(n), b(n - 1), \dots, b(0)$. I was using (7) to extend the range of the computed tables one day when I noticed that some of the values were divisible by 8 and even higher powers of 2. Since this contradicts (iii) of Theorem 1 there had obviously been a miscalculation and I found that I had inadvertently computed not $b(4n)$ but $b(4n) - b(n)$. Further examination showed that $b(4n) - b(n) \equiv 0 \pmod 8$ whenever n is even. This subsequently proved to be a special case of a much stronger result which I published (Churchhouse, 1969) as a conjecture, viz:

CONJECTURE *If* $k \geqslant 1$ *and* $t \equiv 1 \pmod 2$

$$b(2^{2k+2}t) - b(2^{2k}t) \equiv 0 \pmod{2^{3k+2}}$$
$$b(2^{2k+1}t) - b(2^{2k-1}t) \equiv 0 \pmod{2^{3k}}.$$

I was able to prove this conjecture for any given value of k but a general proof did not appear to be so easy. The numerical evidence in support of the conjecture was overwhelming and furthermore it seemed that the congruences are always exact, i.e., that the differences on the left hand sides in the conjecture are always divisible by precisely the power of 2 given on the right hand side and by no higher power.

A few months after I had made the conjecture Rødseth (1970) succeeded in proving it and his proof also reveals that the congruences are indeed exact. I had some of the evidence for $p = 3$, (Churchhouse, 1969), and this showed that the results in this case must be less elegant than in the case $p = 2$. Rødseth (1970) also proved the complete general results for partitions based on powers of a prime $p > 2$.

On the question of the size of $b(n)$ it is not too difficult to prove that

$$b(n) = O(n^{\frac{1}{2}\log_2 n}).$$

In fact, Mahler (1940) proved that if $t_k(n)$ denotes the number of partitions of n as the sum of powers of k then

$$\log t_k(n) \sim \frac{(\log n)^2}{2 \log k}.$$

One final point: the total time required on the SRC Atlas to carry out all the calculations involved in this work was only a matter of seconds. It is however salutary to realise that the most interesting results were discovered because I made a mistake in a hand calculation!

References

Churchhouse, R. F. (1969). *Proc. Cambridge Phil. Soc.* **66,** 371–376.
Hardy, G. H. and Ramanujan, S. (1917). *Proc. London Math. Soc.* **2** (17), 75–115.
Rodseth, Ø. J. (1970). *Proc. Cambridge Phil. Soc.* **68,** 447–453.
Mahler, K. (1940). *J. Lond. Math. Soc.* **15,** 115–123.

Table of Values of the binary partition function

n	$b(n)$	n	$b(n)$
0	1		
2	2	102	10,614
4	4	104	11,514
6	6	106	12,414
8	10	108	13,428
10	14	110	14,442
12	20	112	15,596
14	26	114	16,750
16	36	116	18,044
18	46	118	19,338
20	60	120	20,798
22	74	122	22,258
24	94	124	23,884
26	114	126	25,510
28	140	128	27,338
30	166	130	29,166
32	202	132	31,196
34	238	134	33,226
36	284	136	35,494
38	330	138	37,762
40	390	140	40,268
42	450	142	42,774
44	524	144	45,564
46	598	146	48,354
48	692	148	51,428
50	786	150	54,502
52	900	152	57,906
54	1014	154	61,310
56	1154	156	65,044
58	1294	158	68,778
60	1460	160	72,902
62	1626	162	77,026
64	1828	164	81,540
66	2030	166	86,054
68	2268	168	91,018
70	2506	170	95,982
72	2790	172	101,396
74	3074	174	106,810
76	3404	176	112,748
78	3734	178	118,686
80	4124	180	125,148
82	4514	182	131,610
84	4964	184	138,670
86	5414	186	145,730
88	5938	188	153,388
90	6462	190	161,046
92	7060	192	169,396
94	7658	194	177,746
96	8350	196	186,788
98	9042	198	195,830
100	9828	200	205,658

Multiplanar Partitions

DONALD BURNELL AND LORNE HOUTEN*

Washington State University, Pullman, Washington, U.S.A.

1. Multiplanar Partitions

We define a multiplanar (or solid) partition of a non-negative integer n to be a solution of the Diophantine equation $n = \sum_{i,j,k=0}^{\infty} n_{i,j,k}$ where the $n_{i,j,k}$ are non-negative integers satisfying the inequalities $n_{i,j,k} \geqslant n_{i+1,j,k}$, $n_{i,j,k} \geqslant n_{i,j+1,k}$, and $n_{i,j,k} \geqslant n_{i,j,k+1}$. These partitions were introduced by MacMahon (1916) and were studied numerically by Atkin *et al.* (1967).

In this note we wish to study those multiplanar partitions with the restriction that $n_{i,j,k} > n_{i+1,j,k}$ for all non-zero $n_{i,j,k}$. Examples may be found in Houten (1968).

Define $b(n; f_{11}, \ldots, f_{rs})$ to be the number of such partitions of n with precisely f_{ij} parts on the ith row of the jth plane. It is shown (Houten, 1968) that $b(n; f_{11}, \ldots, f_{rs})$ is uniquely defined by the recursion formula

$$b(n; f_{11}, \ldots, f_{rs}) = \sum_{e_{ij}=0}^{1} b(n - \Sigma f_{ij}; f_{11} - e_{11}, \ldots, f_{rs} - e_{rs})$$

subject to the initial conditions

$$b(n; f_{11}, \ldots, f_{rs}) = \begin{cases} 0 \text{ unless } n \geqslant 0; f_{ij} \geqslant f_{i+1\,j} \geqslant 0; f_{ij} \geqslant f_{i\,j+1} \geqslant 0 \\ 1 \text{ if } n = f_{11} = \ldots = f_{rs} = 0. \end{cases}$$

We use this recursion as a basis for the computing techniques.

2. Computing Techniques and Algorithm

The problem defined in Section 1 is equivalent to the question of computing $f(A, n)$ where $A \in M$ and M is the set of all $n \times n$ matrices such that $A = (a_{ij})$

* The second author is supported in part by NSF Grant GP–8956. The authors wish to thank the Computing Centre at Washington State University for its support and cooperation.

and

(a) $\theta(A) = \sum_{i=1}^{n} \sum_{j=i}^{n} \sigma(a_{ij}) \leqslant n,$

(b) $\sigma(a_{ij}) = \sum_{k=0}^{a_{ij}} k,$

(c) $0 \leqslant a_{i,j+1} \leqslant a_{ij} \leqslant a_{i,j+1} + 1,$

(d) $0 \leqslant a_{i+,1j} \leqslant a_{ij} \leqslant a_{i+1,j} + 1$ and

$$f(A,n) = \sum_{A' \in M(A)} (f(A', n - \rho(A))$$

where

$$\rho(A) = \sum_{i=1}^{n} \sum_{j=1}^{n} a_{ij}, \quad f((0),0) = 1$$

and

$$f(A,k) = 0 \quad \text{for} \quad k < 0$$

and

$$A' \in M(A) \quad \text{iff} \quad a_{ij} - 1 \leqslant a'_{ij} \leqslant a_{ij}.$$

I. One needs some kind of bound on n for a given computation since the technique is to keep all previous A's and $f(A,k)$, $k = 1, 2, \ldots, l$ where l is this previously defined limit. So assume l is predefined and we have decided to limit $\theta(A) \leqslant l$ and $n \leqslant l$. Suppose an algorithm P has been developed such that if p is a partition on the integer n, say $p = (p_1, p_2, \ldots, p_n)$ then P computes the next p' in the 'sequence' by

(1) $p_1 = 1$ then

$$p_1' = n + 1, \quad p_i = 0 \quad (i > 1).$$

(2) $p_1 \neq 1$ then

$$p_i' = p_i \text{ for } i \text{ such that } p_{i+1} \neq 0, 1,$$

$$p_i' = p_i - 1 \text{ for the smallest } i \text{ such that } p_{i+1} = 0, 1, \text{ and}$$

$$p_k' = \min\left(p_{k-1}, n - \sum_{i=1}^{k-1} p_i\right) \quad \text{if} \quad p_k = 0, 1.$$

II. Using algorithm P described in I, compute the set of vectors $P_0, P_1, P_2, \ldots, P_{m'}$ such that

(1) $P_0 = (0, 0, \ldots, 0),$

(2) $P_1 = (1, 0, \ldots, 0),$

(3) If P_i is in the set the P_{i+1} is the first vector generated from P sequentially such that $p_{i+1}^{k+1} + 1 \geqslant p_{i+1}^{k} \geqslant p_{i+1}^{k+1}$ where p_i^k is the kth component in P_i and

$$\sum_{k=1}^{n} \sum_{m=1}^{p_i^k} k \leqslant l.$$

III. Define V_{ijk} for $1 \leqslant i \leqslant m'$, $1 \leqslant j \leqslant l$, $1 \leqslant k \leqslant l$ by

$$V_{ijk} = 1 \quad \text{if} \quad p_i^k = j \quad \text{or} \quad p_i^k = j - 1 \quad \text{for} \quad j \neq 0,$$

$$V_{ijk} = 1 \quad \text{if} \quad p_i^k = 0 = j,$$

$$V_{ijk} = 0 \quad \text{otherwise.}$$

IV. Define $C_i(i) = 0 \quad (i = 1, 2, 3, \ldots, n)$,

$$C_0(0) = 1.$$

V. It is not difficult to show how to list A's in a sequence so that

$$A_0 = (0) \quad \text{and} \quad i \leqslant j \quad \text{if} \quad A_i \in M(A_j).$$

VI. Now $(a_{i1}, a_{i2}, \ldots, a_{in})$ is a partition so A_k can be stored as a sequence

$$m(1, A_k), m(2, A_k), \ldots, m(l, A_k)$$

of integers corresponding to the proper vectors

$$P_{m(1,A_k)}, \ldots, P_{m(l,A_k)}.$$

VII. Now assume that if $i < j$ then $f(A_i, k)$, $k = 0, 1, 2, \ldots, l$ has been computed and saved.

VIII. Now compute $f(A_j, k)$, $k = 0, 1, 2, \ldots, n$, as follows:

(1) Store A by partitions as in VI (remember that $A_j \in M(A_j)$).

(2) Let $n_{k'i} = 1$ if $V(k', a_{i'm}, m) = 1, m = 1, 2, \ldots l$, $n_{k'i} = 0$ otherwise.

(3) $f(A_j, k) = \sum_{A_i \in M(A_j)} f(A_i, k - \theta(A_j))$

where $M(A_j)$, although it is the same set, is now determined by

$$A_i \in M(A_j) \quad \text{if} \quad n_{m(i',A_i)i'} = 1, \text{ for all } i', \; i = 1, 2, \ldots, l.$$

3. Conclusions

The results are discouraging from the point of view of finding a neat product expansion for the generating function as was obtained for the plane case (Gordon and Houten, 1968). These computing techniques have been adapted to compute the lattice functions associated with these partitions. These results will be analyzed in a subsequent paper.

4. Table

Let $b(n)$ be the number of solid partitions of n, distinct along rows.

n	$b(n)$	n	(bn)
1	1	11	2020
2	3	12	3803
3	7	13	7043
4	16	14	12957
5	33	15	23566
6	71	16	42536
7	141	17	76068
8	284	18	135093
9	552	19	238001
10	1067	20	416591

References

Atkin, A. O. L., Bratky, P., MacDonald, I. G. and McKay, J. K. S. (1967). Some computations for *m*-dimensional partitions. *Proc. Cambridge Phil.* **63,** 1097–1100.

Gordon, B. and Houten, L. (1968). Notes on plane partitions I, II. *J. Comb. Theory* **4,** 72–80; 81–89.

Houten, L. (1968). A note on solid partitions. *Acta Arithmetica* **15,** 71–76.

MacMahon, P. A. (1916). "Combinatory Analysis", Vols. I and II. Cambridge University Press.

Some Problems in Number Theory

P. ERDŐS

Academy of Sciences, Budapest, Hungary

In the present paper I discuss some problems in number theory which I have thought about in the last few years; computational techniques can be applied to some of them.

1. On Prime Factors of Consecutive Integers

Let $f(k)$ be the smallest integer with the property that the product of $f(k)$ consecutive integers all greater than k is always divisible by a prime greater than k. A well-known theorem of Sylvester and Schur (see Erdős, 1934) states that $f(k) \leqslant k$. I proved (1955)

$$c_1 \log k \log_2 k \log_4 k/(\log_3 k)^2 < f(k) < c_2 k/\log k,\dagger$$

Recently Ramachandra (1969 and to appear) proved $f(k) < (1 + o(1)) k/\log k$. It seems to me to be very difficult to prove that for all $k > k_0$ we have $f(k) < \pi(k)$, though I have no doubt that the conjecture is true. In fact it seems likely that $f(k)$ is not substantially larger than

$$A_k = \max(p_{r+1} - p_r), \qquad k < p_r < p_{r+1} < 2k.$$

In fact I cannot even disprove $f(k) = A_k$ for all sufficiently large k, though it seems likely that $f(k) > A_k$ for all large k. A well known theorem of Pólya and Störmer states that if $u > u_0(k)$ then $u(u + 1)$ always contains a prime factor greater than k, thus $f(k)$ can be determined in a finite number of steps, and an explicit bound has been given by Lehmer (1964) for the number of necessary steps. It is known (Utz, 1961) that $f(2) = 2, f(3) = f(4) = 3, f(5) = \ldots = f(10) = 4$.

Selfridge and I conjected that if $m \geqslant 2k$ then $\binom{m}{k}$ has a prime factor $\leqslant m/2$, the only exception being $\binom{7}{3}$. This conjecture was recently proved by Earl Ecklund.

† We write $\log \log k = \log_2 k$, etc.

Selfridge and I proved that there is an absolute constant $c > 0$ so that if $m \geqslant 2k$ then $\binom{m}{k}$ always has a prime factor less than m/k^c.

The proof is very simple. Assume first $m \geqslant 2k^{1+c}$, put $l = [k^c] + 1$. It follows from the theorem of Hoheisel–Ingham (see Ingham, 1937) that for sufficiently small $c > 0$ there is a prime p satisfying

$$\frac{m}{l} > p > \frac{m-k}{l} > k.$$

Clearly this prime divides $\binom{m}{k}$ and this proves our assertion if $m \geqslant 2k^{1+c}$. Assume next $m < 2k^{1+c}$. Let

$$s = \left[\frac{3m}{2k}\right] + 1.$$

It follows from the Hoheisel–Ingham theorem that there is a prime p satisfying

$$\frac{m}{s} > p > \frac{m-k}{s-1}.$$

Clearly

$$p \left| \binom{m}{k} \right. \qquad \left(\text{since } \frac{k}{2} < \frac{m-k}{s-1} < p < k \text{ amd } m - k < (s-1)p < sp < m\right)$$

which completes our proof. The simplicity of our proof is caused by the fact that we have not determined c explicitly.

Selfridge and I conjectured that if $m > k^2$ then $\binom{m}{k}$ has a prime factor $\leqslant m/k$; $\binom{7}{3}$ is certainly an exception, and this may be the only one. In connection with this problem we asked: Determine or estimate the smallest integer $g(k)$ so that all prime factors of $\binom{g(k)}{k}$ are greater than k, (it is easy to see that such integers exist).

It is perhaps true that, for $k > k_0(\varepsilon)$ and $m > k^{1+\varepsilon}$, $\binom{m}{k}$ always has a prime factor greater than $k^{1+\varepsilon} - k$. Ramachandra (1969) has some results which point in this direction. More generally let $h(k)$ be the largest integer so that if $m > h(k)$ then $\binom{m}{k}$ always has a prime factor greater than $h(k) - k$.

I am sure that $h(k) > k^c$ for every $c > 0$ and $k > k_0(c)$; $h(k) > ck \log k$ is easy and Ramachandra's result will no doubt give $h(k) > (1 + o(1)) k \log k$. Denote by p_k the least prime greater than $2k$. Faulkner (1966) proved that for $m \geqslant p_k$, $\binom{m}{k}$ always has a prime factor $\geqslant p_k$, except for $\binom{9}{2}$ and $\binom{10}{3}$. Thus $h(k) \geqslant p_k + k$ for $k > 3$.

It is easy to see that $h(2) = 4$, $h(3) = 6$, $h(4) = 16$ (i.e. the product of 4 consecutive integers $\geqslant 13$ always has a prime factor $\geqslant 13$). It is difficult to compute $h(k)$ but by the effectivisation results of Brown this can be done in a bounded number of steps. Lehmer (1963) showed $h(7) \geqslant 43$.

I conjectured that, for every $m \geqslant 2k$, $\binom{m}{k}$ has a divisor d with $m - k < d \leqslant m$. This is easy to see if $k = p^\alpha$. Schinzel (1958) proved that in general it is incorrect, e.g., it is false for $k = 15$, $m = 99125$. He further proved that it is true for all integers $k \leqslant 34$ except 15, 21, 22, 33. Schinzel now conjectures that it is false for all $k > 34$, $k \neq p^\alpha$. This conjecture has been verified for $k < 150$. I proved (see Schinzel, 1958) that my conjecture is false for infinitely many $k \neq p^\alpha$.

In view of the failure of my conjecture one can try to investigate the greatest factor of $\binom{m}{k}$ not greater than m. I would now conjecture that the greatest prime factor $\leqslant m$ of $\binom{m}{k}$ is greater than cm for some $c > 0$. Unfortunately I can prove no non-trivial result.

This question leads me to the following one: Is it true that for every $\varepsilon > 0$ there is a k_0 so that, for $k > k_0$, $k!$ is the product of k integers all greater than $(k/e)(1 - \varepsilon)$. It easily follows from Stirling's formula that if

$$k! = \prod_{i=1}^{k} a_i, \quad a_1 \leqslant \ldots \leqslant a_k,$$

then $a_1 < k/e$, thus our conjecture if true is best possible.

Recently Selfridge and I proved that the product of consecutive integers is never a power (that it is never a square is due to Rigge, 1939); our proof is not quite easy and will be published elsewhere (for a weaker result see Erdős, 1955b). In fact we prove a somewhat stronger result. We prove that for every $l > 1$, $k > 1$ the product $\prod_{i=1}^{k} (m + i)$ contains a prime $p > k$ to an exponent which is not a multiple of l. We conjecture that if $l \geqslant 2$ and $k \geqslant 3$ then $\prod_{i=1}^{k} (m + i)$ contains a prime $p > k$ to the exponent one. The only exception is $48 \cdot 49 \cdot 50$. For $k = 2$ there are infinitely many exceptions. This conjecture if true is very deep.

Put ($p^\alpha \| m$ means $p^\alpha | m$, $p^{\alpha+1} \nmid m$)

$$A_i^{(m)} = \Pi p^\alpha, \quad p^\alpha \| (m + i), \quad p \leqslant k.$$

It is not difficult to prove that for $k > k_0(\varepsilon)$

$$\min_{1 \leqslant i \leqslant k} A_i^{(m)} < (1 + \varepsilon)k.$$

Probably very much more is true, in fact perhaps

$$\lim_{k \to \infty} \frac{1}{k} \max_{0 \leqslant m < \infty} \min_{1 \leqslant i \leqslant k} A_i^{(m)} = 0.$$

2. Covering Congruences

A system of congruences $a_i \pmod{m_i}$, $m_1 < \ldots < m_k$ is called a covering system if every integer satisfies at least one of the congruences $a_i \pmod{m_i}$. I was lead to the problem of covering congruences by a letter of Romanoff who asked if there are infinitely many odd integers not of the form $2^k + p$ (as is well known Romanoff (1934) proved that the lower density of the integers of the form $2^k + p$ is positive).

The simplest covering system is 0 (mod 2), 0 (mod 3), 1 (mod 4), 1 (mod 6), 11 (mod 12) and the system 0 (mod 2), 0 (mod 3), 1 (mod 4), 7 (mod 8), 11 (mod 12), 19 (mod 24) shows (Erdős 1947-51) that the answer to Romanoff's question is positive, in fact there is an arithmetic progression consisting entirely of odd numbers no term of which is of the form $2^k + p$.

The following question seems very difficult: Is it true that to every c there exists a covering system $a_i \pmod{m_i}$ $c \leqslant m_1 < \ldots < m_k$? This is known for $c \leqslant 9$ (see Churchhouse, 1968) but the general case seems very difficult. A positive answer would imply that for every r there is an arithmetic progression no term of which is the sum of a power of 2 and an integer having at most r prime factors.

Schinzel recently investigated the question whether, for fixed r, there is an arithmetic progression no term of which is of the form $2^{k_1} + 2^{k_2} + \ldots + 2^{k_r} + p$; already for $r = 2$ the question seems difficult. Schinzel (1967) recently applied covering congruences to the study of reducibility of polynomials.

There are many further interesting problems on covering congruences, e.g., is there a covering congruence all whose moduli are odd, or is there a covering congruence in which no two moduli divide each other? Schinzel (1967) and Selfridge observed that the two problems are connected.

Call an integer m covering if one can find a covering set whose moduli are all divisors of m; $m = 12$ is clearly the smallest covering integer. Clearly all multiples of a covering integer are again covering. An integer is primitive covering if it is covering but all its divisors are not covering. Clearly we obtain the covering integers by taking the set of all multiples of the primitive covering integers. I can prove using the results in Erdős (1948) that the covering integers have a density. One could try to estimate the number of primitive covering integers not exceeding x.

I expect that for every $c > 0$ there is an m which is not covering and for which $\sigma(m)/m > c$, but I could not prove this (perhaps I overlook a simple idea).

A system of arithmetic progressions $a_i \pmod{m_i}$, $m_1 < \ldots < m_k$ is called disjoint if no integer is in two of them. Denote by $f(x)$ the maximum number of pairwise disjoint arithmetic progressions whose difference does not exceed x. Stein and I conjectured that $f(x) = o(x)$; Szemeredi and I (1968) recently proved this. The sharpest results for $f(x)$ are

$$x \exp(-c_1(\log x \log_2 x)^{\frac{1}{2}}) < f(x) < x(\log x)^{-c_2},$$

perhaps the lower bound is close to the true order of magnitude.

Stein conjectured that if $a_i \pmod{m_i}$, $m_1 < \ldots < m_k$ are k disjoint congruences there is an integer $\leqslant 2^k$ which does not satisfy any of these congruences Selfridge proved this conjecture. I conjectured that if $a_i \pmod{m_i}$, $m_1 < \ldots < m_k$ are any k congruences which are not covering then there is an integer $\leqslant 2^k$ which does not satisfy any of these congruences (Selfridge, Crittenden and Van der Eyden recently proved this conjecture).

It is not hard to see that the density of integers not satisfying any of the disjoint congruences $a_i \pmod{m_i}$, $m_1 < \ldots < m_k$ is $\geqslant 1/2^k$ and that this result is best possible. The same result probably holds for any k congruences which are not covering (Erdős, 1962).

I would like to state one more problem on arithmetic progressions: Let $a_i \pmod{m_i}$, $m_1 \leqslant m_2 \leqslant \ldots$ be an infinite sequence of arithmetic progressions. Is it true that the set of integers not satisfying any of these congruences always has a logarithmic density? Special cases of this conjecture were proved by Davenport and myself (1936 and 1951).

3. Some Problems and Results on the Addition of Residue Classes

Heilbronn and I (1969) proved that if $a_1, \ldots, a_k$, $k \geqslant 3(6p)^{\frac{1}{2}}$ are distinct residues mod p (p prime) then every residue (mod p) can be written in the form

$$\sum_{i=1}^{k} \varepsilon_i a_i, \quad \varepsilon_i = 0 \text{ or } 1.$$

We conjectured that the same holds for $k > 2\sqrt{p}$ and that this is best possible. Olsen (1968) recently proved this conjecture. We further conjectured that the number of distinct residues of the form $a_i + a_j$, $1 \leqslant i < j \leqslant k$, is at least $2k - 3$; as far as I know this conjecture is still unsettled.

Let now m be composite and $a_1, \ldots, a_k$ be k distinct residues mod m. We conjectured (Erdős and Heilbronn, 1969) that if $k > c\sqrt{m}$ then

$$\sum_{i=1}^{k} \varepsilon_i a_i \equiv 0 \pmod{m}, \qquad \varepsilon_i = 0 \text{ or } 1$$

is always solvable (probably $k > \sqrt{2m} + o(\sqrt{m})$ will suffice). Ryavec (1968) proved a slightly weaker result and our conjecture was recently proved by Szemeredi (his paper will appear in *Acta Arithmetica*). Szemeredi's proof works for every abelian group of order m; perhaps the result holds for non-abelian groups too.

Eggleston proved the following result: Let G_m be an abelian group of m elements, $m \leqslant n + k - 1$, $a_1, \ldots, a_n$ are n elements of G_m where at least k of the a's are distinct. Then (e is the unit element of G_m)

$$e = \prod_{i=1}^{k} a_i^{\varepsilon_i}, \quad \varepsilon_i = 0 \text{ or } 1$$

is always solvable.

Eggleston and I conjectured that $m \leqslant n + k - 1$ can be replaced by $m \leqslant n + \binom{k}{2}$; this if true is easily seen to be best possible (it suffices to take G_m to be the additive group mod m and the a's $1, \ldots, k, 1, \ldots, 1$).

We proved this conjecture if $m > m_0(k)$ (unpublished), also we were led to the following question which seems to be of some interset. Let $f(k)$ be the largest integer with the following property; let $a_1, \ldots, a_k$ be k distinct elements of G_m and assume that no product

$$\prod_{i=1}^{k} a_i^{\varepsilon_i}, \quad \varepsilon_i = 0 \text{ or } 1,$$

equals the unit of G_m; then at least $f(k)$ distinct elements of G_m can be represented in the form

$$\prod_{i=1}^{k} a_i^{\varepsilon_i}, \quad \varepsilon_i = 0 \text{ or } 1.$$

We showed $f(2) = 2, f(3) = 5, f(4) = 8, f(k+1) \geqslant f(k) + 2$. Szemeredi showed $f(k) > ck^2$. It does not seem to be easy to determine $f(k)$ or even to give an asymptotic formula for it. These problems can be stated for non-abelian groups too.

4. Miscellaneous Problems, Results and Conjectures

Denote by $\pi(x)$ the number of primes not exceeding x. Is it true that $\pi(x + y) \leqslant \pi(x) + \pi(y))$? This conjecture, if true, is certainly extremely deep. It is not hard to prove for small values of y. I do not know for how large values of y it has been proved and I also do not know for how large values it has been checked.

Following Hardy and Littlewood (1923) put

$$\rho(y) = \limsup_{x=\infty} \left(\pi(x + y) - \pi(x)\right)$$

Probably $\lim_{y=\infty} \rho(y) = \infty$. Hardy and Littlewood conjectured that for $y > y_0$ then $\rho(y) > y/\log y$; this if true is certainly very deep. Using Brun's method they proved $\rho(y) < cy/\log y$ (as far as I know this is the only time they used Brun's method). Denote by $h_m(k)$ the number of integers $m < x \leqslant m + k$ which are not divisible by any prime less than or equal to k. Hardy and Littlewood conjectured that $\rho(k) = \max_m h_m(k)$. It seems probable that $\lim_{y=\infty} \left(\pi(y) - \rho(y)\right) = \infty$.

All these conjectures seem hopeless at present. Perhaps the following questions deserve some investigation. A sequence $m < a_1 < \ldots < a_l \leqslant m + k$ is called complete if $(a_i, a_j) = 1$, $1 \leqslant i < j \leqslant l$, but for every $m < n \leqslant m + k$, $(n, a_j) > 1$ for some $1 \leqslant j \leqslant l$. Denote by $f(m, k)$, respectively, $F(m, k)$ the smallest (largest) value of l. It is easy to see that $\min_m f(m, k) = 2$ $(m = k! - 1)$ but it seems very difficult to determine or give a good estimation for $\max_m f(m, k)$, $\min_m F(m, k)$ or $\max_m F(m, k)$. Clearly all three functions tend to infinity with k, perhaps $\max_m F(m, k) = \pi(k) + 1$ (clearly $\max_m F(m, k) \geqslant \pi(k) + 1$, to see this observe that the $\pi(k) + 1$ integers $k! + 1, k! + p$ [p runs through the primes not exceeding k] are pairwise relatively prime). $F(m, k) < ck/\log k$ trivially follows from Brun's method. For small values of k it is easy to compute all these functions.

One could try to estimate $f(m, k)$ and $F(m, k)$ if both m and k tend to infinity e.g. is it true that if c is a sufficiently large constant then $f(m, (\log m)^c)$ tends to infinity together with m? This question is connected with the problem of the difference of consecutive primes and seems very difficult.

The sharpest known inequality for large differences of consecutive primes is due to Rankin (1938) and states that for infinitely many n we have

$$p_{n+1} - p_n > c \log p_n \log_2 p_n \log_4 p_n / (\log_3 p_n)^2.$$

Denote now by $a_1^{(r)} < a_2^{(r)} < \ldots$ the sequence of integers which have at most r prime factors. I proved (Erdős, 1955c, 1956)

$$\limsup_{k=\infty} (a_{k+1}^{(2)} - a_k^{(2)})/\log k > c;$$

perhaps this inequality holds for every r and perhaps the lim sup is in fact infinite, but I cannot prove this even for $r = 2$.

Let $g(m)$ be the smallest integer so that at least one of the integers m, $m+1, \ldots, m+g(m)$ divides the product of the others. It is easy to see that $g(k!) = k$ and, for $m > k!$, $g(m) > k$. I can prove that for infinitely many m

$$g(m) > \exp\left((\log m)^{\frac{1}{2}-\varepsilon}\right).$$

I have no good upper bound for $g(m)$. $g(m) < c\sqrt{m}$ is easy but probably $g(m) = O(m^{\varepsilon})$ and in fact perhaps $g(m) = O\left(\exp\left((\log m)^{\frac{1}{2}+\varepsilon}\right)\right)$.

Denote by $u_1^{(\varepsilon)} < \ldots < u_s^{(\varepsilon)} \leqslant m$ the integers not exceeding m all whose prime factors are $< m^{\varepsilon}$. $g(m) = O(m^{\varepsilon})$ would follow if we could show $a_{i+1} - a_i = O(m^{\varepsilon})$, but this seems hopeless at present.

Put $f(m) = \sum_{p|m} p$ (this function has recently been investigated from a different point of view by Mohan Lal, 1969). Denote by $F(x)$ the number of distinct integers of the sequence $f(m)$, $1 \leqslant m \leqslant x$. I can prove (unpublished)

$$c_1\, x/\log x \prod_{k=3}^{r} \log_k x < F(x) < c_2\, x/\log x \prod_{k=3}^{r} \log_k x, \tag{4.1}$$

where $1 \leqslant \log_r x \leqslant e$. Analogous questions have been investigated for the functions $\sigma(m)$, $\phi(m)$ and $d(m)$; see Erdős (1935, 1945) and Erdős and Mirsky (1952).

The same function which appears in (4.1) occurs in a completely different question. Let $1 \leqslant a_1 < \ldots < a_k \leqslant x$ be a sequence of integers so that all the sums

$$\sum_{i=1}^{k} \varepsilon_i/a_i, \quad \varepsilon_i = 0 \text{ or } 1,$$

are all different. Put $\max k = f(x)$. Then

$$c_1\, x/\log x \prod_{k=3}^{r} \log_k x < f(x) < c_2\, x/\log x \prod_{k=3}^{r} \log_k x. \tag{4.2}$$

The proof of (4.2) is not published.

Finally I state a conjecture of the 16-year old Hungarian mathematician I. Ruzsa.

Let $f(m)$ be a multiplicative function whose values are elements of a group G. Let g be an element of this group, Is it true that the density of integers m for which $f(m) = g$ always exists? This conjecture if true must be very deep since it would imply the theorem of Wirsing (1967) that every multiplicative function which only assumes the values < 1 has a mean value.

References

Churchhouse, R. F. (1968). "Covering sets and systems of Congruences in Computers in Mathematical Research". pp. 20–36, North Holland.

Davenport, H. and Erdős, P. (1936). On sequences of positive integers, *Acta Arith.* **2,** 147–151 and (1951) *J. Ind. Math. Soc.* **15,** 19–24.

Erdős, P. (1934). On a theorem of Sylvester and Schur. *J. London Math. Soc.* **9,** 282–288.

Erdős P. (1935). On the normal number of prime factors of $p - 1$ and some related problems concerning Euler's ϕ function, *Quarterly J. Math.* **6,** 205–213.

Erdős, P. (1945). Some remarks on Euler's ϕ function and some related problems. *Bull. Amer. Math. Soc.* **51,** 540–544.

Erdős, P. (1947–51). On integers of the form $2^k + p$ and some related problems. *Summa Brasil. Math.* **2,** 113–123.

Erdős, P. (1948). On the density of some sequences of integers. *Bull. Amer. Math. Soc.* **54,** 685–692.

Erdős, P. (1955a). On consecutive integers. *Nieuw Archiev Wiskunde,* **3,** 124–128.

Erdős, P. (1955b). On the product of consecutive integers, III. *Indag. Math.* **17,** 85–90.

Erdős, P. (1955c). *Elemente der Mathematik,* Vol.I., 47.

Erdős, P. (1956). *Elemente der Mathematik*, Vol. II., 86–88.

Erdős, P. (1962). Szamelmeleti megjegyzesek IV. (in Hungarian), *Mat. Lapok* **13,** 241–242.

Erdős, P. and Heilbronn, H. (1969). On the addition of residue classes mod p. *Acta Arith.* **9,** 149–159.

Erdős, P. and Mirsky, L. (1952). The distribution of values of the divisor function $d(m)$. *Proc. London Math. Soc.* **3,** 257–271.

Erdős, P. and Szemeredi, E. (1968). On a problem of Erdős and Stein. *Acta Arith.* **15,** 85–90.

Faulkner, M. (1966). On a theorem of Sylvester and Schur. *J. London Math. Soc.* **41,** 107–110.

Hardy, G. H. and Littlewood, J. E. (1923). Some problems on partitio numerorum III: On the expression of a number as a sum of primes. *Acta Math.* **44,** 1–70.

Ingham, A. E. (1937). On the difference between consecutive primes. *Quarterly J. Math.* **8,** 255–266.

Lal, M. (1969). Iterates of a number theoretic function. *Math of Computation.* **23,** 181–183.

Lehmer, D. H. (1963). Some high speed logic. *Proc. Symp. in Applied Math.* **15,** 141–145.

Lehmer, D. H. (1964). On a problem of Störmer. *Illinois J. Math.* **8,** 57–79.

Olsen, J. E. (1968). An addition theorem modulo p. *J. Combinatorial Theory* **5,** 45–52.

Ramachandra, K. (1969). A note on numbers with a large prime factor. *J. London Math. Soc.* (2), **1,** 303–306.

Rankin, R. A. (1938). The difference between consecutive prime numbers. *J. London Math. Soc.* **13,** 242–247.

Rigge, S. (1939). Über ein diophantisches Problem, 155–160. 9th Scandinavian Maths. Congress.

Romanoff, N. P. (1934). Über einige Sätze der additiven Zahlentheorie, *Math. Annalen,* **109,** 668–678.

Ryavec, C. (1968). On the addition of residue classes modulo *n*, *Pacific J. Math.,* **26,** 367–373.

Schinzel, A. (1958). Sur un probleme de P. Erdős, *Coll. Math.* **5,** 198–204.

Schinzel, A. (1967). Reducibility and irreducibility of polynomials and covering systems of congruences. *Acta Arith.* **13,** 91–101.

Utz, W. (1961). A conjecture of Erdős concerning consecutive integers. *Amer. Math. Monthly.* **68,** 896–697.

Wirsing, E. (1967). Das asymptotische Verhalten von Summen über multiplikative Funktionen. *Acta Math. Acad. Sci. Hung.* **18,** 411–467.

Some Unsolved Problems

RICHARD K. GUY*

University of Calgary, Alberta, Canada

Introduction

The following problems have little in common other than that they are wholly or in part number-theoretic, and that computers have been or could be used to assist in their solution.

1. Sets with Distinct Sums

The set of integers $\{2^i : 0 \leqslant i \leqslant k\}$ has subsets, every pair of which have distinct sums. Erdős (1956a, 1957b, 1961) has asked for the maximum number m of positive integers

$$a_1 < a_2 \ldots < a_m \leqslant 2^k. \tag{1}$$

so that all sums of subsets are distinct. Erdős and L. Moser (1956) have shown that

$$k + 1 \leqslant m \leqslant k + (\tfrac{1}{2} + \varepsilon) \log k \tag{2}$$

where the logarithm is to base 2. Conway and Guy (1968, 1969) have given a sequence, $u_0 = 0$, $u_1 = 1$,

$$u_{n+1} = 2u_n - u_{n-r}, \qquad n \geqslant 1, \tag{3}$$

where r is the nearest integer to $\sqrt{2n}$, from which may be derived a set of $k + 2$ integers

$$\{a_i = u_{k+2} - u_{k+2-i} : 1 \leqslant i \leqslant k + 2\}. \tag{4}$$

They conjecture that this set has subsets with distinct sums for all values of k, and they establish this for $k \leqslant 40$. For $k \geqslant 21$, $u_{k+2} < 2^k$, so this enables the lower bound in (2) to be improved to $k + 2$ for $k \geqslant 21$, since once such a set is found, its cardinality can be increased by doubling the

* Research supported by grant A–4011 of the National Research Council of Canada.

size of its members, and adjoining the member 1. It is desired to give a proof of, or a counter-example to, the statement made concerning the set (4). Erdős offers \$250 for an exact determination of m in (2).

2. No–Three–In–Line Problem

Given an $n \times n$ array of n^2 points of the unit lattice, can $2n$ of them be selected so that no three are in line? This problem goes back, at least in the case $n = 8$, to Dudeney (1917) and Ball (1939). Configurations which solve this problem have been found by various authors, and the numbers of essentially different ones have been checked and extended by Kelly (1967). For $2 \leqslant n \leqslant 10$, they are 1, 1, 4, 5, 11, 22, 57, 51 and 156. Apart from three solutions (for $n = 2, 4$ and 10) with the full symmetry of the square, it is conjectured that there are no solutions with even the symmetry of the rectangle. A computer search has revealed no others with $n \leqslant 32$.

The results quoted indicate a roughly exponential increase in the number of solutions, apart from a natural tendency for more solutions to appear for n even than for n odd. However, Guy and Kelly (1968) have given a heuristic argument which supports the conjecture that for large n, no solutions exist. More precisely they conjecture that for large n, not more than $(2\pi^2/3)^{1/3}\, n \simeq 1{\cdot}87n$ points can be found without there being three in line.

In the other direction, Erdős (Roth, 1951) has indicated that when n is prime, no three points of the n points (i, i^2), $1 \leqslant i \leqslant n$ are collinear, where the second coordinate is to be reduced modulo n. It is clear from the pigeon-hole principle that not more than $2n$ points can be found, but no closer bounds than these appear to be known.

3. Distinct Distances Between Lattice Points

Erdős (1957a) has asked for the largest number k of the n^2 lattice points just considered such that the $\binom{k}{2}$ distances that they determine should be all distinct. A simple counting argument shows that

$$k \leqslant n, \tag{5}$$

and this upper bound can be attained for $n \leqslant 7$ (see Figure 1, for example). However, for large n, the theorem of Landau concerning numbers representable as sums of two squares shows that (5) can be improved to

$$k < c_1\, n(\log n)^{-1/4}. \tag{6}$$

A heuristic argument leads to the conjecture

$$(?) \qquad k < c_2 n^{2/3} (\log n)^{1/6}, \tag{7}$$

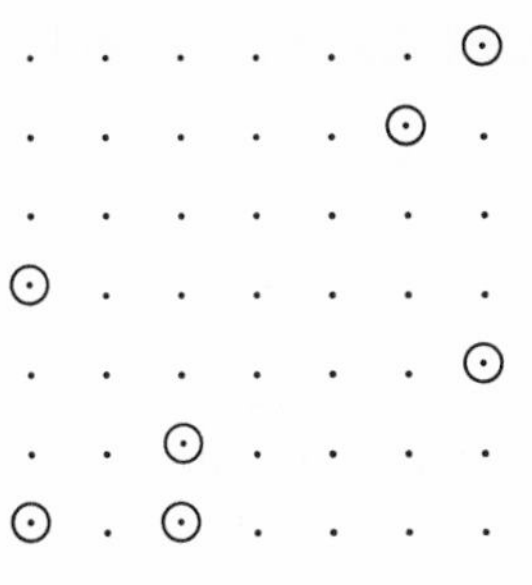

FIGURE 1

while in the opposite direction it has recently been shown by Erdős and Guy (1970) that

$$k > n^{2/3-\varepsilon}. \tag{8}$$

For the corresponding problem in one dimension, the best results are

$$n^{1/2}(1-\varepsilon) < k < n^{1/2} + n^{1/4} + 1.$$

The lower bound is constructed from a Singer (1938) difference set, and the upper bound follows from a recent improvement by Lindström (1969) on the work of Erdős and Turán.

In d dimensions ($d \geqslant 3$) we are unable to improve on (8), while the results corresponding to (6) and (7) are

$$k < c_3 d^{1/2} n,$$

and the conjecture

$$(?) \qquad k < c_4 d^{2/3} n^{2/3} (\log n)^{1/3}.$$

4. The Rhind Papyrus

P. Erdős and E. G. Straus conjectured that

$$\frac{4}{n} = \frac{1}{x} + \frac{1}{y} + \frac{1}{z} \tag{9}$$

could be solved in positive integers for all $n > 1$. There is a good account of this problem by Mordell (1969), where it is shown that the conjecture is true, except possibly in the cases where n is prime and congruent to 1,121, 169, 289, 361 or 529, modulo 840. Several have worked on the problem, including Bernstein (1962), Obláth (1949) Rosati (1954) and Yamamoto (1964); the last named has verified the conjecture for $n < 10^7$. Schinzel (oral communication) has indicated that one can only express

$$\frac{4}{at + b} = \frac{1}{x(t)} + \frac{1}{y(t)} + \frac{1}{z(t)}$$

with $x(t)$, $y(t)$, $z(t)$ integral polynomials in t, provided b is a non-residue of a. Hence the difficulty of the above mentioned cases.

Schinzel has generalized the conjecture by replacing the 4 in (9) by general m, relaxing the condition that y and z be positive, and requiring the truth only for $n > n_m$. That n_m may be greater than m is exemplified by $n_{18} = 23$. Schinzel and Sierpiński (1956, 1957) have verified the latter conjecture for $m \leqslant 18$, and Schinzel later for $m = 19$. Meanwhile Sedláček (1959) had verified it for $19 \leqslant m \leqslant 21$, Palamà (1959) for $19 \leqslant m \leqslant 23$, and Stewart and Webb (1966) from $m < 36$.

5. A Divisor Function Problem

Leo Moser has observed that while $n\phi(n)$ determines n uniquely, where $\phi(n)$ is Euler's totient function, $n\,\sigma(n)$ does not, where $\sigma(n)$ is the sum of the divisors of n. He gives the example

$$m\,\sigma(m) = n\,\sigma(n), \qquad m \neq n, \tag{10}$$

with $m = 12$, $n = 14$. The multiplicativity of $\sigma(n)$ then implies an infinity of solutions of (10), namely $m = 12q$, $n = 14q$, with $(42, q) = 1$. However Moser asks if there is an infinity of primitive solutions, in the sense that m/d, n/d is not a solution for any $d > 1$. His example is the smallest of a set of solutions, $m = 2^{p-1}(2^q - 1)$, $n = 2^{q-1}(2^p - 1)$, where $2^p - 1$ and $2^q - 1$ are distinct Mersenne primes, but only 253 such pairs are so far known.

There are other solutions of (10), for example $m = 2^4.3.5^3.7, n = 2^{11}5^2$, but they seem difficult to construct. Although $(m, n) = 400$ in this example, it is in fact primitive. A more obviously primitive example is $m = 2^5.5$, $n = 3^3.7$. Further solutions are $m = 2^9.5$, $n = 2^3.11.31$; $m = 2^7.3^2.5^2, n = 2^4.5^3.17$ and $m = 2^7.3^2.5^2(2^p - 1), n = 2^{p-1}5^3.17.31$ where $2^p - 1$ is a Mersenne prime other than 3 or 31.

6. Another Divisor Function Problem

The Mersenne primes also feature in the problem of finding solutions of the equation

$$\sigma(q) + \sigma(r) = \sigma(q + r), \tag{11}$$

which was raised by Ramsey (1963). Again one asks if there is an infinity of solutions which are primitive in a sense similar to that defined in the previous problem.

If $q + r$ is prime, the only solution of (10) is $\sigma(1) + \sigma(2) = \sigma(3)$. If $q + r = p^2$, where p is prime, then q (or r) is prime and $r = 2^n k^2$, where n and k are odd integers. The case $k = 1$ leads to solutions when $p = 2^n - 1$ is a Mersenne prime, provided that $q = p^2 - 2^n$ is also prime. Such solutions occur for $n = 2, 3, 5, 7, 13$ and 19. The values of n for which the question is still open are 31, 61, 89, 107, 127, 607, 1279, 4253, 9941, 11213, ... There are no solutions if k contains a factor congruent to 3, modulo 14. For $k = 5$, M. J. T. Guy used Titan to show that there are no solutions, except possibly for $n =$

189 249 501 509 521 573 585 605 621 809
845 861 873 969 ...

For $k = 7$ there are solutions for $n = 1$ and 3:

$$\sigma(5231) + \sigma(2 \,.\, 7^2) = \sigma(73^2),$$

$$\sigma(213977) + \sigma(2^3 \,.\, 7^2) = \sigma(463^2),$$

the next values of n which are in doubt being $n =$

31 33 103 115 121 123 159 169 225 255
259 273 313 355 369 409 435 439 463 483
535 553 561 583 585 625 681 705 759 799
801 841 871 889 895 903 913 915 945 961
979 985 ...

For $k = 11$, $n = 1$ and 13 give the solutions

$$\sigma(24407) + \sigma(2 \,.\, 11^2) = \sigma(157^2),$$

$$\sigma(1410646926617) + \sigma(2^{13} \,.\, 11^2) = \sigma(1187707^2),$$

with $n =$

21 45 57 67 141 145 153 163 177 193
201 211 217 265 273 285 307 333 405 453
481 501 505 513 523 525 537 541 553 571
585 597 613 661 667 685 717 757 811 813
837 865 883 931 933 945 955 981 993 ...

still in doubt. If $k = 13$, no solutions are known, though the values $n =$

53	55	79	91	149	163	175	187	229	277
325	367	373	389	415	437	439	581	583	643
679	703	715	725	763	773	805	823	871	883
895	907	919	943	...					

have not been eliminated as possibilities. Other solutions with $q + r = p^2$ are

$$\sigma(155015849) + \sigma(2^5\,19^2) = \sigma(12451^2),$$

$$\sigma(1193399) + \sigma(2\,.\,5^4) = \sigma(1093^2),$$

$$\sigma(229405235369) + \sigma(2^9\,.\,5^4) = \sigma(478963^2),$$

$$\sigma(2676857975009) + \sigma(2^9\,.\,7^4) = \sigma(1636111^2).$$

For $n = 1$ and k prime, $k = 53, 137, 193$ and 277 give (further) solutions; for $n = 3$, $k = 313$ and 421 give solutions, and for $n = 5$, $k = 97, 107, 131, 149$ and 257 give solutions.

The only solutions so far discovered with $q + r$ equal to the cube of a prime are

$$\sigma(2) + \sigma(6) = \sigma(8),$$

$$\sigma(11638687) + \sigma(2^2\,.\,13\,.\,1123) = \sigma(227^3).$$

There appear to be many more solutions to this problem than to the previous one, but we are unable to prove that there is an infinity of (primitive) solutions.

7. Simpler than Goldbach?

Erdős and Moser (1959) ask if, for each positive integer n, there are integers a, b such that

$$\phi(a) + \phi(b) = 2n.$$

The truth of the Goldbach hypothesis would imply an affirmative answer; can the present question be settled independently?

8. A Problem of Bateman

Let S be the set of odd positive integers k such that $k\,.\,2^n + 1$ is composite for all $n \geqslant 0$. Then Sierpiński (1956) has shown that S has positive asymptotic density. Bateman (1963) asks if S has asymptotic density strictly less than $\frac{1}{2}$. If s is the smallest member of S, Selfridge has shown that $383 \leqslant s \leqslant 78557$. What is the exact value of s?

9. A Conjecture of Chowla

Chowla (1963) has conjectured that if $(a, b) = 1$ and p_n is the nth prime, then there are infinitely many m such that $p_m \equiv p_{m+1} \equiv a$, modulo b. The case for $b = 4$, $a = 1$ follows from the theorem of Littlewood, and the case $b = 4$, $q = 3$ has been settled by Knapowski and Turán (1969).

References

Ball, W. W. (1939). "Mathematical Recreations and Essays", 11th ed. p. 189 (rev. Coxeter, H. S. M.), Macmillan.

Bateman, P. T. (1963). Problem 2, Proc. Number Theory Conf. p. 89. Univ. of Colorado, Boulder.

Bernstein, L. (1962). Zur Lösung der diophantischen Gleichung $\frac{m}{n} = \frac{1}{x} + \frac{1}{y} + \frac{1}{z}$ insbesondere im Fall $m = 4$. *J. reine angew. Math.* **211,** 1–10.

Chowla, S. (1963). Problem 17, Proc. Number Theory Conf. p. 5–24, 93–94, Univ. of Colorado, Boulder.

Conway, J. H. and Guy, R. K. (1968). Sets of natural numbers with distinct sums. *Amer. Math. Soc. Notices.* **15,** 345.

Conway, J. H. and Guy, R. K. (1969). Solution of a problem of P. Erdős. *Colloq. Math.* **20,** 307.

Dudeney, H. E. (1917). "Amusements in Mathematics". pp. 94, 222. Nelson.

Erdős, P. (1956a). "Problems and results in additive number theory". *In* "Colloque sur la théorie des nombres", p. 127–137. Bruxelles, Liège and Paris.

Erdős, P. (1956b). Problem 319. *In* "The New Scottish Book". Wroclaw.

Erdős, P. (1957a). Néhany geometriai problémáról, *Mat. Lapok,* **8,** 86–92.

Erdős, P. (1957b). Problem 209. *Colloq. Math.* **5,** 119.

Erdős, P. (1957c). Problem 11. Some unsolved problems, *Michigan Math. J.* **4,** 291–300.

Erdős, P. (1961). Some unsolved problems. *Publ. Math. Inst. Hungar. Acad. Sci.,* (A), **6,** 221–254.

Erdős, P. and Guy, R. K. (1970). Distinct distances between lattice points. *Elemente der Math.*

Erdős, P. and Moser, L. (1959). Problem 20, p. 340. Report Inst. Theory of Numbers, Univ. of Colorado, Boulder. (1963). Proc. Number Theory Conf. p. 78, Univ. of Colorado, Boulder.

Guy, R. K. and Kelly, P. A. (1968). The no-three-in-line problem. *Canad. Math. Bull.* **11,** 527–531.

Kelly, P. A. (1967). The use of the computer in game theory. Master's thesis. The Univ. of Calgary.

Knapowski, S. and Turán, P. (1969). Über einige Fragen der vergleichenden Prinzahltheorie. *In* "Number Theory and Analysis", (a collection of papers in honour of Edmund Landau, 1877–1938). Plenum Press, New York.

Lindström, B. (1969). An inequality for B_2-sequences. *J. Combinatorial Theory.* **6,** 211–212.

Mordell, L. J. (1969). "Diophantine equations". Academic Press, London and New York.

Obláth, R. (1949). Sur l'équation diophantienne $\frac{4}{n} = \frac{1}{x_1} + \frac{1}{x_2} + \frac{1}{x_3}$. *Mathesis.* **59,** 308–316.

Palamà, G. (1959). Su di una congettura di Schinzel. *Boll. Un. Mat. Ital.* (3). **14,** 82–94.

Rosati, L. A. (1954). Sull'equazione diophantea $\frac{4}{n} = \frac{1}{x_1} + \frac{1}{x_2} + \frac{1}{x_3}$. *Boll. Un. Mat. Ital.* (3). **9,** 59–63.

Roth, K. F. (1951). On a problem of Heilbronn. *J. London Math. Soc.* **26,** 198–204.

Rumsey, M. (1963). Problem. *Eureka* **26,** 12.

Sedláček, J. (1959). Über die Stammbrüche (Czech, Russian and German summaries). *Časopis Pěst Mat.* **84,** 188–197.

Sierpiński, W. (1956). Sur les décompositions de nombres rationnels en fractions primaires. *Mathesis.* **65,** 16–32.

Sierpiński, W. (1957). On the decomposition of rational numbers into unit fractions (Polish). Pánstwowe Wydawnictwo Nauk., Warsaw.

Singer, J. (1938). A theorem in finite projective geometry and some applications to number theory. *Trans. Amer. Math. Soc.* **43,** 377–385.

Stewart, B. M. and Webb, W. A. (1966). Sums of fractions with bounded numerators. *Canad. J. Math.* **18,** 999–1003.

Yamamoto, K. (1964). On a conjecture of Erdős. *Mem. Fac. Sci. Kyushu Univ. Ser.* A, **18,** 166–167.

Languages

P. Barrucand

In the early days of computers only machine languages were used; indeed not really languages, but rather a form of electronic handicraft. Later machine language became "pure", the only work of the programmer being to write the program, punch it onto cards or tapes, and debug it.

Today, for numerical calcutalion, we use almost exclusively languages such as FORTRAN, ALGOL 60, and PL1; and for management and accountancy, languages such as COBOL, Moreover we have now a strange, powerful, but esoteric *theory* of language; a puzzling mixture of mathematical logic, semigroup theory and axiomatices. All these things are, practically and theoretically, excellent; but, strangely, there is some confusion between numerical calculation and "scientific computation" (in fact, computation for physics and chemistry). Thus the usual languages are fairly practical for solving linear systems, etc., which are only a very small application of a logically built language such as ALGOL 60. But what can we say about number theory? Everyone using FORTRAN or ALGOL has seen weak spots; both are maladjusted for many important problems and, at best, not only do we waste a lot of time and money, but also we are compelled to programme in an unpleasant and artificial manner; a clear indication that we are taking the wrong road. In fact, we are like a man attempting to formulate abstract algebra (say) in the vocabulary of Virgil or Pope.

Indeed we have here two different problems: (1) flexible general concepts and (2) flexible atomic software.

1. ALGOL aims to be a "communication language" of algorithms totally independent of special computer opportunities. It uses only "reals" and "integers" (I do not want to say anything on the "boolean" problem). But real and integer are not, for the number theorist, the only arithmetics; why not others? For current routines we need, at the least, arithmetic and algebraic fields (and rings), finite fields, quaternions, also finite rings with divisors of zero, and probably p-adic fields and extensions of these.

Even more obviously, ALGOL and FORTRAN both use "integer division" but give no immediate representation of the *remainder*. But, the result of an integer division is not one number, but an ordered pair (quotient, remainder), and for the number theorist the remainder is often more important than the quotient.

Of coures, we may use, for the determination of this remainder, formulae such as $R = N - (N/D)*D$. But this is exactly the sort of unpleasant writing, uneconomical in operation, which we wish to avoid. In one sense, ALGOL is surely a beautiful and rigorous logical device; in another sense it does not give us the arithmetic we need. Neither does FORTRAN.

I can write in FORTRAN

```
IF (A. GT . 3.7)  Gϕ  Tϕ  31 .
```

But why have we no possibility of using a *simple* language, involving no highly sophisticated processes, no lengthy items, which would enable us to write (using fluent English):

If A and B are coprime, go to 31.

Here A and B are perhaps "ordinary" integers, perhaps integers of a recondite octic field, perhaps (why not?) ideals.

If number theorists had had to build an *ab ovo* language, we should have today not the cut-and-dried real *vs* integer, but one highly flexible arithmetic language. We should have also a library of subroutines in a totally different mood.

In the classical systems, a subroutine is a device enabling us to compute the numerical value of a function: $x = f(y_1, y_2, y_3 \ldots y_{11})$, or generally only $x = f(y_1, y_2)$ or $x = f(y)$. Now let us consider the simple number theoretic function $d(n)$ (number of divisors of n, in Z). What is really necessary is to transform the familiar 2-adic expansion of n into the canonical product $n = p_1{}^{a_1} p_a{}^{a_a} \ldots p_w{}^{a_w}$. Knowing this new representation of n the determination of $d(n)$ is a very simple matter, and also $\sum_{d|n} d^k$ etc...

So what is necessary is not a function-computing subroutine, but one device giving the whole of p and a, for we must know not just one number, but an ordered set of numbers $\{p_1, p_2, p_3 \ldots ; a_1, a_2, a_3 \ldots\}$

So we have found the first requirements: "flexibility of arithmetics" and "flexibility of jump codification".

2. Theoretically, a "communication language" has to ignore the length of a real or integer number: π is a real number, totally distinct from 3 . 14159... to 1000 digits.

Alas for the computer a real number is, in fact, a very special number belonging to the ring generated by adjunction of (generally) $\frac{1}{2}$ to Z. Here the "scientific programmer" finds the unsolved problem of round-off errors. But for the number theorist we have no round-off possibility except in the cases of asymptotic density problems. So if we attempt some research in the field $Q(\sqrt[3]{69})$ we have to consider the fundamental unit $\varepsilon = A + B\sqrt[3]{69} + C\sqrt[3]{69^2}$ where $A = 404, 886, 837, 053, 487, 091, 694, 212, 951, 195, 653, 956, 127$ and not some 40-bits mantissa approximation! In a communication language A is an "integer" but for the computer A is intractable with the usual compilers. A more simple example is given by the values of $p(n)$, the classical partition function. Here we are mostly interested in congruences, for some very long numbers!

So we need multi-length working, and since direct computation with variable length is today too recondite and uneconomical on most machines, we have to consider the splitting of arithmetics into 1, 2, ... n, "precision" (i.e. length).

I will give some practical examples showing the necessity of this.

(a) The possible applications of Jacobi's algorithm (generalization of con- inued fractions) to the study of cubic (or perhaps more generally n^{ic}) units and fields;

(b) The determination of many quadratic units (and associated class-numbers);

(c) Let

$$\left(\sum_{n=-\infty}^{+\infty} q^{n^2}\right)^k = \sum_{n=0}^{\infty} R_k(n)q^n;$$

what can we say of $R_k(n)$ if k is rational, but not integral? From a theoretical point of view I have built an algorithm enabling us to compute easily many coefficients, but practically—with the classical double precision floating-point (96 bits!) the algorithm in this case loses quickly every stability showing a destructive accumulation of round-off errors;

(d) The extensive study of the distribution of the bits of 2-adic irrational, or binary expansion or irrationals.

On the other hand we have also the problem of the storage of extensive information. This implies not so much multi-length as mini-length. D. H. Lehmer has shown how to teach a computer combinatorial tricks. I give a new example.

(e) Let $$b(n) = \begin{cases} 1 \Leftrightarrow n = a^4 + b^4 + c^4 + d^4 \\ 0 \Leftrightarrow n \neq a^4 + b^4 + c^4 + d^4 \end{cases}$$

($n, a, b, c, d \in Z$). The direct computation is easy, but the key problem is the storage. For this we must consider not whole words, but the actual binary bits. Using these and observing

$$\forall n, \forall m : 5 \leqslant m \leqslant 15, n \equiv m(16) \Rightarrow b(n) = 0,$$

we can, in a word of 45 bits store the whole of $b(n)$ for 144 consecutive n, enabling us to store in 32,000 words of a CDC 3600 a table of $b(n)$ from 1 to 4,608,000!

Naturally we can do this within a FORTRAN programme, but only by trickery, using the special properties of a special computer (and in this case we lose the properties of a universal communication language) or else by writing a nonsensical sequence of divisions and multiplications, an offence against common sense.

(f) Even more striking is the case of rules such as $n^{(p-1)/k} \equiv 1 \pmod p \Leftrightarrow n$ is a $k - ic$ residue mod p [here n is an integer, p a prime $\equiv 1 \pmod k$]. This fantastic power is, *modulo p*, exceedingly quick to determine, using each bit of the 2-adic expansion of p.

(g) Very similar is the case of the determination of $\sqrt{m}$, $\sqrt[3]{m}$ etc. modulo p.

So we have found two new requirements, the easy possibility of using multi-length and mini-length arithmetic, and direct access to each bit.

Each machine language works well with specific problems, and gives a quick and easy technique for numerical study. It seems that, in the theory of languages, we have a sad gap; no theory of such languages exist. We have full theories for polysynthetic, or high agglutinating or flexional ones such as ALGOL and FORTRAN (as in linguistics ancient Greek or Navajo) but not for monosyllabic languages such as the so called "AUTOCODES" (as Chinese or English).

Now, machine languages are not totally artificial, so "logical intersection" is really a true binary operation on 2-adic numbers and not only a "boolean" one, and one binary operation is algebraically as worthy as another—shifting is also a basic facility and we must remain dissatisfied with languages giving no straightforward possibility to programme this. I do not want to dwell on this—but I suggest that we have something to say, not only on the software, but also on the hardware of new machines.

A very quick, and technically perfect computer using heavily floating point arithmetic may be, from a logical point of view, less perfect than a slower one, but given a good integer arithmetic, then generally, the less we use floating point, the better the device.

As it is of the greatest interest to be able to use each facility of each machine language, the most reasonable solution seems to be to have the possibilities of "quotation", i.e. some simple process for mixing highly synthetic language and "autocode". The actual systems are, from this point of view, extremely recondite and almost unserviceable. Naturally such mixed language is not a "universal communication device" and *it must not be so.*

The reading of a journal such as "Mathematics of Computation" shows a continuous rise of interest in number theoretic problems. This enables us to demand, either a language fully shaped for our special needs and aims, or (and this is perhaps better and easier) the ability to transform slightly the existing languages, a kind of "do it yourself". This implies the ability to know exactly the structure of the compiler and the exact localization of every code. I can see no reason to consider such structures as "top secret".

I wish to remark also that as between numerical analysis and number theory, there is no unbridgeable gap. Some of our problems have applications in numerical analysis. Thus a key problem in numerical integration is: knowing a moment sequence c_m determine the sequence of associated orthogonal polynomials. This may be done using Hankel determinants, or some ingenious device such as the QD- algorithm. Now what is needed in order to do this without serious rounding errors leading quickly to a breakdown? If every c_m is rational (or more generally algebraic) we need rational (algebraic) arithmetic, and no floating point. Many differential equations may also be solved thus. Conversely the fascinating study of the zero distribution of the zeta-functions and L functions is a problem typical of numerical analysis.

On another point I wonder if it would not be wise to mingle communication algorithm language (a man-to-man language) and computer-programmer language: classical mathematical language is already an algorithmic language. Thus the easy transition to "formula translator" (FORTRAN). In number theory, communication algorithm language is badly missing. So the methods used for the determination of cubic units by people such as Markov, Reid, Wolfe, Selmer, or Voronoi, are exceedingly perplexing. And the process used by D. H. Lehmer for his magnificent machine theorem about cubic residuacity is not too easy to transform into a fully written programme.

The reasons for this gap are probably (1) ALGOL is maladjusted to our problems; (2) we have not begun building some new language. This duty is, today, indispensable if we want to progress; exceedingly tedious things such as ideal determination, cubic and biquadratic forms composition, should be easy, but in practice they are not, owing to this lack of language opportunities.

On the other hand, even if our problems are not of paramount economic importance (racing cars do not have the economic importance of lorries), they are of great logical interest, and they must be exceedingly illuminating for computer-engineers and makers: an incentive for technical progress. So, the building of (a) an algorithm communication language (between number theorists), and (b) a fully computer-programmer language, should be of the greatest interest for scientific research and also, I think, for computer progress.

Author Index

The numbers in *italics* refer to the pages where references are listed in full. Absence of this page number indicates a general reference used in the chapter.

D

E

F

G

H

T

U

V

W

Y

Z